Water Politics and Development Cooperation

Deutsches Institut
für Entwicklungspolitik

German Development
Institute

d·i·e

Waltina Scheumann · Susanne Neubert ·
Martin Kipping (Eds.)

Water Politics and Development Cooperation

Local Power Plays and Global Governance

Springer

Editors

Dr. Waltina Scheumann
Helmholtz-Zentrum for
Umweltforschung (UFZ)
Abt. Umwelt und
Planungsrecht
Permoserstr. 15
04318 Leipzig
Germany

Dr. Susanne Neubert
Deutsches Institut for
Entwicklungspolitik (DIE)
Tulpenfeld 6
53113 Bonn
Germany
susanne.neubert@die-gdi.de

Martin Kipping
Deutsche Botschaft Kabul
Einsatz I. Ktgt.
ISAF AFG, Feldpost
64298 Darmstadt
Germany

ISBN 978-3-642-09541-2 e-ISBN 978-3-540-76707-7

Cover design: deblik, Berlin

Printed on acid-free paper

9 8 7 6 5 4 3 2 1

springer.com

The world has the technology, the finance and the human capacity to remove the blight of water insecurity from millions of lives. Lacking are the political will and vision needed to apply these resources for the public good.

UNDP, Human Development Report 2006. Beyond scarcity: power, poverty and the global water crisis, New York, Oxford 2006, p. 27

Contents

List of Contributors

Franz-Josef Batz is a senior advisor with GTZ. He is heading a team of water policy advisors supporting BMZ's water sector desk. Dr Batz holds a PhD in Agricultural Economics. He has been the coordinator of the International Conference on Freshwater. e-mail: Franz-Josef.Batz@gtz.de

Anne Cramer is a graduate of business and economics studies from University of Regensburg Germany. She has worked at the World Bank in Washington DC in strategic communications for water sector reform programs. Since 2003, she has been working as an advisor for dam-related issues to GTZ and BMZ. Ms Cramer is now working as a freelance consultant. e-mail: Anne.Cramer@gtz.de

Michael Fink is in charge of GTZ's advisory project on sustainable dams. This includes appraisals of individual dams worldwide on behalf of BMZ as well as steering pilot initiatives towards a more sustainable planning and management of dams. Prior to working for GTZ, Mr Fink studied urban and regional planning in Germany and Ghana. e-mail: michael.fink@gtz.de

Volkmar Hartje is professor of environmental economics within the Planning Department of the University of Technology Berlin. His current research focuses on the international dimensions of water policy and the economics of water and nature protection policies. Prior to joining the University, he has been working with KfW in Frankfurt and as a research fellow at the Science Center Berlin. e-mail: hartje@imup.tu-berlin.de

Elke Herrfahrdt-Pähle is an economist and a research officer in the environment department of DIE since 2003. Her fields of research include natural resources management, water management and adaptation to climate change. e-mail: Elke.Herrfahrdt@die-gdi.de

Lena Horlemann holds a master's degree in Southeast Asian Studies from Bonn University Germany. 2005–2006 she worked with DIE on global water management issues. Since July 2007, she has been participating in the DIE and InWEnt (Capacity Building International) program "Managing Global Governance". Ms Horlemann is preparing her PhD on environmental NGOs within the global governance process. e-mail: Lena.Horlemann@web.de

Martin Kipping has studied political science at Free University Berlin and the Institut d'Etudes Politiques (IEP) in Paris. After post-graduate training at DIE Bonn, he has been working as water sector desk officer at BMZ. Mr Kipping is currently seconded to the German Embassy in Kabul, Afghanistan, as its 1st secretary for development cooperation and economic affairs. e-mail: martin.kipping@gmx.net

Thomas Kluge studied Jurisprudence at University of Frankfurt/Main. He holds his post doctoral lecture qualification and readership at University GH Kassel. Dr Kluge is head of the research area "water and sustainable planning" at the Institute for Social-Ecological Research (ISOE) in Frankfurt. e-mail: kluge@isoe.de

Matthias Krause is an economist at DIE Bonn. He has several years of research and working experience in Latin America. His main research activities lie within the fields of private sector participation in infrastructure as well as in regulation and the political economy of infrastructure in developing countries. Matthias.Krause@die-gdi.de

Stefan Lindemann is a political scientist and research fellow at the German Advisory Council on the Environment (SRU). Mr Lindemann is a PhD candidate in Development Studies at the London School of Economics and Political Science (LSE). Water-related issues are among his main areas of research. S.Lindemann@lse.ac.uk

Manfred Matz is a water engineer by profession and works currently as a project director at the Stockholm International Water Institute (SIWI) in Sweden. He has worked in various African and Latin-American countries in different water related sectors, such as erosion control, irrigation, rural and urban water supply and water sector reforms. His current engagements are in water governance, strategies and policies. e-mail: manfred.matz@siwi.org

Peter P. Mollinga is senior researcher at ZEF in Bonn. He is convener of SaciWATERs (South Asia Consortium for Interdisciplinary Water Resources Studies), based in Hyderabad, India. Dr Mollinga has worked on irrigation management and irrigation reform, and more generally on the politics of water. He is presently involved in land and water management related research in Uzbekistan, Afghanistan and India. e-mail: pmollinga@hotmail.com

Susanne Neubert is senior researcher at DIE, Bonn since 1997. She studied Environmental Sciences and Agricultural Engineering and received her PhD from Humboldt University Berlin in Agricultural Economics. Her main fields of research are water resources management, poverty oriented water and irrigation policies as well as strategic virtual water trade. Her regional focus is Africa. e-mail: Susanne.Neubert@die-gdi.de

Lena Partzsch is a researcher at Helmholtz Centre for Environmental Research (UFZ), Leipzig. She has been a researcher at Wuppertal Institute for Climate, Environment and Energy and received her PhD from the Environmental Policy Research Centre, Department of Political and Social Science, Free University of Berlin, analyzing the EU Water Initiative as a new global governance approach. e-mail: lena.partzsch@ufz.de

Danuta Sacher studied geography, sociology and communication sciences. Since 2004 she has been heading the Division "Policy and Campaigns" at the NGO "Bread for the World" in Stuttgart. Before, she was regional director for Central America of an US human rights organization (WOLA) and of the Project Counseling Service for Latin American Refugees. Ms Sacher is initiator of the "Bread for the World" campaign "Water is a Human Right". d.sacher@brot-fuer-die-welt.de

Ulrich Scheele is currently working with the Regional Planning and Environmental Research Group (ARSU Ltd.), a private research institute. Until 2006 Dr Scheele has been assistant professor for economic policy at University of Oldenburg. His main research interests include water policy and water management, infrastructure planning, privatization and regulation of network industries. e-mail: scheele@uni-oldenburg.de

Waltina Scheumann holds her Master in political science and a PhD in engineering. She has been assistant professor at the Chair in Environmental and Land Economics, University of Technology Berlin. Dr Scheumann later worked as senior researcher at the Helmholtz Centre for Environmental Research (UFZ), Leipzig, and is presently leading a research project at DIE. e-mail: scheumann@die-gdi.de

Maria Schnurr holds a university degree in social and communication sciences with a focus on international environment and development politics. She is currently preparing her PhD at the Future Research Department of Volkswagen AG. e-mail: hmschnurr@web.de

Imme Scholz is sociologist and received her PhD from Free University of Berlin. She is working as a senior researcher at DIE since 1992. At DIE, she has been heading the environment department since 2002. Between 1999 and 2002, Dr Scholz worked as policy advisor in the Pilot Programme for the Protection of Brazilian Rainforests (PPG7), Brazil. Her work focuses on climate change and sustainable development, global environmental governance and forestry issues. e-mail: Imme.Scholz@die-gdi.de

Insa Theesfeld is an agricultural economist, specialized in institutional economics and resource economics. She is research fellow in Humboldt University Berlin and Leibniz Institute of Agricultural Development in Central and Eastern Europe, Halle. Her focus is on natural resources management in Central and Eastern Europe, and, in particular, on the relationships of formal policy and local institutional changes in water resources management. e-mail: insa.theesfeld@agrar.hu-berlin.de

Michael Windfuhr is political scientist. Since 2006, he has been Human Rights Director for the Social Service Agency of the Protestant Church in Germany (Diakonisches Werk). Since 1988 he has been involved in the international human rights organisation FIAN (Food First Information and Action Network), being its Secretary General for several years. His publications focus on economic, social and cultural rights, particularly on the right to adequate food. e-mail: m.windfuhr@diakonie-human-rights.org

List of Abbreviations

ADB	Asian Development Bank
ADE	Agences de l'eau
ACIAPA	Associação Comercial, Industrial, Agropecuária e de Profissionais de Altamira
A-EUWI	EU Water Initiative for Africa
AMCOW	African Ministers' Council on Water
BCM	billion cubic metres
BGR	Bundesanstalt für Geowissenschaften und Rohstoffe (German Federal Institute for Geosciences and Natural Resources)
BGRM	Bureau de Recherches Géologiques et Minières (France)
BMZ	Bundesministerium für wirtschaftliche Zusammenarbeit und Entwicklung (German Federal Ministry for Economic Cooperation and Development)
BOT	Build-Operate-Transfer Schemes
BSP	Bulgarian Socialist Party
BWDB	Bangladesh Water Development Board
CATHALAC	Centro del Agua del Trópico Húmedo para América Latina y el Caribe (Water Centre for the Humid Tropics of Latin America and the Caribbean)
CBD	Convention on Biological Diversity
CEDARE	Centre for Environment and Development for the Arab Region and Europe
CESCR	UN-Committee on Economic, Social and Cultural Rights
CGG	Commission on Global Governance
CF	Constituição Federal (Federal Constitution, Brasil)
CFC	Chlorofluorocarbon
CIED	Centro de Investigación, Educación y Desarrolló (Centre for Research, Education and Development)
CIM	Centrum für internationale Migration und Entwicklung (German Centre for International Migration and Development)

CITES	Convention on International Trade in Endangered Species of Wild Fauna and Flora
CMWs	Community Mobilizing Workers
CODE	Committee on Development Effectiveness (World Bank)
COHERE	Centre on Housing Rights and Evictions
COIAB	Coordenação das Organizações Indígenas da Amazônia Brasileira (Organization of Indigenous Peoples of the Amazon)
CONAMA	Conselho Nacional do Meio Ambiente (National Environ mental Council, Brasil)
CRA	Comisión de Regulación de Agua Potable y Saneamiento Básico (Regulatory Commission for Water and Sanitation)
CREE	Commission de la régulation de l'électricité et de l'eau du Mali
CSD	Commission on Sustainable Development
CSO	Civil Society Organization
CWR	crop water requirement
DAC	Development Assistance Committee (OECD)
DANE	Departamento Administrativo Nacional de Estadística (National Statistics Department)
DDP	Dams and Development Project (under UNEP)
ded	Deutscher Entwicklungsdienst (German Development Service)
DfID	Department for International Development
DIE	Deutsches Institut für Entwicklungspolitik (German Development Institute)
DiFu	Deutsches Institut für Urbanistik (German Institute of Urban Affairs)
DNH	Direction National de l'Hydraulique (Mali)
DNP	Departamento Nacional de Planeación
DRC	Democratic Republic of Congo
DSI	Devlet Su Isleri (State Hydraulic Works, Turkey)
DVGW	Deutsche Vereinigung des Gas- und Wasserfaches e.V.
DWA	Deutsche Vereinigung für Wasserwirtschaft, Abwasser und Abfall e.V.
DWM	Department of Water Management (Kyrgyz Republic)
ECOSOC	Economic and Social Council (UN)
EBRD	European Bank for Reconstruction and Development
ECA	Export Credit Agency
ECGD	Export Credits Guarantee Department, United Kingdom
Ecosan	ecological sanitation
EDM	Energie du Mali
EECCA-EUWI	EU Water Initiative for Eastern Europe, Caucasus and Central Asia
EIA	Environmental Impact Assessment
EIB	European Investment Bank
EMWIS	Euro-Mediterranean Information System on the know-how in the Water sector

ENCOP	Environment and Conflict Project (Swiss research project)
ERG	Export Risk Guarantee Switzerland
ESC-Rights	Economic, Social and Cultural Rights
ESSD	Environmentally and Socially Sustainable Development
EU	European Union
EUWI	European Union Water Initiative
FAN	Freshwater Action Network
FAN-CA	Freshwater Action Network - Centro América
FAO	Food and Agricultural Organization of the UN
FAP	Flood Action Plan
FDI	Foreign Direct Investment
FIAN	Food First Information and Action Network
GAP RDA	Güney Dogu Anadolu Projesi (Regional Development Administration)
GAP	Güney Dogu Anadolu Projesi (Southeastern Anatolia Project)
GARWSP	General Authority for Water and Sanitation Projects (Yemen)
GATS	General Agreement on Trade in Services
GDI	German Development Institute
GDP	Gross Domestic Product
GEF	Global Environment Facility
GKKE	Gemeinsame Konferenz Kirche und Entwicklung
GIS	Geographic Information System
GTZ	Deutsche Gesellschaft für technische Zusammenarbeit
GWh	Gigawatt hour
GWP	Global Water Partnership
GWP-Med	GWP-Mediterranean
HDI	Human Development Index
HDR	Human Development Report
HEPP	Hydroelectric Power Plant
HIC	Habitat International Coalition
HIV	Human Immunodeficiency Virus
IAP	Infrastructure Action Plan
IBAMA	Instituto Brasileiro do Meio Ambiente e dos Recursos Naturais Renováveis (Federal Environmental Authority, Brasil)
IBRD	International Bank for Reconstruction and Development (World Bank)
ICESCR	International Covenant on Economic, Social and Cultural Rights
ICG	International Crisis Group
ICID	International Commission on Irrigation and Drainage
ICJ	International Commission of Jurists
ICLEI	International Council for Local Environmental Initiatives
ICOLD	International Commission on Large Dams
IDA	International Development Association
IDV	Individuality
IFC	International Finance Corporation

IFI	International Finance Institution
IGOs	Intergovernmental Organizations
IHA	International Hydropower Association
IMT	Irrigation Management Transfer
IME	Mediterranean Water Institute
INSFOPAL	Instituto Nacional de Fomento Municipal (National Institute for District Promotion)
InWEnt	Internationale Weiterbildung und Entwicklung (Capacity Building International, Germany)
IPAM	Instituto de Pesquisa Ambiental da Amazônia
IPCC	Intergovernmental Panel on Climate Change
IRD	Institut de recherche pour le développement
IRN	International Rivers Network
ISA	Instituto Socioambiental (Brasil)
ISC	Irrigation System Company
ISF	Irrigation Service Fees
IUCN	International Union of the Conservation of Nature (The World Conservation Union)
IWRM	Integrated Water Resource Management
JBIC	Japan Bank for International Cooperation
JMP	Joint Monitoring Programme (WHO & UNICEF)
JPOI	Johannesburg Plan of Implementation
KfW	KfW Development Bank (German financial cooperation)
LA-EUWI	EU Water Initiative for Latin America
LANBO/RELOC	Latin America Network of Basin Organisations/Red Latinoamericana de Organizaciones de Cuenca
LCs	Local Water Supply and Sanitation Corporations
LDCs	Least Developed Countries
LMP	Living Mekong Programme
LOUAL	Law for Ownership and Use of Agricultural Land
LWSCA	Local Water & Sanitation Corporation Aden Governorate
MAI	Ministry of Agriculture and Irrigation (Yemen)
MAS	Masculinity
MAVDT	Ministerio de Ambiente, Vivienda y Desarrollo Territorial (Columbia)
MAWMPI	Ministry of Agriculture, Water Management and Processing Industry (Kyrgyz Republic)
MDB	Multilateral Development Bank
MED-EUWI	EU Water Initiative for the Mediterranean
MedWet	Mediterranean Wetlands Initiative
MDG	Millennium Development Goal
MDTX	Movimento em Defesa da Transamazônica e Xingu
MENA	Middle East and North Africa
MENBO	Mediterranean Network of Basin Organisations
MMA	Ministério do Meio Ambiente (Brasil)

MME	Ministério de Minas e Energia (Brasil)
MRF	Movement for Rights and Freedom (Bulgaria)
MStF	Multi-Stakeholder Forum
MW	Megawatt
MWE	Ministry of Water and Environment (Yemen)
NEPAD	New Partnership for Africa's Development
NGO	Non-Governmental Organization
NICs	Newly Industrializing Countries
NWRA	National Water Resources Authority (Yemen)
NWSA	National Water and Sanitation Authority (Yemen)
NWSSIP	National Water Sector Strategy and Investment Program (Yemen)
ODA	Official Development Assistance
ODI	Overseas Development Institute
OECD	Organisation for Economic Cooperation and Development
OECD-DAC	OECD Development Assistance Committee
OED	Operations Evaluation Department (World Bank)
OHCHR	Office of the High Commission of Human Rights
O&M	Operation and Maintenance
OPEC	Organization of the Petroleum Exporting Countries
OPS	Organización Panamericana de la Salud (Pan American Health Organization)
PDRY	People's Democratic Republic of Yemen
PDI	Power Distance Index
PPP	Public Private Partnership
PRS	Poverty Reduction Strategy
PRSP	Poverty Reduction Strategy Paper
Pseau	Programme Solidarité Eau
PSP	Private Sector Participation
PT	Partido dos Trabalhadores (Workers' Party, Brasil)
PTP	Provincial Towns Programme (Yemen)
RAP	Resettlement Action Plan
ROR	Rate of Return
SADC	Southern African Development Community
SAGE	Schémas d'aménagement et de gestion des eaux
SAM	Sustainable Asset Management Group
SECTAM	Secretaria Executiva de Ciência, Tecnologia e Meio Ambiente (Environmental Authority of Pará state, Brasil)
SEF	Stiftung für Entwicklung und Frieden (Bonn)
SISNAMA	Sistema Nacional do Meio Ambiente (National Environmental System, Brasil)
SMEs	Small and Medium Enterprises
SNM	Simeon II National Movement (Bulgaria)
SSPD	Superintendencia de Servicios Públicos Domiciliarios (Superintendency for Public Utility Services)

SWAp	Sector Wide Approach
SWAT	World Bank Sanitation & Wastewater Advisory Team
TA	Technical Assistance
ToR	Terms of Reference
UAI	Uncertainty Avoidance
UBS	Union Bank of Switzerland
UDF	Union of Democratic Forces (Bulgaria)
UDHR	Universal Declaration of Human Rights
UFZ	Umweltforschungszentrum Leipzig-Halle
UK	United Kingdom
UNCBD	UN Convention on Biodiversity
UNCCD	UN Convention to Combat Desertification
UNCED	United Nations Conference on Environment and Development
UNDP	United Nations Development Programme
UNECE	United Nation Economic Commission for Europe
UNEP	United Nations Environment Programme
UNESCAP	United Nations Economic and Social Commission for Asia and the Pacific
UNESCO	United Nations Educational, Scientific and Cultural Organization
UNFCCC	United Nations Framework Convention on Climate Change
UFPA	Universidade Federal do Pará (Federal University of Pará, Brasil)
UN-HABITAT	United Nations Human Settlements Programme
UNICEF	United Nations International Children's Fund
UNSGAB	United Nations Secretary General's Advisory Board on Water and Sanitation
US AID	United States Agency for International Development
US	United States (of America)
USBR	United States Bureau of Reclamation
USP	Utility Support Programme
UWV	Unit Water Value
VWC	Virtual Water Content
WAFED	Water and Energy Users' Federation (of Nepal)
WAJ	Water Authority of Jordan
WASH	Water, Sanitation and Hygiene campaign of the WSSCC
WBGU	Wissenschaftlicher Beirat der Bundesregierung Globale Umweltveränderungen
WCD	World Commission on Dams
WDM	World Development Movement
WEED	World Economy, Ecology & Development
WFD	Water Framework Directive
WHO	World Health Organization
WPDC	Water Power & Dam Construction
WRI	World Resources Institute

WRM	Water Resources Management
WRMG	Water Resources Management Group
WRSS	Water Resources Sector Strategy
WSS	Water Supply and Sanitation
WSSCC	Water Supply and Sanitation Collaborative Council
WSSD	World Summit on Sustainable Development
WSU	Water User Association Support Units
WUA	Water Users Association
WUO	Water User Organization
WWAP	World Water Assessment Programme
WWC	World Water Council
WWF	World Wide Fund For Nature
YAR	Yemen Arab Republic
YR	Yemeni Riyal
ZEF	Zentrum für Entwicklungsforschung (Center for Development Research, Bonn)

Introduction

This statement in the most recent Human Development Report articulates a conviction that has increasingly gained ground in the water community over recent years: the key challenge in the water sector is not a lack of water, knowledge, financial resources or technology. In general, it is the political sphere that determines whether or not water problems are solved, whether or not people have access to drinking water, irrigation water and sanitation, whether our natural resource base is developed sustainable or overexploited, and whether new challenges for the water sector – such as adaptation to climate change – will be tackled or not. Politics (the process of decision-making of groups of people, involving the authoritative allocation of e.g. resources), the actors, their interests and interactions determine whether progress is made or hindered. The outcome of water politics is then reflected in water policies, the substantive outcome of the political interplay in terms of regulations, action programs or spending priorities of the various public or private entities concerned.

The importance of the political sphere for understanding and solving water sector problems is the basic rationale of this book. It is not the first time that the Dialogues on Water have touched upon water politics and policies. But these Dialogues, unlike earlier ones, focus on the political processes of policy formulation and the strategic behavior of the actors involved. The chapters assembled in this book analyze debates and investigate water politics and policies at the international, national and local level, each considering different aspects or different elements of policy formulation and implementation processes from a variety of disciplinary backgrounds. They examine policies that result from power plays of state and non-state actors alike over water resources and modalities of water service delivery and as a function of their respective means of bringing influence to bear. In line with the general focus of Dialogues on Water, specific attention is devoted to the implications for development cooperation.

As regards epistemic approaches, the book allows for a variety of perspectives: some chapters follow a constructivist line, elaborating on how global norms on water-related issues evolve and how international debates influence them. These chapters consider whether and how global norms evolve, become effective and are adhered to at the national level and discuss the role played and means used by the major actors – the World Bank, the European Union, bilateral donors, national elites and civil society groups – in influencing ideas and concepts and their translation into policies. Other chapters adopt a political economy or public choice perspective, adhering to a rational actor approach. They explain policies more in terms of the self-interest and power of the actors involved and their means of exercising influence.

The book consists of the chapters given at the Fifth Dialogues on Water, which took place at the German Development Institute (DIE) in Bonn in October 2005. Since 2001 the Dialogues have been held bi-annually as a joint initiative of the senior researchers Susanne Neubert and Waltina Scheumann in collaboration with partners from the BMZ and/or such German development cooperation implementing agencies as GTZ. On this occasion, the editors were joined by Martin Kipping, the then water sector desk officer at the Federal Ministry for Economic Cooperation and Development (BMZ). The Dialogues on Water serve to enhance communication among water researchers and water practitioners in the development cooperation field, the aim being to strengthen the link between the academic and practitioners. In this way, research questions can assume a more practice-oriented form and – conversely – scientific research results can be absorbed into policy formulation and practical development cooperation.

The Structure of the Book

The first chapter acts as a conceptual introduction. In his article *Water policy – water politics. Social engineering and strategic action in water sector reform* **Peter P. Mollinga** maps the "politics of water" as a field of research. He argues that water control should be conceived as a politically contested arena. Two regulative principles are relevant to the mapping of that contest: first, distinguishing different levels of water politics as relatively autonomous areas of interaction, and second, identifying issue-networks that encompass processes of contest within or across levels. Water politics is divided into four different areas: the everyday politics of water control, the politics of national water policy, inter-state hydropolitics and the global politics of water. These four areas can be distinguished by their different space and time scales, their different combinations of actors, the different problems they face, their different modes of contest and the different sets of institutional arrangements in which they are located. Some of the most interesting and important questions in water policy and politics concern the links between and across domains with respect to certain issues or questions. Among the plethora of issue-networks that constitute concrete water politics and policy practices, the chapter focuses on two main "sticking points" in present-day water policy reform processes: first, the internalization of "new concerns", notably the environment and human development, into the professional practice of mainstream water sector organizations and second, the transformation of state-centered water resource policy processes into society-centered policy processes. The chapter contains a critique of the dominant social-engineering approaches to institutional transformation and argues that, unless a confident political strategic action approach to institutional transformation is taken, deadlock in water-sector reforms may continue for some time.

Part I: Global Norms and National Policies

Part I of the book focuses on an issue that illustrates the actual impact of ideas, debates and conceptual development: the on-going debate on large dams. As the debates and disagreements over large dams continued, major international actors set up the World Commission on Dams (WCD) in 1998, which undertook a thorough analysis of the positive and negative impacts of large dams and developed recommendations for improving planning procedures. The WCD report published in 2000 met with a mixed response. Most stakeholders agreed with the values and principles underlying the WCD recommendations, while others, such as financial institutions and industrial associations, criticized the planning guidelines put forward by the WCD as too far-reaching and imprecise. That said, the WCD has had a remarkable impact.

Michael Fink and **Anne Cramer** report on experience five years after the WCD published its recommendations. They argue that the recommendations had to remain fairly abstract to be universally applicable, but need to be operationalized and translated into the specific context of a region, country or dam project before they can be used to improve the planning of future dams or the management of existing ones. The challenge of adapting the WCD recommendations for practical use has been tackled by various institutions and stakeholders in a wide range of thematic and geographical contexts. After more than five years, it has become possible to take stock of the experience thus gained. The chapter examines two practical attempts to implement some of the WCD recommendations. It concentrates on the lessons learned from both positive and negative outcomes and identifies critical factors responsible for success or failure. The "Constructive Dialogue on Dams and Development" in Nepal and the World Wildlife Fund initiative on "Environment criteria for hydropower development in the Greater Mekong Region" serve as examples of how to address the translation of the WCD recommendations into practice and how to appreciate their significance and applicability today.

Waltina Scheumann considers how the global norms and recommendations developed by the World Commission on Dams influence decision-makers and decision-making. The development of norms for internationally highly controversial large dams differs from the paths traditionally followed in international policy formation, as in the establishment of international environmental regimes, where nation states are the decisive actors: the WCD members were selected for their personal abilities and to reflect regional diversity, expertise and stakeholder perspectives and do not represent states. This process of global norm development has been welcomed by many as a prototypical example of how trisectoral networks (including the governmental and private sector and civil society groups) can help overcome stalemate in highly conflict-ridden policy arenas. On the other hand, independent assessments and analyses have pointed to the fact that the consensus achieved within the Commission is not reflected in a broader stakeholder consensus. A case study from Turkey (the Ilisu Dam on the Tigris River) traces how respective pressure was exerted on the Turkish government, one of the governments that sharply rejected the WCD's guidelines.

Imme Scholz is interested in how global environmental governance influences national water policies. Even without a global convention on the right to water, national water policies and politics are already heavily influenced by global environmental governance. These processes are based on simultaneous interventions by multiple actors at the local, national and global levels. The existence of other legally binding global environmental regimes (e.g. the conventions on climate change, biodiversity and measures to combat desertification), global concepts (e.g. IWRM) and approaches supported by the UN forges many links between national water policies on the one hand and global policies seeking environmental sustainability on the other. Scholz first asks whether it is necessary to adopt a specific global convention for each environmental problem area or whether the potential links between existing conventions and water policies are strong enough to ensure greater sustainability in water management at national and local level. She goes on to ask how life can be breathed into global regimes or conventions by public and civil actors to make them relevant in practice. Global regimes can be enhanced if national actors integrate them into their cognitive, administrative and political structures and strategies. In support of this argument, the chapter presents a case study on the policy processes and conflicts associated with the construction of a new dam and hydroelectric plant in Brazil's Eastern Amazon (Belo Monte), demonstrating how national water policies and local water-related politics are already permeated by global governance elements.

Maria Schnurr takes a normative approach to the concept of global water governance. She argues that the concept of governance can address the dynamics, complexity and interdependencies of current water-related problems. The wide range of institutions, programs and action plans constitutes a barrier to coherent, efficient action, resulting in ever increasing implementation gaps in water politics. Applying the principles of governance – cooperation, coordination, common values, integration of decision-making levels and subject matters – rather than deepening hierarchical structures could lead to a "Global Water Governance" architecture, which might guarantee the more efficient use of human and financial resources, thus helping to close implementation gaps, especially in the case of the Millennium Development Goals (MDGs). According to Schnurr, the change to multi-level, polycentric global water governance would entail structural reforms, foremost among them the strengthening of UN-Water and the establishment of binding global rules through the addition to existing water conventions of provisions on water supply and sanitation, for example, accompanied by corresponding measures at the national level, especially regarding accountability and good governance. While the path to a "Global Water Governance" architecture is not without its obstacles, research could provide programmatic support for progress in this direction.

Part II: Critical Debates Revisited

This part of the book undertakes a critical reassessment of their practical implications.

Susanne Neubert analyses the vigorous debate on virtual water trade within the water community with the aim of identifying genuine controversies and points of agreement. To differentiate and carry forward the debate and to identify the need for further research, the most relevant arguments are examined and reassessed from a multidisciplinary perspective. Against the background of the water crisis, strategic virtual water trade has the potential to help save enormous amounts of water where it is scarce. However, the specific implications of virtual water trade, such as the likely adverse social consequences for the virtual-water- importing countries, also need to be considered. The chapter concludes that the debate on virtual water trade is not yet sufficiently focused on the fact that an increase in water productivity is not enough in itself to protect water resources against overuse. There is, rather, a need for intelligent water management strategies in which multiple aspects, e.g. the opportunity costs of alternative uses and ecological sustainability, are considered against the background of the spatial-temporal nature of water resources. Provided this approach is adopted, strategic virtual water trade has the potential to become an element of IWRM strategies, particularly in water-scarce middle-income countries.

Danuta Sacher and **Michael Windfuhr** analyze the debate on "water as a human right" and its implications for development assistance. The human rights approach is increasingly attracting attention as not only an ethical, but also a legal framework for the prioritization of water and sanitation as well as an expanded set of tools for policy-makers and civil society groups. The chapter summarizes the current state of the debate on "water as a human right" and discusses policies and instruments for attaining the right to water. The authors also analyze current initiatives to implement the rights-based approach in development and national water policies in the wake of changing development paradigms in the last three decades.

In his chapter, **Manfred Matz** argues that water management is closely associated with cultural aspects which are usually neglected in the IWRM concept. Culture comprises aspects of human interaction, social organization and adaptation. Matz endorses Geert Hofstede's conclusion that cultural differences translate into political reality. Culture should therefore be considered a significant part of a country's legal, administrative and political system. Consequently, the chapter questions the universal validity of some elements of IWRM, giving as an example differences in water governance in France and Germany. Cultural differences have given birth to country-specific systems: while France has embarked on a largely participatory and monetary (dis)incentive-based approach, Germany's system is almost entirely based on approval or disapproval. This difference of approach to water management also tends to be reflected in the development advisory services in the water sector that France and Germany provide for their respective partner countries, leading to culturally biased "blue-print" approaches. The two developing countries Mali and Jordan are given as examples of the difficulties encountered when approaches that do not reflect cultural-political realities of the partner countries are adopted. The chapter therefore calls for greater cultural sensitivity on the part of water experts.

Part III: Politics of Water Supply and Sanitation

This part of the book is devoted to the water sub-sector that generally dominates the water debate: the provision of water supply and sanitation (WSS).

Ulrich Scheele and **Thomas Kluge** suggest that more private-sector participation (PSP) in the provision of WSS is needed if the MDGs are to be achieved. More PSP should therefore become or remain a key element of donors' strategies. However, the privatization of WSS has given rise to a very emotional and heated debate between fundamentally opposed positions: water as a human right vs. water as a commercial good. The chapter argues that water supply and sanitation are classic cases of monopoly situations. Successful PSP therefore requires strong public regulation. Developing countries, however, often lack the necessary institutional capacity or appropriate governance structures for regulation. PSP projects to date have therefore had rather mixed results, with project failures often largely due to a lack of regulation. As a result, PSP strategies have increasingly come under pressure, which has led to waning interest on the part of large international corporations in developing countries' water and sanitation sectors. However, new corporations – including some in developing countries themselves – are entering the market, and promising new opportunities for public–private partnerships are emerging.

Franz-Josef Batz argues that a substantial increase of investment in WSS is required if the MDGs are to be achieved. Funds for investment need to be mobilized from all sources. However, mobilizing local resources is the key to sustainable financing. Tariffs should be one of the sources of finance. In sub-Saharan Africa, however, tariffs do not as a rule cover operation and maintenance costs, let alone total costs. Utility performance and regulation are also generally weak. This situation leads to inadequate investment, poor coverage and poor service quality, without providing incentives for private investment in the water sector. On the other hand, public investment does not fill the gap, either. African governments allocate on average less than 1 percent of their annual budgets to the water sector. The reasons for this include low state revenues due to ineffective tax systems and the low priority given to WSS. Despite this, local financial and capital markets should be in a position to channel funds into the water sector on a demand-driven basis. However, financial systems in sub-Saharan Africa are extremely weak. In most African countries capital markets are inadequately developed or do not exist at all. Batz argues that sectoral reforms are imperative if the financial gap in WSS is to be closed. They include the reform of the water sector as such and also of the financial markets and public administration. Sectoral experts and policy-makers thus need to broaden their view and to work across sectors if sustainable financing of WSS is to be achieved. Development cooperation can assist in this task, but will never be sufficient on its own.

Matthias Krause claims that adequate domestic WSS services are closely related to the availability of adequate governance structures. During the last decade, the academic and political debate on WSS reform has focused on the polemic issue of private vs. public ownership of service providers. Krause argues that this focus is too narrow to explain successes or failures in the delivery of WSS. The chapter therefore shifts the emphasis to the importance of governance for service provision. Achieving broad access to good-quality and low-cost services presupposes a complex mix of poverty- and efficiency-oriented WSS policies. Political and administrative governance structures should therefore have a major influence on the poverty orientation and efficiency of service delivery. The empirical part of the chapter includes a case study on Colombia. It explores the hypothesis that weaknesses in WSS policies can be attributed to weaknesses in governance, focusing on (i) the institutional articulation of the roles of WSS policy-making, regulation and service delivery and on (ii) the poverty-orientation of subsidization policy. With respect to the former, empirical evidence shows that the reluctance of politicians to grant full independence to regulators and public service providers has hampered improvements in service delivery efficiency. This reluctance can be plausibly ascribed to the fact that politicians would forfeit control of tariff-setting and fund allocation and so lose an important means of winning elections and maintaining clientelistic networks. As far as the second issue is concerned, there is evidence of discrimination against the rural poor, who have the greatest need of safe WSS services, and this is due to their low degree of organization and their lack of the political voice that is necessary if subsidization policy is to be reformed and subsidies focused on needy households.

Part IV: Power Plays in Irrigation Reforms

This part of the book concerns the sub-sector that consumes most water: agriculture. Globally, around 70 percent and in many developing countries more than 90 percent of all water resources extracted from natural bodies of water is consumed in agriculture. Agriculture is thus the sector where political struggles over quantitative water allocation are at their most virulent.

Insa Theesfeld analyses the extent to which state actors and international donors intervene in Bulgaria's irrigation sector by enforcing legislation and implementing development projects. The process of designing national irrigation sector policy reform and, in particular, its implementation in post-communist Bulgaria are regarded as having been shaped by the various holders of political power. The formal devolution-oriented reform of Bulgaria's irrigation sector is compared to the de facto concentration of power in the hands of the state authorities. This is done by describing the official objectives and the actual introduction of four subsequent innovations in Bulgaria's irrigation sector policy: the World Bank project to set up water-user organizations, the Bulgarian Water Law, the Water User Association Act to facilitate the organization of water-user associations and the latest bills amending the Water User Association Act. Examples confirm that no more than pseudo-devolution has taken place: state authorities have actually increased their decision-making power by determining the way regulations are implemented. In this way, individual actors are able to extract private short-term rents from the system. Aspects of public choice theory of institutional change help to explain the stages of Bulgaria's water sector reform as a function of different periods in the holding of political power. The sequence in this respect shows how political actors develop and implement policies beneficial to their own clientele, as they vie for the votes of the rural electorate. With this analysis, Theesfeld's chapter provides evidence of both political and economic strategies for undermining reforms and concludes that the commitment of political leaders is a major determinant of effective devolution in natural resources management.

Water sector reforms in the Kyrgyz Republic are the subject of **Elke Herrfahrdt-Pähle**'s chapter, which focuses specifically on irrigation. For the Kyrgyz Republic, which has been committed to the reform of its water governance since independence in 1991, IWRM has clearly been the leading concept underlying reform efforts since 2001. Despite relatively good conditions for change, the water sector is lagging behind in the implementation of reforms. The analysis of the various actors involved reveals a discrepancy between the actual demand for change and the donor-driven design of reforms by discussing both the new Water Code and the resistance of many actors to the changes and organizational restructuring it entails. The author argues that the reform program has been induced largely by the government's financial constraints, and that the government responded to such conceptual innovations as IWRM solely to gain access to external funds. The reform process, which paid little attention to informal rules, led to the co-existence of incompatible new and old water institutions and new and old formal water management organizations.

Part V: Development Cooperation

While most of the chapters in this book touch upon aspects of development cooperation, the last part addresses this field in even greater depth:

Volkmar Hartje analyses the World Bank's role in water sector policy reforms. The World Bank is certainly the most powerful donor agency to endorse the concept of Integrated Water Resources Management (IWRM) and provides substantial financial support for its application. It has undertaken two major efforts – in the 1990s and between 2001 and 2005 – to translate the concept into a set of operational guidelines. On the basis of these new management policies, the Bank hoped to guide client countries' water policies towards IWRM or at least to influence them systematically in this respect. Hartje assesses the coherence and effectiveness of the Bank's use of IWRM as a water policy paradigm, the process and the effectiveness of translating the concept into practical Bank policy as well as the process and effects of and the limits to its translation into policy reforms in client countries. The chapter employs an institutional economics analytical framework to analyze the enhancing and restricting factors in the diffusion process.

Martin Kipping contends that water is one of the keys to poverty reduction and the achievement of the Millennium Development Goals (MDGs), particularly those concerning health and environmental sustainability. However, 1.1 billion people continue to lack access to safe drinking water, and 2.6 billion still have no access to sanitation facilities. While the MDG for drinking water is likely to be achieved, the sanitation target will be missed unless a much greater effort is made in this area. Germany is the third largest bilateral donor in the water sector, the regions on which it focuses being the Middle East and Africa. The German Federal Ministry for Economic Cooperation and Development (BMZ) and its implementing agencies are steadily working on further improvements to their approaches and strategies in the context of the global effort to increase aid effectiveness. Six challenges deserve particular attention in the water sector related activities of German development cooperation: (1) mainstreaming the concept of Integrated Water Resources Management; (2) addressing water management in agriculture; (3) taking questions of water governance and water politics seriously; (4) prioritizing waste water management, sanitation and hygiene; (5) ensuring the sustainability of large-scale hydro-infrastructure; and (6) increasing funds for development cooperation in the water sector through innovative funding mechanisms. As it takes these challenges seriously, German development cooperation is committed to further improving its water sector activities. The probable reorganization of BMZ's implementing agencies would make a significant contribution to this effort.

Stefan Lindemann discusses the challenges of development assistance to water service delivery in fragile states and particularly Germany's involvement in Yemen. As fragile states are either unable or unwilling to provide water services for the majority of their people, especially the poor, they now account for about a third of the people in the world who do not have sustainable access to safe drinking water. Western donors are increasingly recognizing the specific challenge of inadequate (water) service delivery in fragile environments and seeking guidance on how to deliver services in fragile states more effectively. The Yemeni case comes closest to a fragility scenario of (enduring) recovery in which a relatively stable government is in place and basic state functions are slowly being established. Here, the water sector is of crucial importance since Yemen is among the countries in the world with the least water: while it has recently made important progress in the institutional and organizational consolidation of the water sector, its performance in terms of water policy development and implementation is still weak and overall structures remain largely unsustainable. German development cooperation has taken up this challenge by devising a "multi-level strategy" that successfully combines support for sectoral reform at the macro and meso level with the creation of decentralized and commercialized service utilities at the micro level. Specific "lessons learned" from German donor involvement in the Yemeni water sector include the need for context sensitivity, state-building through intervention at different levels, dialogue and participation, conflict prevention, alignment with local priorities and donor coordination.

Lena Partzsch asks whether the EU Water Initiative (EUWI) is an innovative form of development aid, with partners from the private sector and from civil society organizations involved. She examines how non-state actors, and especially women, have been considered in EU water policies in the past. Special emphasis is placed on the EU Water Framework Directive and on the Communication from the European Commission on water strategies for developing countries. Against this background, it is possible to gauge the extent to which the EUWI can be considered innovative. There is evidence to show that, in fact, the partnership builds on both internal and external EU water strategies, but goes further than both in terms of stakeholder participation. The second part of her chapter elaborates on theoretical assumptions underlying the new partnership approach and develops an analytical framework to enable the effectiveness and legitimacy of including non-state actors in the EUWI to be examined. Partzsch argues that, on the one hand, the EUWI adopts an innovative approach to the coordination of state actors and "new" private-sector and civil-society actors in different areas of policy. On the other hand, she believes the partnership must be considered non-innovative because it is de facto dominated by European actors, actors in the partner countries being underrepresented.

To sum up, this book sets out a multitude of approaches and perspectives for the analysis of the political drivers and implications of water-related decisions. The editors have avoided forcing the authors to accept a single concept of what the "politics of water" might be or how they should be analyzed, let alone adopt a common normative stance. The intention of this book is not to narrow down the political analysis of water sector issues prematurely, but rather to open up and encourage new lines of thought. The editors share the conviction that the book will play its intended role if it leads to an even greater number of more elaborate and conceptually innovative contributions that consider "the political" as the key source of challenges and solutions in the water sector.

Bonn/Kabul
September 2007

Waltina Scheumann
Susanne Neubert
Martin Kipping

Water Policy – Water Politics

Social Engineering and Strategic Action in Water Sector Reform

Peter P. Mollinga

Abstract The contribution maps the 'politics of water' as a field of research. Water control is understood as *politically contested resource use*. Contestation is mapped along two axes: (1) different levels or domains of water politics; (2) issue-networks encompassing processes of contestation within or across levels and domains. The four domains are: the everyday politics of water control, the politics of national water policy, inter-state hydropolitics, and the global politics of water. These have different space and time scales, are populated by different configurations of main actors, have different types of issues as their subject matter, involve different modes of contestation and take place within different sets of institutional arrangements. Some of the most important questions in water policy and water politics involve the interlinkages across domains, around certain issues. Among the plethora of issue-networks of concrete water politics policy, the chapter focuses on two main 'sticking points' in present-day water policy reform processes. (1) The internalization of 'new concerns', notably environment and human development, into the mainstream water sector organizations' professional practice, and (2) the transformation of state-centered water resources policy processes into society-centered policy processes. The chapter provides a critique of the dominant social engineering approaches to institutional transformation, and argues that unless a self-consciously political strategic action approach to institutional transformation is taken, the deadlock in water sector reform may continue for some time.

1 Introduction

The objective of this chapter is to map the 'politics of water' as a field of research. Such mapping logically has two parts. The first is an explanation of what is meant by politics and what could be the overall conceptual approach for analyzing the politics

Peter P. Mollinga
e-mail: pmollinga@hotmail.com

W. Scheumann et al. (eds.), *Water Politics and Development Cooperation*,
doi: 10.1007/978-3-540-76707-7_1, © Springer-Verlag Berlin Heidelberg 2008

of water – the formal part of the mapping. The second part of the mapping is the substantive dimension: what are the concrete issues and questions around which research on water politics could be organised? While the first part can have a single answer, the approach one prefers to take, the second is an in principle endless list of relevant and interesting topics for concrete investigation, each with their own specific conceptual and methodological demands. Selection within that list follows primarily, at least in this chapter, from an assessment of what are pertinent policy questions in (a certain part of) the real world of water resources management.[1] This chapter focuses on the issue of water sector reform in developing and transition countries, particularly the reform of the public organisations that manage agricultural water. Agriculture is the dominant form of water use in most developing and transition countries, and changes in water resources management towards a more 'integrated' approach require quite fundamental changes in how agricultural water management is done. The need for a more integrated approach to water resources management is taken as the context for the argumentation in this chapter, though 'integration' is by no means a clear, single 'thing', but a contested concept.[2]

Given this demarcation, I summarise the two main concerns and research foci regarding the politics of water that this chapter wants to elaborate as follows.

1. The internalization of 'new concerns', notably environment and human development, into the mainstream water sector organisations' professional practice.
2. The transformation of state-centered water resources policy processes into society-centered policy processes.[3]

The suggestion is that these are two crucial questions for those interested in furthering reform in the water sector, because they are 'sticking points': issues that hold up the reform process, where there is a need for new analysis to inform strategic action.

The mapping exercise is organised in three sections. The first presents a framework for water politics analysis (Sect. 3), followed by two sections that discuss the two main foci mentioned above (Sects. 4 and 5). Preceding these three sections is

[1] The alternative approach for defining concrete research foci would be from an academic starting point: the pursuit of certain theoretical or methodological interests with the water resources domain as the area of enquiry. Water resources *management* is the generic term used in this chapter as the broadest reference to all activities related to water governance, management (in the narrow sense), use, finance, and other aspects.

[2] Policy and research statements arguing the case for integration abound. Cf. for instance GWP (2000), Rogers and Hall (2003), and the websites of the International Water Management Institute, www.iwmi.org, and the Comprehensive Assessment of Water Management in Agriculture, http://www.iwmi.cgiar.org/Assessment/; http://www.bradford.ac.uk/acad/bcid/seminar/water/ for the 'Challenging the global water consensus' conference/seminar series and Mollinga et al (2006) for critical engagements. In the European Union water policy context 'adaptive water management' is a concept that seems to be gaining currency, see for instance www.newater.info

[3] In a recent volume on the politics of irrigation reform, Mollinga and Bolding (2004, 302–306) suggest the following three research fields as in need of more research attention: (1) the resilience of irrigation bureaucracies, (2) the role of international development funding agencies, and (3) the capture of irrigation reform policy in implementation. The two foci of this chapter are formulated at a higher level of abstraction, and include the three fields just mentioned, and more.

an explanation of why 'politics' should be given special attention in the first place (Sect. 2). The main argument of the chapter is summarised in the concluding Sect. 6.

2 Why Emphasize *Politics*?

We still, in 2007, live in an era in which it is necessary to explain that water resources management is an inherently political process. Ten years ago politics and the political were anathema in most circles of the water policy discourse.[4] As discussed below, the social engineering paradigm reigned largely unquestioned. The rise of the theme of (good) governance brought in politics into the mainstream development discourse through the backdoor. When talking governance, good or bad, and associated ideas like accountability, transparency and legitimacy, it is rather difficult not to acknowledge that such processes and relations have political dimensions, and to stay confined within an instrumentalist perspective. As governance is about the exercise of authority and allocation of rights and resources, the issue of social power, generally excluded from social engineering rationales, becomes difficult to avoid.[5] Nevertheless, it remains very difficult for those holding positions

[4] This statement derives from participation in policy related discussions on water management by the author since the early 1990s. However, in past years the politics word seems to have acquired some acceptability. On 25 February 2004 a double session on 'Driving the Political Economy of Reform' took place as part of the World Bank Water Week, the yearly gathering of World Bank staff and partners in Washington, DC (see http://www.worldbank.org/watsan/waterweek2004, session 14). This was the first time that the political dimensions of water/irrigation reform processes were given such explicit space in the Water Week event. On 26 and 27 February 2004 the World Water Council (WWC) launched a 'Water and Politics' initiative by organizing a workshop in Marseille – the site of its headquarters (cf. http://www.worldwatercouncil.org/water_politics/index.html). In the corporate sector, the RWE Thames Water company emphasizes the importance of water politics on its website (http://www.thames-water.com/TW/division/en_gb/content/General/General_000101.jsp?SECT=General_000101). Cf. Merrey et al. (2006) for further discussion.

[5] Cf. Hoebink (2006) for an interesting discussion of how the concept of (good) governance was taken up in West European bilateral development assistance programs in recent years. In the global water discourse, the moment of 'closure' for establishing governance as a core theme seems to have been the Bonn Freshwater Conference in 2001 and the Johannesburg Summit of Sustainable Development in 2002. Much quoted is the phrase 'The world water crisis is a crisis of governance – not one of scarcity' from the *No Water No Future* speech at the Summit by the Prince of Orange. Jenkins (2001) argues that 'governance' as used in the mainstream international development discourse of the international development funding agencies tends to become a 'technical' issue: it, as it were, depoliticizes the understanding of politics. I do agree with much of Jenkins' criticism of the global (good) governance agenda, but from the perspective of water sector reform, I find the acceptance of the importance of the issue of governance, after the acceptance of 'management' in the 1970s to improve upon 'operation', a step forward. Governance in this context refers to the allocation of rights (rights to water and technology, decision-making rights) and resources (water itself, but also maintenance and investment funds for instance), and thus brings in issues of interest groups and social power more forcefully than the notion of management, which has been the leading concept for the past decades in water policy reform discussions, tends to do. Hence my phrasing of 'politics through the backdoor'. There are other discursive trajectories leading to acknowledgement of the relations of social power. The most notable one is the participation discourse that often

of authority, to acknowledge that the social fabric can not be understood and transformed without questioning existing social relations of power, and that, therefore, an explicit, scientific and professional discussion on social power and politics is desirable at all levels of decision making. As contributions like those of Ferguson and Harriss have argued, there may be compelling reasons for governments and other actors to depoliticise debates on development, reasons located in the way instrumental reason, which actively claims to exclude 'politics', assists in reproducing state power and legitimacy, as well as the reproduction of development assistance programmes (Ferguson 1994; Harriss 2001; also see Scott 1997).

Another indication that 'politics' is not the most likely topic for detailed investigation within water resources management studies, is that despite the strong emphasis on 'rent-seeking' in the neo-liberal development discourse of the 1990s (cf. Repetto 1986) there is virtually no follow-up research on Wade's seminal paper on 'the system of administrative and political corruption' (Wade 1982). In most seminars and workshops on irrigation reform that the author has attended in India and elsewhere in the past 15 years, the issue was not even mentioned – except sometimes in the corridors.[6] Though there are domains of water politics that are well researched (see below), the importance of the political dimensions of water resources management still needs to be consolidated in global and national water policy discourses.

The English language, and Dutch, the other language that I commonly use, have two separate words for policy and politics. Till I started preparing this chapter for a workshop on Wasserpolitik it had never occurred to me that this might be different in other languages. In German the word for both is *Politik*, in French it is *politique*, in Italian *politica*, in Spanish *política*, in Finnish *politiikka*[7] and in Russian политика (politika).[8] This use of a single word may reinforce that for most people 'politics' is often first and primarily associated with official, state politics, as practiced in the polity. In a dictionary definition, politics is 'the art and science of directing and administering states and other political units' (The New Collins Concise English Dictionary 1982, 877). State governance is the substance of politics in this perspective. Politics is, however, a much broader term. In the same *lemma* in the

started from populist and instrumentalist perspectives but has produced the notion of 'empowerment' as a much more political understanding of 'involvement of stakeholders' (cf. Scoones and Thompson 1994).

[6] The most striking example of this I found is a meeting convened in Delhi in February 2003 that brought together a large number of Indian water researchers and NGO water practitioners and activists, as well as government representatives to discuss the need for an India-wide dialogue on the massive 'interlinking of rivers' plan that shot to prominence in 2002 through a Supreme Court Order and a Presidential speech. This eminent collection of water experts managed to conduct the meeting without a single mention of the rent-seeking issue, while arguably lobbies for larger investment in water infrastructure are partly informed by this interest. At the global scale the issue of corruption in the water sector seems to slowly become a more acceptable topic. The Stockholm Water Symposium has had sessions on the topic in recent years.

[7] Though for 'policy', a newer word, the plural form is sometimes used, *politiikat* (personal communication Erja Hänninen).

[8] These are the languages for which I checked with colleagues. I thank Rutgerd Boelens, Tommaso Trevisani, Darya Zavgorodnyaya, Waltina Scheumann, Fabian Scholtes, François Molle and Erja Hänninen for their reflections on this issue.

dictionary quoted, politics is also defined as 'the complex or aggregate of relationships of men [sic!] in society, especially those relationships involving authority or power', 'any activity concerned with the acquisition of power' and 'manoeuvres or factors leading up to or influencing (something)'. Politics is a dimension or quality of many social processes, that is, all social processes in which interests of individuals or groups are mediated. This is, of course, conceptually well established in the social science literature, but needs to be incorporated into the analysis of water issues more systematically that it has been so far.[9]

For water, the basic idea that informs the approach presented in this chapter is that water control should be conceived as politically contested resource use. In this formulation water control is the subject matter of water management. It is something that humans have done since time immemorial. Any human intervention in the hydrological cycle that intentionally affects the time and/or spatial characteristics of water availability and/or its qualities, is a form of water control.[10] Water control has three dimensions: a technical/physical, an organisational/managerial, and a socio-economic and regulatory. These generic categories refer to, respectively, the manipulation of the physical flow and quality of water, the guiding of the human behavior that is part of water use, and the socio-economic, legal, administrative and other structures in which water management is embedded and that constitute conditions and constraints for management and regulation (cf. Bolding et al. 1995 and Mollinga 2003 for detailed discussion of the water control concept). These three categories coincide, for the case of irrigation, with three literatures that each use the category, but are largely separate: the engineering literature, the literature on irrigation management, and the literature on irrigation as part of the broader development process (see Mollinga 2003 for references). Each of these literatures abstracts from water control in a different way, highlighting a different dimension of it. An important implication of this understanding of water control is that it constitutes the case for interdisciplinarity as a necessary requirement for comprehensive understanding of water management, but this point is not pursued in this chapter.[11]

The focus of this chapter is on the second part of the formulation, the politically contested resource use that water control is. Contestation is another generic category. It is meant to refer to a range of interaction patterns in water management,

[9] A much quoted treatment of social power I find very helpful is Lukes (2005). On 'politics' cf. publications like Lasswell (1936), Leftwich (1984), and many others.

[10] My usage of the term 'control' in this manner has been found problematic by some. In critical perspectives 'control' tends to a 'bad thing', associated with the excessive and arrogant desire or mastery over nature by humankind. As an actual description of what humans do with water, water guidance, direction or regulation would be better, as intervention in the hydrological cycle is basically that. However, all three terms are awkward and confusing as general categories, and I therefore stick to water control till a better term becomes available. In Ostrom's (1990) framework one would speak of 'water use and control' as she distinguishes two categories of rights: use rights and control rights.

[11] On the premise that the three dimensions of water control are internally related, that is constitute each other. As such, water control can be regarded as a 'boundary concept' that creates space for different perspectives to discursively meet for more comprehensive understandings of water resources management.

including negotiation and struggle, and also less explicit and longer term disputations. The idea is to convey that there tends to be something at stake in water management, and that different individuals or groups involved have different interests. This is not meant to be a general theoretical statement, but an empirical one. The approach is interested to analyse those situations where water management is an issue. Because the societal issues around water management are proliferating, this seems to be a relevant perspective. The addition of the adjective political to contestation is simply meant to highlight that there is a political aspect to contestation and thus to water control.[12] As soon as the political would be a self-evident property of water control, it would become unnecessary to give it special emphasis: the adjective could be dropped and contestation understood to include it. How the political aspect of water control can be elaborated is discussed in the following section.

3 The Politics of Water: A Framework

Ubiquitous water politics is an assembly of domains and issues. Depending on the purpose of analysis and/or action, the structure of this sociotechnical practice can be represented in different ways. Two ordering principles are relevant for the purposes of this chapter: first, a distinction of different levels of water politics as relatively autonomous domains of interaction, and second, the identification of issue-networks encompassing processes of contestation within or across levels.

3.1 Levels and Issue Networks

'Levels' is a problematic metaphor for describing social structure (cf. Kitching 1988), but it has empirical relevance in the case of water resources. The social dynamics of water control play out at different geographical levels – where geography has to be understood in a combined physical-spatial and socio-political sense.[13]

The hydrological and hydraulic behavior of water has geographical boundaries in basins, in aquifers, and in human-created water control systems. Thereby, the social processes of water control are spatially situated and interlinked. This is obvious for basins, where very visibly the direction, magnitude, and timing of surface water flow constitute socio-geographical patterns, in terms of, for instance, settlement, mobility, and land-use patterns. The same applies to aquifers, though these are less visible because underground. Basins, aquifers, and water control systems together with

[12] This does not intend to suggest that water control can be *reduced* to its political nature, that is, that water control is *only* political or that its political aspect *determines* all other aspects. How and how strongly the mediation of actors' interests and the social relations of power shape the different properties and dimensions of water control processes is an empirical question, though my starting assumption is that it is always present and often important.

[13] For a discussion of the concept of space, see for instance Massey (1999).

climate (rainfall and snowmelt patterns particularly) and landscape characteristics shape quantity, quality and timing of water availability, and thus provide the fluid grid of human existence. Floods and droughts are among the phenomena that regularly bring home the message that these systems may have relatively clear and stable boundaries, but exhibit dynamic and sometimes erratic behavior.

The institutions and organisations created by societies for accomplishing water control also have spatial reference, though the reach of these by no means necessarily coincides with the physical boundaries of water control practices. The organising principle is socio-political space rather than physiographical order. Two link-concepts are, for instance, 'territory' and 'jurisdiction'. These are hybrid concepts that unite geographical and socio-political extent.

Water politics is organised in four different domains.[14] In each domain the 'main stake' or subject matter of the interaction processes is different.

1. *The everyday politics of water control*
   Everyday politics is a phrase coined by Kerkvliet (1991). Regarding water it refers to contestation of day-to-day water use and management. In many cases everyday politics is a relatively small scale phenomenon, including, for instance, how access to local groundwater markets is negotiated between community members, how maintenance obligations connected to water rights are enforced in a farmer-managed irrigation system, and many other examples. However, the management of a big reservoir distributing stored water to canals and areas hundreds of kilometres away from the dam is also 'local' in the sense of being a concrete, situated water use and management practice, with an everyday politics associated with it, for instance focused on the negotiation of gate settings and discharge monitoring, determining how much is released to whom at what time.
2. *The politics of national water policy*
   Politics of policy is a phrase coined by Grindle (1977, and subsequent work). It refers to the contested nature of policy processes. In the water resources domain I use it to refer to policy processes at the level of sovereign states, or states within a federation. The concept is a critique of linear views of policy formulation and implementation (Hill 1997), and aims to 'demythologise planned intervention' (Long and van der Ploeg 1989). The idea is that water policies, like other policies, are negotiated and re-negotiated in all phases or stages and at all levels, and are often transformed on their way from formulation to implementation. The political contestation of water policies takes place within state apparatuses, but also in the interaction of state institutions with the groups directly and indirectly affected by the policies.[15]

[14] These can be seen as (territorial/jurisdictional) levels, (action) arenas, semi-autonomous fields, domains of interaction etc. depending on one's purpose and focus of analysis. I settled for the general term 'domain of interaction'.

[15] An example of an approach addressing the issue how societal interest groups influence policy formulation and implementation is Sabatier's work on advocacy coalitions (Sabatier 1988). Mooij and de Vos (2003) is an annotated bibliography of policy process literature.

3. *Inter-state hydropolitics*
   Hydropolitics is a phrase that has been coined in the literature on international water conflicts, notably those in the Middle East (cf. Waterbury 1979; Ohlsson 1995). It there refers primarily to conflicts and negotiation processes between sovereign states on water allocation and distribution, particularly in relation to transboundary rivers or aquifers. Turton and Henwood (2002) propose to broaden the term to encompass all water politics, but I prefer to use it in its original meaning, including inter-state water conflicts in federal political setups. Hydropolitics is the part of water politics that has been well researched and documented, perhaps because it is a very public phenomenon, with sometimes high stakes and geopolitical relevance, and an interesting case for international relations studies (Zeitoun and Warner 2006).
4. *The global politics of water*
   Rather than being a phrase coined for long-existing practices, the global politics of water refers to a relatively new phenomenon: the recently, in the 1990s, invigorated international level of water discourse, policy and tentative regulation. The global politics of water contains several processes. These include the institutions and organisations set up as a follow up of the 1992 Dublin and Rio international conferences on water, environment and development, notably the World Water Forums, the World Water Council (WWC) and the Global Water Partnership (GWP). The GWP has become the international social carrier of the IWRM concept. The WWC has played an important role in the advocacy in recent years for more investment in water infrastructure. Another component of the global water politics is the World Commission on Dams' process, triggered by large political controversies around the effects of large dam building. A third component is the process related to the World Trade Organisation negotiations regarding water, notably around the issue of the privatisation of water and water service provision. A fourth relates to global advocacy for access to water as a human right (cf. Klaphake and Scheumann in this volume.)

The four domains can be distinguished because they have different space and time scales, are populated by different configurations of main actors, have different types of issues as their subject matter, involve different modes of contestation and take place within different sets of institutional arrangements.

The categorization of four domains of water politics provides a generic classification of the major action arenas, to use Ostrom's term (1995), of water control. However, the domains are only semi-autonomous fields, to use Falk-Moore's phrase (Falk-Moore 1973). Some of the most interesting and important questions in water policy and water politics involve the interlinkages between or across domains, for instance the abrogation of national sovereignty in water policy making by international development agencies and banks, the translation of global politics and policy ideas into national approaches and local impacts, and vice versa, and others. For instance the large dams issue is an example how local and national politics can be taken to the global level (cf. Klaphake and Scheumann in this volume).

All four domains and their interlinkages are hybrid spaces in which 'things happen' regarding water control. However, the vessels have no content as yet. The substance of contested water control is located in what I would call issue networks or problemsheds.[16] These are configurations of actors, social relations and practices around certain subjects that matter. Issue networks have concrete bones of contention, actors involved in shaping the issues and their impacts, while applying their perceptions, pursuing their interests by implementing their strategies, while mobilising the variety of their resources, with certain transformative (and reproductive) outcomes. Societal concerns concretely configure actors and interests around essential themes and topics – essential from the perspectives of the actors involved. The issue and sub-issues around which an issue network or problemshed has emerged may (be) play(ed) out in or across one or more domains. Issue networks or problemsheds give life to the structural landscape of water control contestation (and constitute it recursively[17]). Issue networks or problemsheds are sometimes stable and persistent, but often adaptive and dynamic, even transient.

3.2 Choosing a Focus

There is a plethora of issues and problems, issue networks and problemsheds. All are important in their own right. However, taking an overall look at reform in the water sector, several general observations can be made.

The boom in infrastructure investment in the water sector for irrigation/drainage, hydropower and flood control in the second half of the 20th century strengthened a quantitative 'harnessing' approach to water resources development. Allan (2006) has called this the phase of the pursuance of a 'hydraulic mission' by water resources agencies and by societies. For irrigation, for instance, this meant a supply enhancement approach, allocating available 'dependable flow' in rivers to new irrigation systems for boosting agricultural production. The approach was set in 'planned development' discourses of different varieties, with agricultural growth and national food security being the main drivers. The organisations responsible for creating and managing the infrastructure systems were predominantly populated by civil engineers. In countries where irrigation is very important for the national economy (like for instance in Pakistan, India, Indonesia, Mexico, Egypt, China, and several other countries), these have become very powerful organisations that occupy a large part of water resources public policy space.

This 'harnessing' approach to water resources development has been very successful in some respects. In a country like India there seems to be little doubt that the creation of large areas of surface irrigation systems in the 1950s–1970s has helped to achieve national food security and contributed to overall economic growth significantly.[18] Nevertheless, already in the 1960s, and prominently in the 1970s,

[16] I have taken the term 'issue network' from the literature on policy processes, and 'problemshed' from Viessman (1998); see also Merrey et al. (2006).

[17] Or rather, develop it in a cyclic process of morphogenesis (cf. Archer 1995).

[18] This point is not uncontested. For India, see for instance Dhawan (1988), Sengupta (1985), and Nadkarni (1984). Another question to be asked is whether there would have been other ways to

the 'underutilisation' of the irrigation systems created started to be noticed and discussed. Even when making perhaps significant contributions, the systems performed below expectation.[19] This challenge was addressed with a social engineering approach, to match the civil engineering approach that created the systems.[20] The 1970s and 1980s saw the rise of irrigation management policies and programmes, and research, that in a highly instrumentalist manner tried to 'get the institutions right'. The first focus was on law and order approaches to water management ('scientific' as against 'politicised' water management in the Indian discourse) focused on the implementation of the operational requirements of the systems, quickly followed by a focus on organising farmers in water users associations or other forms of cooperative management. The most recent proposal for the social engineering of water management forms is the establishment of river basin organisations (cf. Merrey et al. 2006, and Shah et al. 2006 for a summary and an extended discussion).

Though the insight that local water management in large-scale systems depended very much on the effectiveness of management at higher levels of the system, was articulated early on in the irrigation management policy discourse (Wade and Chambers 1980), it would take till the late 1990s till reform of irrigation bureaucracies became a serious item on the international policy agenda, with the idea of irrigation management transfer (IMT) linking the on-farm water management ideas of the 1970s and early 1980s, with the bureaucratic reform emphasis that emerged in the 1990s.[21] The 1990s saw an interlude in which much faith was put in engineering water markets and water rights, on the idea that tradable water rights would enhance both allocative efficiency and water use efficiency. Though such programmes, except in the case of Chile, were rarely introduced in a radical fashion, discursively they represented, and continue to represent, though with less support than before,

spend the budgets involved that could have achieved better results. I do not enter into discussion on this issue in this chapter. Historically, the choice for the development trajectory as just sketched has been made in many places, and we have to face the impacts and consequences – whatever they are.

[19] For India, some of the relevant documents that mark the emergence of this growing awareness are GOI/PC/PEO (1965), GOI/MOIP (1972) and GOAP (1982).

[20] For a more detailed critique of the social engineering paradigm in agricultural water management see Merrey et al. (2006).

[21] For India, I have suggested in Mollinga (2003) that 'pushing' the water management issue to the farmers level initially was not so much based on a lack of understanding of water management dynamics, but a quite conscious effort to position such interventions outside the domain of the Irrigation Department. The focus on the farm level should thus, in part at least, be seen as an effort of irrigation bureaucracies to reproduce and defend the orientations of their organizations: infrastructure creation and operation by hierarchically organized centralized civil engineering bureaucracies. The positioning of discussions on 'participation' almost exclusively at the local level, is another instance of this perspective. I have argued elsewhere that participatory irrigation *governance* is the core issue rather than participatory irrigation *management* (in the narrow sense of management), as participatory governance would be about sharing power (over allocation of rights and resources, over inclusion/exclusion and rule making; cf. Ostrom's (1990) 'constitutional choice' and 'collective choice' levels of rule making and enforcement *vs.* the 'operational' rule making and enforcement, while participatory management can easily be conceived in a 'technical' manner, aimed at enhancing 'system performance'.

the economic variant of social engineering 'institutional fix' approaches to irrigation management.

The achievements of management reform programmes in the irrigation sector have generally been very limited, with some exceptions, like for instance the Mexican and Turkish reforms of the early 1990s (on Mexico cf. Kloezen 2002; Rap 2004). Instrumentalist approaches to management transformation have proven to be not very effective in reshaping the institutional fabric and dynamics of the water sector. In Merrey et al. (2006) it has been argued that social engineering approaches need to be replaced by self-conscious concepts of strategic action to be able to engage effectively with the dynamics of inherently political water policy processes.

Instrumentalism is not just a civil engineering predilection. It also fits the concerns of bureaucratic and political decision-makers. These prefer simple, single or limited point, standardized solutions to policy problems, rather than emphasis on context-specificity, and the open-endedness, complexity and uncertainty of intervention processes. Instrumentalism is a general feature of prescriptive, linear approaches to policy making and implementation. In the literature on policy processes this linear, prescriptive approach to planning has been fundamentally criticized (Hill 1997), the real world being complex and non-linear. However, as a policy format, instrumentalism is very tenacious. It may be hypothesised that this tenacity has to do with the political and practical purposes it serves, in the water sector enhanced by the technical instrumentalism of the dominant professional group in the sector, i.e. the civil engineers.[22]

The instrumentalist engineers-bureaucrats configuration in the water sector has been challenged in different ways. The internal challenge of acknowledgement within the sector of the under-performance of existing water infrastructure mentioned already is one such challenge, and a problem that has not been willing to go away, despite a series of models, toolboxes and identification of best practices to be emulated. More profound challenges have come from outside the sector. The controversies around large dams have brought two issues to the fore: firstly, the displacement of people living in the areas to be submerged by new dam reservoirs, and the problems associated with resettlement and rehabilitation of these groups, and secondly, the negative ecological consequences of dams. Also in a broader sense the negative environmental 'externalities' of large-scale water infrastructure development have gained prominence in public debates: the effects on coastal and freshwater fisheries by changes in river and flood plain hydrology, the waterlogging and salinisation problems related to large-scale irrigation, the polluting effects of high external input agriculture (nutrients and pesticides/herbicides), the effects of reduced or increased sediment deposition in reservoirs, river alignments and delta formation, and several others. Challenges have also come from the budget/financial front: governments have become less willing and/or able, for different domestically

[22] This phrasing homogenizes the categories of bureaucrats and civil engineers far too much of course for concrete analysis. Discussing the issue of 'internalization' below, a more nuanced position will be presented. The statement made here should not be read as an empirical generalization regarding these categories of people, but as a description of the 'structural configuration of dispositions' within which social engineering is practiced and contested.

and internationally generated reasons, to invest the same amounts of money in large-scale water infrastructure as in the past, and support the rehabilitation and maintenance of the infrastructure without higher user contributions. Yet another challenge to the system is that which calls for more democratic forms of resource management, as part of overall decentralisation and democratisation agendas and processes.[23]

This rather mixed bag of pressures on the water sector together constitutes a demand for reform. The sector is being asked to change its ways in the management of water, and move towards approaches that incorporate a broader set of concerns and objectives than was the case in the past. As part of the global politics of water, international dialogues and assessments are held on how the environmental, human development and other broader concerns could be incorporated in water management more effectively.[24]

However, despite an enormous amount of text produced in global and national policy discourses on the need for and desirable features of 'integrated approaches', reality on the ground seems quite far removed form such ideas in many cases. Methodologically the 'integrated' approaches place emphasis on 'adaptive management' and 'social learning' (cf. Pahl-Wostl 2002 for discussion of these concepts in the European context). In many cases entrenchment, polarisation, defensiveness and even a refusal to learn seem to be more characteristic features of the water bureaucracies and the policy processes they are involved in. There are very few cases indeed where environmental, equity and democracy concerns have effectively translated into new objectives and activities and new styles of management. The only front at which there has perhaps been noticeable change is that of the financial sustainability of the water infrastructure, notably the level of financial contributions by users, but even in this area very little has changed in the past 10–15 years in many cases.

As a result of all the 'bad press' for large-scale water infrastructure development the investment of institutions like the World Bank in the sector were at a historical low in 2000–2003. However, that trend seems to have been reversed, and in recent years investments have grown substantially again (see Fig. 1).

How this trend change should be interpreted is not fully clear: as the result of successful lobby work of the pro-infrastructure investment network (following the Camdessus report and on the wings of the Millennium Development Goals achievement priority), as the result of a recognition within the World Bank that the water/agricultural sector is finally acknowledging environmental and other concerns as part of its core business, as the result of effective pressure by recipient countries, as a response to a perceived world food shortage, or something completely different?[25]

---

[23] In South Africa post-apartheid social and political reform objectives translated into water sector reform. In Indonesia the decentralization drive following the fall of the Suharto regime had important implications for water sector reform.

[24] For detailed argumentation and evidence I refer to the website of the *Comprehensive Assessment of Water Management in Agriculture* (http://www.iwmi.cgiar.org/Assessment).

[25] Another reason for reduced investment in irrigation development commonly suggested by economists is the lower need for increasing cereal production at the global scale, as evident in surpluses existing at the world market level (Rosegrant and Svendsen 1993). In the lobby for the

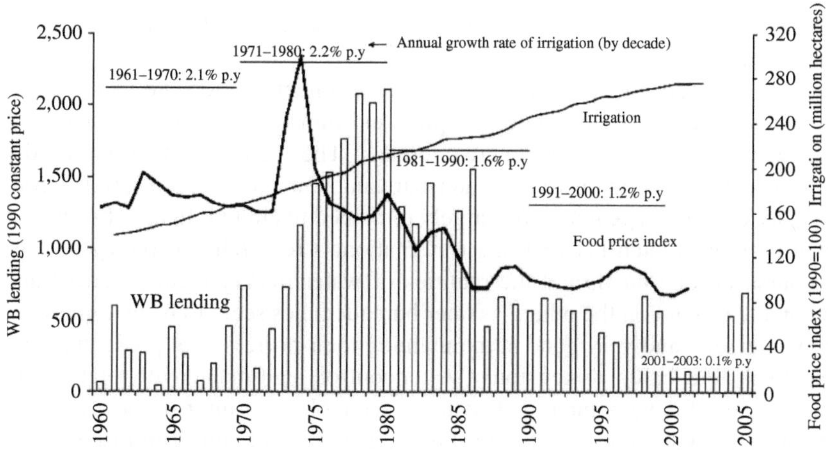

Fig. 1 World Bank lending (bars) for irrigation and drainage, area under irrigation and world food price index (of 1990 constant USS)
Source: Faures et al. (2007, in print) based on World Bank and FAO data

My concluding assessment is that since the early 1990s, first slowly, then with more momentum, new international and national policy discourses have been created around the need for 'integrated approaches', with IWRM, Integrated Water Resources Management, as the main brand name.[26] The discourses not only involve debates and documents, but also a new set of institutions and organisations, as noted above. Quite a few of the ideas regarding IWRM are drawn from developed country experiences (notably Australia, the USA and the European Union[27]), and in this sense a discourse is being imposed on developing and transition countries that may be partly misdirected and that may find little local resonance.

However, the need for 'integrated approaches' in developing countries does not only emanate from global policy imposition or assimilation. Real 'integration' issues are proliferating on the ground and are increasingly finding their way into national and regional policy debates. Examples are the increasing problems regarding water quality (water pollution), and the increasing pressure of cities and industries to move water out of agriculture, both frequently leading to conflicting situations. Such problems are set against a background of more and more basins reaching closure, liberalising and urbanising economies, and a continuation of large-scale rural poverty. The IWRM discourse provides a space in which such issues can be more

recent increase in investment the need to feed the increasing world population has also played a role. Evidencing of such analyses/statements would involve looking at the details of water and agricultural policy decision-making processes, something that has hardly been done.

[26] As suggested above, there are also other currents in global water politics. How these different strands in the discourse articulate, supporting or contradicting each other, is not the subject of this chapter.

[27] With a tendency to glorify these experiences, like the Murray-Darling case example, and the European Water Framework Directive policy approach. Problems in the 'home' of implementing these concepts are often lost in translation to other places.

legitimately raised than in the earlier sectoral and productivist discourses on (agricultural) water management and development.[28]

However, this said, the second part of my assessment of the state of the art as regards 'integrated approaches' in developing and transition countries is that very little has been achieved so far in concrete terms.[29] The 'new discourse' has hardly been internalised by the mainstream water bureaucracies: their organisational concerns and styles of management have largely remained unaltered. In my observation there is a large mismatch between the (international) research and policy notions and discourse on one hand, and the realities of water management practices and projects on the ground on the other.[30] Water bureaucracies seem to be extremely resistant to change,[31] and very good at maintaining their professional pre-occupation with the physical/technical dimension of water control, legitimated with reference to modernist views of development that focus on technological improvement enhancing economic growth. Associated with this are hierarchical and centralised expert populated organisational structures and styles (for evidence cf. the case studies in Mollinga and Bolding 2004).

[28] In that sense IWRM's relevance as a concept should primarily be understood as a boundary concept in international, national, regional and local policy debates on the future of water management, rather than a definite approach or model. Its 'looseness' is its strength, as that provides discursive space (and policy and political space if institutionally consolidated) based on a minimum agreement that there is some need of bringing concerns together that used to be separately treated or ignored. Complaints that the concept is vague and should be operationalized are, therefore, in my view partly misdirected. Closing the discursive space at the general, abstract level by some authoritative definition/operationalization would be counter-productive. What is relevant is how the concept is concretely and dynamically deployed in specific contexts, and through which kind of process this happens.

[29] I do not discuss the issue how much has been achieved in the context of developed nations. Uncritical success stories are, I feel, unwarranted, even when there is a lot of interesting experience to learn from. Water control is not only contested in developing and transition countries. Jaspers (2003) (quoted in Bruin et al. 2006) identifies the following five criteria (called 'triggers for change' in the chapter) along which the level of IWRM can be assessed: (1) water management based on hydrological boundaries; (2) integration of quantity, quality and ecological issues for both surface and subsurface water; (3) stakeholder participation in decision-making; (4) cost recovery, and (5) subsidiarity. These criteria are perhaps debatable, but they do refer to important aspects of more 'integrated' forms of water management. Most countries would score low on at least 2, 3 and 5, and many also on the other two criteria.

[30] This observation follows from the difficulty of finding other than very modest examples of really existing IWRM, and my long term study of the Indian water policy and practice situation.

[31] I say extremely, because my impression is that they may be more so than other technical state organizations, like agriculture or forestry, though I cannot substantiate this. The impression derives from the observation that the irrigation/energy/flood part of the water sector seems to be a late comer compared to agriculture and forestry bureaucracies as regards participatory approaches and poverty alleviation programs. My hypothesis is that the explanation of this lies in the strongly disciplinary population of the organizations (mainly civil engineers and hydrologists), and the high status accorded to water engineering organizations and their staff in the second half of the 20th century as (literally) nation-builders. This, however, is an untested hypothesis. A factor suggested to me by Roel Slootweg (personal communication) is the absence of a strong private sector presence as opposed to for example the energy sector.

From this assessment I derive two main issues for water politics research that I want to discuss in more detail in this chapter: (1) how does/can it happen that water bureaucracies make environmental and human development (poverty) concerns part of their 'core business', and (2) how does/can it happen that water bureaucracies adopt or get engaged in more participatory planning processes, in which different actors involved in issue networks can co-shape policy formulation and implementation?

4 Internalizing New Concerns: Environment and Human Development

To understand the process of internalization of concerns like ecology and human development (poverty particularly) into water bureaucracies' mandates, policies and professional orientation, it is instructive to look at cases where such internalization has happened, and where the process has been documented from a socio-political perspective. Well documented cases are only available for the internalization of ecological concerns. This is not so surprising, as environmental concerns have been the major trigger for water sector transformation in the West European, North American and Australian contexts, where the documented cases come from.[32]

4.1 Case 1: Environment and Flood Protection in the Netherlands

In Dutch water management there has been an 'ecological turn' in water management in response to the 'environmentalist wave' of the 1970s (Disco 2002). With environmental critiques of modernisation proliferating, the Dutch 'had to face up to the fact that ecological damage was a precondition of survival and prosperity', with the one-third of the country that is below sea level being the location of the most important economic activities and the habitat of most of the population. A massive national flood protection plan (the Delta Works) had been started after a damaging flood in 1953; by the 1970s the ecological damage of closing of large parts of the unique brackish estuarine delta system had gained recognition in the polity. The

[32] To my knowledge there is no systematic review of the internalization of environmental/ecological issues by water bureaucracies. For reasons of space I only briefly discuss a Dutch and a USA case. For the processes around restoring environmental flows in the Murray-Darling basin in Australia, see for instance, www.mdbc.gov.au/nrm/water_issues/environmental_flows, and www.rivermurray.sa.gov.au/major/water_flow.html, and numerous other sites and publications. Other cases that would be interesting to explore are the impact of the Flood Action Plan (FAP) related debates and controversies in Bangladesh, which has brought ecology, fisheries and livelihoods concerns more into the mainstream of water resources policy and the professional orientation of the Bangladesh Water Development Board (BWDB; www.bwdb.gov.bd/). In the USA and Australia rights and livelihoods concerns of American Indians and Aboriginal peoples, respectively, have played an important role as well in rethinking approaches to water management.

issue played out around a large infrastructural intervention, the closing of the Oosterschelde, a deep and elongated maritime intrusion. The closure would reduce the length of coastline to be protected against severe storms and high floods enormously and thus provided highly increased safety to the inhabitants of that part of the country. It would also destroy an ecosystem, and the shellfish industry that depended on it. The 'old' regime of water management exclusively emphasised 'safety'; the 'new' regime tried to find a balance between safety and other concerns, notably 'ecology'. A long-drawn political struggle ensued, technologically resolved by the design of a permeable storm surge barrier that can close off the 8 km wide opening of the Oosterschelde by lowering a series of gates in the case of dangerous flood levels, but normally lets through most of the tidal fluctuation, and thus preserves most of functions of the original ecosystem. First resisted heavily and considered technically impossible, the storm surge barrier is now the pride of Dutch hydraulic civil engineering. The dominance of the civil engineering profession in water resources development and planning reduced, at the cost of increased influence of ecologists and biologists. Institutionally, the decision-making involved a process of civil society agitation, government committees, and discussion in/by political parties in parliament. The event triggered a still ongoing process of reconsidering the basic premises of water management policy, and the expertise needed for it. In the 1990s 2 years with very strong inland flood threats, and some flooding, through extraordinarily high river discharges, provided another important trigger for such reconsideration. There is an ongoing, contested, process of ecological modernisation taking place.[33]

4.2 Case 2: Environment, Rationality and Submergence in the United States of America

The USA was once the largest and leading dam builder in the world; it is now systematically decommissioning dams, and regenerating river regimes. Like the process in the Netherlands described above, this was, and is, a politically highly contested process, with the rethinking of the 'hydraulic mission' beginning in the 1970s with the environmental movement. How environmental concerns got internalised into the United States Bureau of Reclamation (USBR) is described by Espeland (1998). The case she discusses, playing out from about 1980, is the planned building of the Orme Dam, part of the Central Arizona Project to bring Colorado River water to Arizona's desert. The dam would submerge most of the Fort McDowell reservation of the Yavapai, an Indian community. The Yavapai, despite their poverty, rejected the financial offer of the agency to buy their land, for many reasons, including their belief that it was not their right to sell ancestral land. The aspect highlighted here of this sophisticated case study of modern water politics in an arid

[33] Other sources on the *Oosterschelde* case and changes in the orientation of Dutch and European water management policy and practice are Lintsen (2002), Bijker (2002), Kuks (2005) and Bressers and Kuks (2005).

region, the nature of state power, and many other things,[34] is the internalization of environmental concerns into the USBR through the process of conducting environmental impact studies, including a valuation of the different functions/ecosystems services of the river system, and the effects of the dam on that. Espeland describes how a 'New Guard' of professional environmental and social scientists, hired in response to external pressure on the agency and with the political wind of the Carter administration's pro-environment policies in the back, enters the USBR bureaucracy, managed to become a powerful force in that bureaucracy changed planning procedures for dam building, and brokered the Orme Dam controversy successfully, with the rational decision models that they developed as the main instrument. This valuation instrument became the main tool for negotiating the trade-offs in the dam building. The major problem in this exercise was that of commensuration: to express different values of different costs and benefits in a single vocabulary – that of price – to be able to compare alternative plans. Espeland discusses these problems in detail, showing that despite the attraction of such instruments and methods as seemingly objective, the tool involves a process of social construction of what constitutes a value, and which values get included and excluded. Despite this, the tool was instrumental in taking the decision – not to build the dam in this case – and also boosted the status of the 'New Guard' facilitating a reinvention of the agency into an more environment-sensitive and participatory direction. However, the Yavapai were of the opinion that they had won the case for the wrong reasons – their view of things had not found, and perhaps could not find, a place in this rational decision-making tool.

A number of things can be learnt from these experiences:

1. How non-replicable, that is situation-specific they are. Each has its own specific and unique trajectory.
2. However, what they do have in common is firstly, that the pressure for policy transformation was to a large extent based in broader changes in society in terms of consciousness of and support for environmental political agendas, and secondly, partly related to this, a weakening of the agricultural constituency as a political force was an important enabling factor. It might be argued that if this is the condition for transformation, then transformations in developing and transition

[34] Espeland (1998, xi) describes the theoretical thrust of her case study as follows. '(...) in trying to analyze the complex set of conditions that produced this outcome [the decision not to build the dam, PPM], I came to reengage some old and formidable questions: the uneasy and sometimes volatile, relation between instrumental reason and substantive values; the conditions that propel commensuration – the transformation of qualities into quantities – and the difference that this makes for how we create and unmake boundaries, attach ourselves to categories, and negotiate identities; the consequences attendant on different modes of valuing; and the capacity of ordinary citizens to participate in decisions that affect their lives, especially when these are brokered by powerful bureaucracies. In the end I came to see this decision as a forum for analyzing competing concepts of rationality and how these shape our understanding of political participation. The debate about a dam site became, for me, a theoretical site to consider the politics of rationality in relation to democratic practice; a place to examine relations between our ideas for how to be rational, how to do politics, and what sort of people count as political actors.'

countries are still a long way off.[35] I do not think that this is a fully correct inference. What the examples show in a more general sense, is that bureaucratic transformation depends on sufficient political clout/momentum in the society at large for sustained pressure. What the coalitions that exert such sustained pressure (should) look like is not necessarily the same across countries. Though agriculture is a very prominent concern in most developing countries, the relative balance of forces among sectors/constituencies is not unchanging. Globalization and urbanization processes may imply new political priorities in which agriculture's predominance declines. Each situation requires its own strategic analysis of the socio-political dynamics through which change processes are happening or could happen. What the cases suggest is that broader socio-political processes are likely to have to provide the setting for water sector reform (rather than water sector reform being internally generated in and by the sector), and that qualitative change like the described does not happen easily and overnight. Furthermore, we may not want to accept the agriculture-environment and environment-poverty oppositions that inform much discourse on this matter. When the issue is primarily conceived as a trade-off, we miss out on options for productive and remunerative agricultural production that is ecologically sustainable and does generate employment.[36]

3. In both cases new professional groups, with new environmental expertise, were inducted into the water bureaucracies. This provided an important consolidation of the 'new concern' in the business as usual of the water bureaucracy.[37]
4. We also learn that a new approach addressing new concerns may provide new technical challenges to engineers. This seems to be how the incorporation of new concerns in the knowledge systems of the civil engineering and hydrology disciplines happens: by having to solve problems forced upon the discipline, rather than by a process of abstract 'increased awareness' and then self-redesign of technology.
5. Finally, the outcome of the processes is ambiguous. In the Dutch case the ecological concern was incorporated in an eco-modernist approach, and left several issues unaddressed. In the US-case the American Indians thought they had won the case for the wrong reasons. The changes are not revolutionary 'world-view' changes, but gradual processes of partial, but qualitative, transformation.

All five elements seem to have relevance in the context of developing and transition countries. The first two points imply a need for situation-specific strategic coalition- and alliance-building around certain issues to advocate policy and institutional transformation. Regarding irrigation reform in India, Bottrall in the early 1990s opined that there could possibly be a coalition for irrigation reform.

---

[35] Cf. the environmental Kuznets curve argument discussed in Shah et al. (2006).

[36] This issue is as relevant for Western/European contexts as it is for that of developing countries. For a concept of sustainable development along these lines developed in Maharashtra, India, cf. Datye (1997), and Paranjape and Joy (1995).

[37] This is in line with arguments about the importance of 'strategic groups' in knowledge system management (Menkhoff and Evers 2005).

> Those currently opposed to the status quo, or with good reasons to oppose it, include finance ministries (concerned about ID's [Irrigation Departments, PPM] never-ending demands on public funds); politicians and their constituents in regions disadvantaged by present patterns of water development (either through direct damage, as in waterlogged areas, or through long neglect, as in tank areas); environmental action groups; local issue-based groups (such as opponents of state water policies in Maharashtra); and non-agricultural water users, including urban domestic and industrial users, who suffer from the absence of efficient methods of inter-sectoral water allocation. (Bottrall 1992, 244).

This alignment of interests has not happened so far, but the question to be answered remains the same: which, in Sabatier's (1988) terms, advocacy coalitions can further policy and institutional transformation? Also the fifth point is relevant in this context: the process is never completed.[38]

The third and fourth points suggest that the habitus of technical professionals should be taken very seriously. At a practical level, reform of education and training programmes is a necessary long-term investment.[39] However, perhaps more important or effective may be the enrolment of technical professionals in the solution of the new technical challenges that an environment and poverty focus generates.

Despite this sketch of potential transformation options and avenues, the empirical observation has to be that most water bureaucracies have internalised the issues of environment and poverty only to a very limited extent, if not outright resisted it. The types of explanation for the resistance of water bureaucracies to societal demands to adopt new, 'integrated' or otherwise alternative approaches to water management can be classified in three types.

1. The 'vested interests' explanation, strongly popularised in international policy circles by Repetto's 'skimming the water' perspective (Repetto 1986).
2. Explanations focusing on the institutional characteristics of water bureaucracies, and their resultant 'inertia' and 'rigidity'. Such states may be due to the inheritance of colonial administrative structures (cf. for India for instance Kaviraj 1997; on Uzbekistan for instance Yalcin and Mollinga 2007).
3. What I tentatively call 'knowledge system' explanations, that is explanations that look beyond direct day-to-day concerns and practices of technocrats, but try to read the 'mental maps' of technocrats. Very little analysis of this exists for the water resources sector.

Where explanations one and two require changes in governance and management structures and styles of water organisations, the third addresses the professional identity of water bureaucrats most directly.

---

[38] This despite the suggestion strongly embedded in development assistance and government programs alike, that (fixed duration) projects are the desirable instrument for solving such problems.

[39] Efforts to establish and support more 'integrated' water resources engineering education and training programs include the *Centro Agua* project in Bolivia, the *WATERnet* project in Southern Africa, and the *Crossing Boundaries* project in South Asia, see: http://www.centroagua.org/, http://www.waternetonline.ihe.nl/, and http://www.saciwaters.org/crossing_boundaries.htm

5 State-Centered, Society-Centered and Donor-Centered Water Policy Processes

The second major 'sticking point' in water sector reform seems to be the unwillingness of water bureaucracies to share power with other interest groups, that is, adopt more inclusive policy formulation and implementation processes. Policy is used here as a generic term that includes all forms of state planning and public policy formulation and implementation for water resources management.

A short discussion of the Indian state of Andhra Pradesh's irrigation reform process can illustrate the power-sharing issue. The Andhra Pradesh Farmer Management of Irrigation Systems Act of 1997 is the largest scale effort at irrigation reform so far in India. The Act states that the Irrigation Department that used to have full jurisdiction over irrigation management will be transformed into a service providing agency, while for the management of the large-scale systems management bodies elected by and composed of irrigators (or more precisely, users of irrigated land) will be formed. At the local level these would be Water Users Associations, at the secondary canal level these would be Distributary Committees, and each system would have a Project Committee. The first two bodies were established through state-wide elections in 1997. The third, project level committee has not been established till 2005. The explanation for this is resistance to the idea on both the Irrigation Department side and on the side of politicians (parliamentarians). The main issues that Project Committees would deal with would be allocation of water at system level, and the allocation of the funds for executing larger physical works in the system. It would be a governing body setting the rules of the game for water distribution and physical maintenance of the system. If established with such authority, the Project Committees would become very powerful bodies at a regional scale (the large-scale systems usually spread across several districts). There would be serious competition in terms of resource brokerage with parliamentarians in their constituencies, and the domain of authority of the Irrigation Department would be seriously curtailed. This, so far, has been a bridge too far for the irrigation reform to allow irrigators/farmers this level of control/governance power, despite strong political support for the process from the Chief Minister of the state in the first phase of the process. A recent development is that the Act has been amended in such a way that the envisaged shift to a governance and management system with strong irrigator participation has become more unlikely. In Grindle's terminology, irrigation policy, governance and management in the Andhra Pradesh case is highly state-centered, and efforts at making it more society-centered will meet with resistance from the state itself, both its administrative and political sections.[40]

Grindle (1999) discusses two major approaches to the analysis of policy processes, as summarised in Table 1.

[40] Cf. Nikku 2006 for an analysis of the Andhra Pradesh irrigation reform process.

Table 1 Rational choice and comparative sociological approaches to the analysis of policy processes

	Approaches based on economic frameworks (rational choice theory)	Comparative sociological approaches (comparative institutionalism)
Why and how are politicians interested in shaping policy change?	Rational choice explanations of politicians behavior in a 'political market' with votes as the currency and access to public resources as benefits Capture of politicians by interest groups and rent seeking 'Context' is a strategic decision making arena	Strong emphasis on institutions and collectivities, rather than individual choice ('statecraft' as theme) Larger role of contingency Conflict over policy is the 'normal stuff' of politics; emphasis on social interaction in economic, social and policy arenas in relation to social power 'Context' is a complex environment with history, shaping perspectives, references and values
How do political institutions affect the choices made by politicians?	Institutions are strategic arenas for individual choice	Institutions have histories, which shapes preferences, orientations, values, and strategies of collective actors
How are new institutions created or transformed?	Intrigued by the creation of new institutions that constrain the power of politicians. The behavior of reformist politicians: how long-term interest and short-term interest relate. Transaction costs in political life to explain change Principal-agent problems; role of institutional designers	Criticize apolitical explanations of institutional change; new institutions are the result of historically embedded conflicts about the distribution of power and benefits in society
What are the consequences of new rules of the game for economic and political interaction?	Consequences generate new strategies for achieving first order preferences, towards a new equilibrium	More dynamic approach: institutional change creates new sources of conflict, new claims for resources, new spaces for contestation

Source: based on Grindle (1999, 3–11)

Grindle (1999, 11) also observes that adherents of the two schools 'have been outspokenly harsh about the other':

> Those who favor the elegance and parsimony of economic models of political behavior accuse comparative institutionalists of avoiding rigorous theory and scientific methodology and of producing primarily descriptive studies. Those who work from within the sociological tradition retort that economic models produce political banalities and historically inaccurate analyses that ignore empirical evidence.

Grindle is of the view that both schools are deficient in important ways. The 'political economy of public policy' perspective has been developed particularly, though not exclusively, with reference to European and North American policy processes. It carries several biases as a result, one being strong assumptions about societal groups actively contesting government policy and thus being involved in policy formulation. Grindle shows that developing countries may be characterised by state-centered policy processes, while these analytical approaches are society-centered. In developing and transitional countries policy may be generated primarily in elite (government) circles. Also the institutional setting of developing and transitional countries may be very unstable, and institutional and policy evolution a different process as a result. Another strong assumption is the sovereignty of the voter in electoral processes, which may not apply elsewhere. Such differences suggest a general point in terms of analytical approach: that the study of the process of policy needs to be contextualised historically and geographically.[41] She concludes her review and assessment with a call for more 'grounded' research on actual processes of institutional transformation while 'seeking to stretch theoretical models' (1999, 21).

Formulated in this terminology, a major challenge facing the water sector is to transform the highly state-centered policy processes that exist in many countries into more society-centered processes. It could also be observed that the global IWRM-focused water policy discourse is highly normative in its emphasis on establishing society-centered governance and management structures.[42] If Grindle's analysis of the bias in existing frameworks of policy analysis is correct, and I think it is, then we don't have much conceptual grasp on existing state-centered water policy processes, and the normative proclamations of the global water resources discourse become somewhat vacuous as a result.[43]

I propose that to the two categories, or ideal types, of state-centered and society centered policy processes a third category needs to be added which could be tentatively called 'donor-centered' policy processes. The notions of both state and society centrism implicitly assume the existence of sovereign states within which these processes occur. As Jenkins (2002) discusses, the sovereignty of developing countries has been undermined or otherwise become weak or eroded in two senses. Firstly, since the introduction of structural adjustment policies in the 1980s, there has been increasing influence of development funding agencies on not just the content of policies but the policy making and implementation structure as a whole

[41] Grindle (1999, 17) also identifies three factors that are, in her view, under-explained: leadership, the role of ideas, and successful policies. She makes no reference to the literature on policy discourses and policy narratives, which in my view have a lot to say on the power of ideas in policy processes. She does make the observation that 'ideas may be important means through which international actors become players in domestic policy debates.' This seems to be very applicable to the water resources domain, and is perhaps what the global politics of water is largely about.

[42] And in Jenkins' (2001) terms, employs highly 'sanitized' understandings of civil society that are idealistic and unrealistic.

[43] This argument also implies that, for instance, the European and USA literature on changing water policy regimes in these relatively society-centered regions may not be of extremely great value for analyzing the relatively state-centered situations in many developing and transition countries, and neither for the category of 'donor-centered' policy processes proposed below. Our analysis of the institutional transformation in the Uzbekistan context seems to confirm this (Yalcin and Mollinga 2007). Society-centered policy processes seem to be the exception rather than the rule.

('good governance reforms'). Secondly, the actual capacity to govern that is part of the concept of sovereignty may or may not exist for a variety of reasons (cf. the debate on 'failed states'). No country is without external influence on its policies and policy institutions, but the degree and impact of that influence varies strongly. When external influence on policy making becomes dominant as compared to the domestic generation of policy dynamics, one could perhaps speak of a 'donor-centered' policy process.

Small countries with small economies are, logically, more vulnerable to 'donor centrism'. One indicator that might provide a signpost to donor-centered policy regimes is the percentage of the government (development) budget that is financed from loans and grants of international development funding agencies. In South Asia, Bangladesh's, Nepal's and Sri Lanka's water policy formulation and prioritisation have in periods been very strongly donor-influenced (though this has not necessarily translated into donor influence on policy implementation), but the much smaller country of Bhutan has avoided such influence. India is too large an economy and country to be very prone to donor centrism, as became clear in the unrolling of the controversy around the Sardar Sardovar project, where many observers opine that if the World Bank had not withdrawn its funding for the project itself, they would probably have been invited to do that by the Indian government. Pakistan is a case of strong donor influence, but its geo-political significance implies that it can relatively easily get away with poor or non-implementation of conditionalities for development loans (cf. VanderVelde and Tirmizi 2004). Concrete analysis of such water resources policy regimes would undoubtedly produce a more refined categorisation than that of society, state and donor centered regimes.[44] This discussion amounts to a call for the undertaking of such comparative institutional and policy regime analysis, with an open mind as regards the conceptual frameworks that might be most useful in such an endeavor.

6 Summary and Conclusion

This chapter has emphasised the importance of squarely addressing the political dimension of water resources use, management and governance (water control). The development of technical infrastructure for surface irrigation, flood control and hydropower generation is the 'core business' of (government) water agencies in many countries, and definitive of the professional identity of the civil engineers and hydrologists that staff these organisations – the social carriers of the 'hydraulic mission' (Allan 2006). Calling these organizations hydrocracies (Rap 2004) expresses their dominant role in water policy making and implementation in those countries where water control is a strategic dimension of development. Addressing the politi-

[44] For instance, India provides an interesting case in that it exhibits the paradoxical situation of strongly state-centered policy processes in the water resources sector, set in an overall very vibrant and long-standing democratic regime, with a very active civil society (cf. Mollinga 2004 for discussion of the paradox, and a first attempt to explain it; also see the Indian case studies in Mollinga et al 2006).

cal dimensions of water resources management self-consciously and analytically is a relatively recent phenomenon, at least in the 'dominant discourse' (Allan 2006). It has partly come in 'through the backdoor' as a consequence of the emphasis on 'governance' in the global development policy debates. Whatever the currents and fashions in global water resources discourse are, the relevance of the issue of water politics follows from the concrete contestations over and around water resources in many places around the world. On that front it can be safely stated that water in many cases, though not necessarily, is a highly contested resource, and that in all likelihood conflicts and disputes over water will increase in the coming decades. The objective base for emphasising attention to the political dimension of water resources management thus lies in the day-to-day reality of its use, and the effects and impacts of that.

The framework presented divides the politics of water into four domains of interaction: the everyday politics of water, the politics of national water policy, hydropolitics and global water politics. These are different domains in the sense that different configurations of actors populate them and interact around different subject matters and issues. However, these domains are only semi-autonomous fields, and some of the most important questions regarding water resources management span across and through them.

After this formal conceptualising of the field of water politics, the second step in elaborating it as a research field is to identify the issues and problems, and their issue networks and problemsheds. This can be done in many different ways, depending on the purpose of the study and the characteristics of the setting in which it will take place.

The present chapter identifies two major issues to be the 'sticking points' in the unrolling of water sector reform processes in situations where powerful hydrocracies dominate the water resources terrain. It is observed that in most of the larger irrigation, flood control and hydropower countries, there is a deadlock in water sector reform. The two sticking points identified are the following.

1. The internalization of the 'new' environmental, poverty and democratic governance related concerns into the professional identities, approaches, and institutional frameworks of water resources professionals and their organisations seems to happen rarely and with great difficulty. Existing identities and approaches, characterised by a focus on increasing production, supply enhancement, 'harnessing' of water resources, and with a social engineering (Merrey et al. 2006) approach to policy implementation, are tenaciously defended against societal pressures to rethink and change the paradigm.
2. The 'unlocking' of the process of institutional and organisational transformation of hydrocracies, in order to establish more balanced an productive relationships between water managers and water users is the second sticking point. This is about changing the social relations of power between the different actors involved in water resources management, with or without the internalization of the concerns mentioned under point 1. These power relations are always implicit in reform, if not explicitly contested, but rarely self-consciously addressed and analysed as part of a 'strategic action' perspective on institutional transformation. Despite emphasis on issues like governance and accountability in the wa-

ter policy discourse, most reform initiatives remain set in conventional 'policy as prescription' (Mackintosh 1992) frameworks, often with a 'single solution' approach, be this water users associations, water markets, public-private partnerships or river basin organisations (Merrey et al. 2006).

The main observations on the first theme of internalization of 'new' concerns are the following.

1. Internalization processes are highly situation specific (they have a history, and are therefore particular and path-dependent), and they are non-linear. They involve a lot of strategic action with inherently uncertain outcomes.
2. The internalization processes that have taken place and have been documented, seem to have been primarily driven by broader socio-political processes and forces, rather than having emerged form internal dynamics and learning within the sector.
3. New professional groups being the carriers of 'new' concerns and approaches are playing an important role in internalization processes.
4. 'New' concerns, notably environmental ones, provide technical challenges for water professionals, which may be instrumental in inducing internalization and transformation processes.
5. Outcomes of internalization and transformation processes tend to be ambiguous. The route towards a 'paradigm shift' is not likely to be a revolutionary 'big bang' one, but more likely to be a gradual, step-by-step, backwards and forwards, process.

The main point raised regarding the second 'sticking point' is the appropriateness of the analytical frameworks for policy process analysis, largely developed as 'society-centered' approaches suited for democratic regimes with some level of organised and regularised public contestation and shaping of public policy. In many developing countries, but not only there, and more significantly for this chapter, in many national water sectors, policy making and implementation processes are highly 'state-centered'. To understand the policy dynamics in such settings different policy analysis frameworks are needed. A third category of situations or policy regimes are those where international development (financing) agencies have a strong role in national water policy making, to the point that national sovereignty is put into question. For such cases one would need a 'donor-centered' framework of analysis. This is admittedly a very crude typology of policy regimes, and mainly meant to 'open up' the debate on appropriate approaches, and a call for comparative research on water policy regimes and transformation processes.

It is evident throughout the chapter that the author's driving concern for unpacking the politics of water is the desire to contribute to a paradigm shift in water resources management. Such a shift would involve a transformation of the policy regime dominated by technocratic, social engineering disposed hydrocracies implementing their 'hydraulic missions', towards an inclusive, polycentric system of water governance and management having a focus on sustainable human development addressing the complex mix of economic growth, welfare, equity, sustainability and democracy concerns. In terms of the typology of policy regimes – how does one move from state- and donor-centered into the direction of society-centered situations? This a strongly normative and political driver to which I gladly plead guilty. The emphasis on 'politics' is not only the addition of another extra dimension to the list of already recognised dimensions of water resources management, nor does it only have implications for the diversity of policy analysis frameworks that need to be deployed. It is a standpoint that critiques the dominant social engineering approaches to institutional transformation, and which states that unless a self-consciously political strategic action approach to institutional transformation is taken, the deadlock in water sector reform may continue for some time.

References

Allan, JA (2006) IWRM: the new sanctioned discourse? In: Mollinga PP, Dixit A, Athukorala K (eds) IWRM in South Asia: global theory, emerging practice and local needs, Water in South Asia Series 1, Sage, New Delhi, 38–63

Archer M (1995) Realist social theory: the morphogenetic approach, Cambridge University Press, Cambridge

Bijker, WE (2002) The Oosterschelde storm surge barrier. A test case for Dutch water technology, management, and politics. In: Technology and Culture 43, July, 569–584

Bolding A, Mollinga PP, van Straaten K (1995) Modules for modernisation: colonial irrigation in India and the technological dimension of agrarian change. In: Journal of Development Studies 31 (6), 805–844

Bottrall A (1992) Fits and misfits over time and space: technologies and institutions of water development for South Asian agriculture. In: Contemporary South Asia, 1 (2), 227–247

Bressers HTA, Kuks SMM (2005) Integrated regimes on sustainable use of natural resources: a multiple case study analysis. Paper for the conference on "Sustainable water management: comparing perspectives from Australia, Europe and the United States", National Europe Centre, The Australian National University, Canberra, Australia, 15–16 September

Bruin EFLM de, Jaspers FGW, Gupta J (2006) The EU Water Framework Directive: Challenges for institutional implementation. In: Vermaat J, Bouwer L, Turner K, Salomons W (eds) Managing European coasts. Past, present and future, Springer, Berlin, 153–171

Datye KR (1997) (assisted by Paranjape S, Joy KJ) Banking on biomass. A new strategy for sustainable prosperity based on renewable energy and dispersed industrialisation. Environment and Development Series, Centre for Environment Education, Ahmedabad

Dhawan BD (1988) Irrigation in India's agricultural development, productivity, stability, equity, Sage, New Delhi

Disco C (2002) Remaking "nature": the ecological turn in Dutch water management. In: Science, Technology and Human Values 27 (2), 206–235

Espeland WN (1998) The struggle or water, politics, rationality, and identity in the American Southwest, University of Chicago Press, Chicago

Falk-Moore S (1973) Law and social change: The semi-autonomous social field as an appropriate subject of study. In: Law & Society Review 7 (4), 719–746

Faures JM, Svendsen M, Turral H (2007, in print) Reinventing irrigation, Chapter 9 of Molden D (ed) Water for food, water for life, Earthscan, London

Ferguson J (1994) The anti-politics machine. "Development", depoliticization, and bureaucratic power in Lesotho. University of Minnesota Press, Minneapolis, London

GOAP (Government of Andhra Pradesh) (1982) Report of the commission for irrigation utilisation. Vol. I & II. Hyderabad

GOI/MOIP (Government of India/Ministry of Irrigation and Power) (1972) Report of the irrigation commission, Vol. I, New Delhi
GOI/PC/PEO (Government of India/Planning Commission/Programme Evaluation Organisation) (1965) Evaluation of major irrigation projects - some case studies
Grindle MS (1977) Bureaucrats, politicians and peasants in Mexico. A case study in public policy, University of California Press, Berkeley
Grindle, MS (1999) In quest of the political: the political economy of development policy making, CID Working Paper No. 17, Center for International Development at Harvard University
GWP (2000) Integrated water resources management. GWP Technical Committee Background Paper 4, Stockholm
Harriss J (2001) Depoliticizing development: The World Bank and social capital, New Delhi, Leftword
Hill M (1997) The policy process. A Reader, 2nd edn, Pearson Prentice Hall, London
Hoebink P (2006) European donors and 'good governance': conditions or goal? In: European Journal of Development Research 18 (1), 131–161
Jaspers FGW (2003) Institutional arrangements for integrated river basin management. In: Water Policy, 5, 77–90
Jenkins R (2001) Mistaking "governance" for "politics": foreign aid, democracy and the construction of civil society. In: Kaviraj S, Khilnani S (eds) Civil society: history and possibilities, Cambridge University Press, Cambridge, 250–268
Jenkins R (2002) The emergence of the governance agenda: sovereignty, neo-liberal bias and the politics of international development. In: Desai V, Potter R (eds) The companion to development studies, Oxford University Press, New York, 485–488
Kaviraj S (1997) The modern state in India. In: Doornbos M, Kaviraj S (eds) Dynamics of state formation. India and Europe compared, Indo-Dutch Studies on Development Alternatives 19, Sage, New Delhi, 225–250
Kerkvliet BJT (1991) Everyday politics in the Philippines. Class and status relations in a Central Luzon village, University of California Press, Berkeley
Kitching Gavin (1988) Karl Marx and the philosophy of praxis, Routledge, London
Kloezen WH (2002) Accounting for water. Institutional viability and impacts of market-oriented irrigation interventions in Central Mexico, PhD thesis, Wageningen University
Kuks SMM (2005) The evolution of national water regimes in Europe. Transitions in water rights and water policies. Paper for the conference on "Sustainable water management: comparing perspectives from Australia, Europe and the United States" 15–16 September 2005 at the National Museum of Australia, Canberra, Australia. Hosted by the National Europe Centre at the Australian National University, Canberra, Australia
Lasswell HD (1936) Politics: who gets what, when, how, McGraw-Hill, New York
Leftwich A (ed) (1984) What is politics? The activity and its study, Basil Blackwell, Oxford
Lintsen H (2002) Two centuries of central water management in the Netherlands. In: Technology and Culture, 43, (3), 549–568
Long N, van der Ploeg JD (1989) Demythologizing planned intervention: an actor perspective. In: Sociologia Ruralis 29 (3–4), 226–249
Lukes S (2005) Power: a radical view, 2nd edn, Palgrave Macmillan
Mackintosh M (1992) Introduction. In: Wuyts M, Mackintosh M, Hewitt T (eds) Development policy and public action, Oxford University Press, Oxford in association with the Open University, Milton Keynes, 1–9
Massey D (1999) Philosophy and politics of spatiality: some considerations. In: Power-Geometries and the politics of space-time, University of Heidelberg, Department of Geography, Heidelberg, 27–42
Menkhoff T, Evers HD (2005) Strategic groups in a knowledge society: knowledge elites as drivers of biotechnology development in Singapore, ZEF Department of Political and Cultural Change Working Paper Series 7, Bonn
Merrey D, Meinzen-Dick R, Mollinga PP, Karar E (2006) Policy and institutional reform processes for sustainable agricultural water management: the art of the possible. Chapter 5. In: David

Molden (ed) Water for food, water for life. A comprehensive assessment of water management in agriculture, Earthscan, London, 193–231
Mollinga PP (2003) On the waterfront. Water distribution, technology and agrarian change on a South Indian canal irrigation system, Wageningen University Water Resources Series. Hyderabad, India: Orient Longman (orig. 1998)
Mollinga PP (2004) Sleeping with the Enemy: Dichotomies and Polarisation in Indian Policy Debates on the Environmental and Social Effects of Irrigation, Nr. 11. Center for Development Research Department of Political and Cultural Change Research Group Natural Resources and Social Dynamics, Working Paper Series 11, Bonn
Mollinga PP, Bolding A (eds) (2004) The politics of irrigation reform. Contested policy formulation and implementation in Asia, Africa and Latin America, Global Environmental Governance series, Ashgate
Mollinga PP, Dixit A, Athukorala K (eds) (2006) Integrated water resources management. Global theory, emerging practice and local needs, Water in South Asia Series 1, Sage, New Delhi
Mooij J, de Vos V (2003) Policy processes: an annotated bibliography on policy processes, with particular emphasis on India, Working Paper 221, Overseas Development Institute, London
Nadkarni MV (1984) Irrigation and rural development – A sceptical view. In: Economic and Political Weekly, Review of Agriculture 19 (26), A67–A73
Nikku BR (2006) The politics of policy. Participatory irrigation management in Andhra Pradesh, Wageningen University (PhD thesis)
Ohlsson L (1995) Hydropolitics: conflicts over water as a development constraint, Zed Books, London
Ostrom E (1990) Governing the commons. The evolution of institutions for collective action, Cambridge University Press, New York
Ostrom E (1995) Understanding institutional diversity, Princeton University Press, Princeton, New Jersey
Pahl-Wostl C (2002) Towards sustainability in the water Sector – the importance of human actors and processes of social learning. In: Aquatic Sciences 64, 394–411
Paranjape S, Joy KJ (1995) Sustainable technology. Making the Sardar Sarovar project viable. A comprehensive proposal to modify the project for greater equity and ecological sustainability. Environment and Development Series. Centre for Environment Education, Ahmedabad
Rap E (2004) The success of a policy model. Irrigation management transfer in Mexico, Wageningen University (PhD thesis)
Repetto R (1986) Skimming the water: rent seeking and the performance of public irrigation systems, World Resources Institute, Washington, DC
Rogers P, Hall AW (2003) Effective Water Governance. TEC Background Papers No. 7. GWP, Stockholm. http://www.gwpforum.org/servlet/PSP?iNodeID=215&itemID=197. Cited 3 Jul 2007
Rosegrant MW, Svendsen M (1993) Asian food production in the 1990s: irrigation investment and management policy. In: Food Policy, 18 (1), 13–32
Sabatier PA (1988) Am advocacy coalition framework of policy change and the role of policy-oriented learning therein. In: Policy Sciences, 21 (2–3), 129–168
Scoones I, Thompson J (eds) (1994) Beyond Farmer First, IT Publications, London
Scott, JC (1997) Seeing like a state: how certain schemes to improve the human condition have failed, Yale University Press
Sengupta N (1985) Irrigation: traditional vs. modern. In: Economic and Political Weekly, 20 (45, 46 & 47), 1919–1938
Shah T, Makin I, Sakthivadivel R (2006) Limits to leapfrogging: issues in transposing successful river basin management institutions in the developing world. In: Mollinga PP, Dixit A, Athukorala K (eds) IWRM in South Asia: global theory, emerging practice and local needs, Water in South Asia Series 1, Sage, New Delhi, 109–144
Turton A, Henwood R (eds) (2002) Hydropolitics in the Developing World. A Southern African Perspective, AWIRA, University of Pretoria. www.internationalwaterlaw.org/Articles/hydropolitics_book.pdf. http://waterlaw.org/articles/hydropolitics_book.pdf. Cited 3 July 2007

VanderVelde E, Tirmizi J (2004) Irrigation policy reforms in Pakistan: who's getting the process right?' In: Mollinga PP, Bolding A (eds) The politics of irrigation reform. Contested policy formulation and implementation in Asia, Africa and Latin America, Global Environmental Governance series, Ashgate, Aldershot, 207–239

Viessman W (1998) Water policies for the future: an introduction. In: Water resources update, Universities Council on Water Resources, 111, 4–7

Wade R (1982) The system of administrative and political corruption: Canal irrigation in South India. In: Journal of Development Studies, 18 (3), 287–328

Wade R, Chambers R (1980) Managing the main system: canal irrigation's blind spot. In: Economic and Political Weekly, 15 (39), A107–A112

Waterbury J (1979) Hydropolitics of the Nile valley, Syracuse University Press, New York

Yalcin R, Mollinga PP (2007) Institutional Transformation in Uzbekistan's Agricultural and Water Resources Administration: The Creation of a New Bureaucracy. Center for Development Research Department of Political and Cultural Change Research Group Natural Resources and Social Dynamics. Working Paper Series 22, Bonn

Zeitoun M, Warner J (2006) Hydro-hegemony – a framework for analysis of trans-boundary water conflicts. In: Water Policy 8 (5), 435–460

Part I
Global Norms and National Policies

Towards Implementation of the World Commission on Dams Recommendations
Experiences and Reflections After 5 Years

Michael Fink and Anne Cramer[1]

Abstract From 1998 to 2000, the multi-stakeholder World Commission on Dams (WCD) developed recommendations on improved planning procedures to overcome continuous debates and conflicts on large dams. The publication of the WCD report met a mixed response. Most stakeholders agreed to the underlying values of the WCD recommendations, but criticized the planning guidelines brought forward as too far-reaching and imprecise. Nevertheless, the WCD had a remarkable impact on dam policies.

The WCD recommendations, which necessarily had to remain quite abstract to be universally applicable, need to be operationalized before they can be applied in practice. This challenge has been tackled by different institutions and stakeholders in a wide range of thematic and geographical contexts. After more than 5 years, it has become possible to take stock of such experiences. The chapter examines two concrete attempts from Nepal and the Mekong Basin to implement parts of the WCD recommendations by conducting national level dialogues and by engaging in long-term planning. The lessons learnt, both through positive and negative outcomes, are useful to improve WCD implementation approaches in future and understand the significance and applicability of WCD recommendation today.

1 Introduction

In the light of continuous debates and conflicts on large dams, major actors established the World Commission on Dams (WCD) in 1998. The Commission undertook an extensive analysis of positive and negative development impacts of large

Michael Fink
e-mail: michael.fink@gtz.de

Anne Cramer

[1] With contributions from Martin Geiger, Elke Nickel and Maricla Costa.

W. Scheumann et al. (eds.), *Water Politics and Development Cooperation*,
doi: 10.1007/978-3-540-76707-7_2, © Springer-Verlag Berlin Heidelberg 2008

dams, developed recommendations on improved planning procedures and proposed better-integrated decision-making processes.

The publication of the WCD report *Dams and Development: a new framework for decision-making* in 2000 (WCD 2000) met with a mixed response. Most stakeholders agreed to the values and principles underlying the WCD recommendations. Others, such as financing institutions and industry associations, criticized the planning guidelines brought forward by the WCD as too far-reaching and imprecise. These institutions therefore decided not to embrace the WCD recommendations. Nevertheless, the WCD had a remarkable impact. The Asian Development Bank (ADB), for example, is introducing some of the Commission's suggestions into their own safeguard policies. The World Bank will probably do the same as it is currently in the regular process of reviewing its safeguard policies. However, although the Bank is generally backing the principles contained in the report, its position is diverging from the report in relation to a few points that are considered "not practical and would virtually preclude the construction of any dam" (World Bank 2004a, 38; 2001a, b, c). Additionally, the international dam industry presented their own sustainability guidelines in 2005, addressing many of the concerns also voiced by the WCD report (IHA 2006).

The WCD recommendations, which necessarily had to remain quite abstract to be generally applicable, need to be operationalized and translated into the specific context of a region, country or dam project before they can be used for improved planning of future dams or management of existing dams. The challenge of adapting the WCD recommendations for practical use has been tackled by different institutions and stakeholders in a wide range of thematic and geographical contexts.

This chapter takes stock of first experiences made in such activities by examining two concrete attempts to implement parts of the WCD recommendations: The *Constructive Dialogue on Dams and Development* in Nepal and the World Wildlife Fund (WWF) initiative on *Environment criteria for hydropower development in the Greater Mekong Region*. The chapter concentrates on lessons learned, through the analysis of both positive and negative outcomes, and attempts to identify critical factors responsible for success or failure. These results in turn can be useful to improve WCD implementation approaches in future.

1.1 Background: The Debate on Dams

Like any major infrastructure, large dams[2] have a high impact on the natural and social environment in their vicinity. They are often located in remote upstream areas featuring an intact water catchment area or beneficial topological conditions

[2] According to the definition of the International Commission of Large Dams adopted by the WCD, large dams are considered those with a height of more than 15 meters, or 5–15 meters high with a reservoir volume of more than 3 million m^3.

for hydropower. While sparsely populated, these areas are often home to poor or indigenous populations and exhibit a great variety of flora and fauna.

Conflicts are inherent to the functions of large dams. They deliver a multitude of considerable benefits in terms of water and energy services (hydropower, flood protection, water for irrigation or household use) on a regional or national level. Depending on the quality of the planning process, dams might also incur serious negative environmental and social impacts, which are borne disproportionately by the population and the environment at the site of the dam. The contrast of national or regional macro-economic benefits versus locally concentrated negative impacts often translates into conflicts of interest and confrontational attitudes. The work of the Environment and Conflict Project (ENCOP) for example has demonstrated how diverging interests over natural resources can represent the fundament of conflicts that were only explained so far as ethnic or religious (Bächler and Spillmann 1996a, b, c).

The debate about large dams is very complex and touches the key questions of development and how it should happen. Proponents and opponents of dams range from the local to the global level and are motivated by many different concerns, such as national development options, business interests, political influence, considerations for the environment and indigenous rights. Dams, similarly to any large infrastructure development, are prestige projects and their development is quite often a matter of politics and political decisions and not based only on technical and scientific considerations. Additionally, as environmental and social sciences have recently started a process of integration, dams are increasingly seen not only through the rather narrow lenses of technical engineering, but are better studied in terms of their impacts on people and the environment. These are found to be more complex and to have longer lasting effects than previously thought. Matters as the cumulated effect of a multitude of dams in a given river basin or the chances of success for an alternative livelihood program, for example, are still not fully understood and subject to scientific and political debate.

As dam building accelerated in the 1960s and 1970s, opposition to dams became more widespread and organized. While communities facing forced resettlement strongly resisted dam building worldwide from the 1980s, the debate acquired a global stage. The global anti-dam movement was sparked by prominent conflicts such as the local opposition to the Sadar Sarovar projects in India.

Despite efforts made by planners and operators to improve the outcomes of dam projects by paying more attention to environmental and social mitigation measures, the debate on dams continued unabated. Instead, the question became more controversial throughout the 1990s, dividing pro- and anti-dam interest groups. During the 1990s, financing institutions became more and more reluctant to fund large dam projects, leading to a decrease in the number of large dams constructed from about 5,000 in the 1980s to about 2,000 in the 1990s (WCD 2000). Major actors in the dam industry as well as representatives of opposing groups saw an urgent need to defuse the dam confrontation.

1.2 The World Commission on Dams

In 1997, the World Conservation Union (IUCN) and the World Bank organized a workshop aimed at discussing some of the controversial issues associated with large dams. Although serious differences concerning benefits and problems deriving from dams emerged, the debate has nevertheless been perceived as constructive. One major result of the workshop was to propose to set up the World Commission on Dams, which was conceived to work as an independent body, without being attached to any organization.

The two objectives of the WCD were "to review the development effectiveness of large dams and assess alternatives for water resources and energy development, and to develop internationally acceptable criteria, guidelines and standards, where appropriate, for the planning, design, appraisal, construction, operation, monitoring and decommissioning of dams" (WCD 2000, 28). Twelve commissioners were selected in negotiations between key stakeholder groups to represent the full spectrum of dam stakeholders. The commissioners came from environmental groups, industry, indigenous peoples' organizations, public administrations, scientific bodies and development assistance agencies. They however did not formally represent a constituency, but were invited to contribute their knowledge and experience independently. This decision was consciously made to enable the process to reach tangible progress on the content in short time, even if the risk had to be accepted that the outcome might not be acceptable to all stakeholders.

In order to fulfill its mandate, the WCD had scheduled an ambitious work plan that aimed to include as many stakeholders as possible. The core elements of the WCD approach were: (a) a balanced multi-stakeholder representation of the full range of dam interests and extensive consultations with other stakeholders; (b) keeping the process lean through working through a small group of commissioners to represent all interests; (c) consensus orientation of the process; and (d) a mandate to reach an outcome within a limited, and rather short, period of 2 years. Within this process, the WCD commissioned and assessed eight detailed case studies on large dams, a crosscheck survey of 125 large dams, which provided data for quantitative analysis, 17 thematic reviews dedicated to a specific topic respectively, and four regional consultations. Additionally, governments, NGOs, financial institutions, private sector representatives and affected communities were invited to present their view on all aspects of the dam debate and discuss the issues raised. The WCD asked all stakeholders to send in their views and concerns in the form of submissions through its web site. In total, 947 submissions were received and evaluated for the analysis. From the analysis, the Commissioners developed recommendations in the form of "core values", "strategic priorities", "policy principles" and "guidelines". Both the results of the analysis and the recommendations were published in a final report, which was released under the title *Dams and Development – A New Framework for Decision-Making* in November 2000 (WCD 2000).

The recommendations developed by the WCD are meant to guide the planning process of future dams. In most cases, the seven strategic priorities are used to structure the messages of WCD. They consist of: (i) gaining public acceptance;

(ii) comprehensive options assessment; (iii) addressing existing dams; (iv) sustaining rivers and livelihoods; (v) recognizing entitlements and sharing benefits; (vi) ensuring compliance; and (vii) sharing rivers for peace, development and security (WCD 2000, 214).

Once the WCD report was published, the commission had fulfilled its mandate and ceased to exist. Despite the efforts undertaken to include the views of all stakeholders, the WCD report was greeted with mixed reactions. The WCD website[3] alone lists almost 40 official reactions that were submitted. Almost all stakeholders agreed to the basic findings (especially the strategic priorities) that were put forward by the WCD while in the view of many, the specific planning and implementation guidelines went too far.

Environmental and social organizations greeted the report as a step ahead towards more sustainable planning and implementation of dams, and directed their criticism to areas where the recommendations could have been even more progressive in their opinion. The response from financial institutions and industry representatives however was quite negative. From their point of view, especially the guidelines were seen as unrealistic and unworkable, overloading the planning process with complicated issues and in effect prohibiting any further development of dam projects. The WCD recommendations were proposed as a voluntary commitment and are not equipped with enforcing mechanisms.

Notwithstanding the different reactions of all stakeholders participating in the debate, the effects of the commission's work are remarkable and continue to have an impact today. The WCD report is used as a benchmark by a wide range of stakeholders, including both dam supporters and opponents.

The WCD can be considered as an exercise in global policy for a very complex and contentious issue. WCD-related implementation efforts are therefore very diverse. When launching the WCD report in 2000, Kader Asmal, the WCD Chairman and then South African water minister, said:

> We have told our story. You can walk away from the WCD report, if you so choose, or turn your backs on the controversial situation which gave rise to the WCD in the first place, and which the WCD report can, if used, help resolve. One only needs to see it not as another crisis but as a sudden opportunity. What happens next is up to you. (Kader Asmal, cited in Baur 2001, 29).

Since then, a variety of stakeholders have implemented a multitude of follow-up activities. Several development organizations, for example, jointly supported the formation of a global Dams and Development Programme (DDP) which is currently hosted by UNEP. Activities of DDP have included promoting multi-stakeholder dialogue at national, regional and global levels and producing non-prescriptive tools to help decision-makers in applying the WCD recommendations.

More than 5 years on, it has become possible to take stock of the first experiences made with the WCD recommendations and extract lessons learned in order to improve WCD implementation approaches in the future.

[3] www.dams.org

1.3 WCD and German Development Cooperation

The German Ministry for Economic Cooperation and Development (BMZ) is one of the international actors that have fully endorsed the WCD recommendations as guidance for all dam related activities (BMZ 2006). They are seen as a viable and useful frame for action with the capacity of ensuring the sustainability of future dam projects. BMZ's commitment also binds its implementing agencies, such as the Gesellschaft für Technische Zusammenarbeit (GTZ) and the KfW Development Bank. These institutions implement the position of German development cooperation on WCD (GTZ and KfW 2004). Dams remain an option for German development policy to advance sustainable energy and water projects. Future dam projects, however, must meet the WCD recommendations and provide a real benefit both for the country concerned and for the directly affected population. German development cooperation works towards implementing the WCD recommendations both at international and national levels through GTZ's project "Implementation of the WCD recommendations".

The project is primarily advising BMZ on dam-related matters, streamlining the WCD approach in projects and liaising with national and international dam stakeholders. Additionally, it supports various efforts related to the implementation of the WCD recommendations as well as the continuing global dialogue on dams. In particular, GTZ encourages national level multi-stakeholders dialogues on dams and sustainability, aimed at building consensus among country stakeholders on institutional and legal frameworks. Among other purposes, the GTZ project aims to study concrete examples on how the issues raised by the WCD can be pragmatically addressed, to document these approaches and to widely disseminate them among dam stakeholders.

This chapter examines two implementation efforts that have been or are currently supported by GTZ. The authors have been involved directly in both cases that represent two different approaches towards the sustainability of hydropower development inspired by the WCD report. The first case analyses the Nepalese national dialogue on sustainable dams that has taken place since 2003. While many countries are in the process of undertaking dialogues on dams, the case of Nepal has already seen the completion of the initial stages and thus offers the possibility of evaluating studies and various reports to extract lessons to be used for potential further dialogue activities. The second example presents a WWF-led initiative to establish practical tools for hydropower site selection in the Mekong region. It represents a landmark for collaboration between stakeholders in sustainable hydropower and river basin approaches in the area of option assessment.

2 "Constructive Dialogue on Dams and Development in Nepal"

Through its project on the implementation of the WCD recommendations, GTZ has supported stakeholders in Nepal to undertake a constructive dialogue on dams and development. The first phase of the dialogue was launched in early 2003 by several

Nepalese institutions and experts previously involved in the World Commission on Dams. These experts continued to be involved in dam issues and are currently members of DDP. In January 2003 a steering committee was formed by representatives from various institutions.

In November 2004, a scoping report was presented to the Nepalese Government in a consultation meeting in order to provide a comprehensive review of existing laws and regulations related to dams in Nepalese policies and discuss guidelines towards the implementation of WCD recommendations.

It was agreed that in-depth discussions would have been held on four out of the seven strategic priorities identified by WCD: "Gaining Public Acceptance", "Ensuring Compliance", "Recognizing Entitlements and Sharing Benefits" and "Comprehensive Options Assessment". GTZ supported IUCN Nepal as the lead organization in the process and the implementation of this second phase of the dialogue process started in January 2005. The results of the discussions on the four selected strategic priorities have been published in four reports at the end of 2005 (Siwakoti and Shrestha 2005; Singh et al. 2005; Dixit and Basnet 2005; Pokharel 2005).

2.1 Limitations of this Case Study

Reviewing a process such as the Nepalese dams dialogue poses an important and yet difficult challenge to the supporting institutions and organizers trying to document and facilitate the process. Practices like national dialogues defy traditional standards of measurement because the factors they attempt to influence are partly abstract and recognizable only in the mid- and/or long-term perspective. On the one hand, specific activities such as the dissemination of the WCD report, the engagement of all stakeholders with emphasis on those not previously involved, or facilitating the flow and availability of information are relatively easy to monitor. On the other hand, the overall impact of a dialogue process on improving dam policies in a given country is much more difficult to judge. It is important to recognize some major limitations in measuring the effectiveness and impact of external assistance to a dialogue process. These include:

- The difficulty to establish causality between the dialogue process and the possible outcomes due to the immeasurable impact of external variables;
- The challenges faced in capturing data and information on the dialogue process and outcomes;
- The high costs of making comprehensive assessments;
- The existence of potential unintended impacts which are by nature not captured in the scope of the evaluation.

In addition, further obstacles in assessing the dialogue in Nepal were posed by the unfavorable political context as well as problems relating to ineffective communication with the relevant actors caused by inadequate telephone and internet facilities.

The evaluation was conducted through a desk review of project documents, a literature survey and discussions with selected experts.

2.2 Nepalese Context

Nepal is one of the least developed countries in the world, with a population of about 22 million, growing with an annual rate of 2.5 percent. According to recent estimates, about 42 percent of the Nepalese population lives below the poverty line. Nepal is and has been an agrarian economy with over 80 percent of the people still dependent on agriculture for their livelihoods.

The country is particularly rich in water resources. The gradient provided by the mountainous topography, the monsoon rains and the Himalayan-fed rivers offer a great opportunity for hydropower development. The hydropower potential in Nepal is estimated to be 83,000 megawatts (MW), of which 42,000 MW are economically feasible at present. A substantial proportion of the potential is based on reservoir projects, rather than run-of-river power stations. At the end of 2002, the national grid supplied electricity to an estimated 33 percent of Nepal's population. An additional 7 percent had access to electricity generated from alternative energy sources like micro-hydro and solar (ADB 2006). Only 18 percent of the Nepali population has access to electricity services from the national grid (Dixit et al. 2004).

Therefore, Nepal is keen to encourage hydropower development. Private-sector investment in hydropower has gradually increased since 1992 (ADB 2006). However, considering the poor infrastructure development in many mountainous areas, the construction of large hydropower dams in this geologically unstable area is particularly complex.

Additionally, factors that have so far contributed to the low level of hydropower development, such as the country's political instability, are now changing after an agreement with the Maoists and the ceasefire has been signed with the major political parties of Nepal in September 2005. Conditions for hydropower development have thus improved significantly, also in relation to the possibility of exporting electricity, for example to India.

In this context of a high development potential, but equally high interests and risks at stake, a dialogue process can make considerable contributions.

2.3 Initiation, Objectives and Outcomes of the Nepalese Dialogue on Dams

A request for financing the second phase of the national dialogue on Dams and Development was submitted to GTZ by IUCN Nepal, who facilitated already the first phase of the national dialogue process (GTZ 2004). IUCN Nepal has coordinated the Dialogue Task Force representing the government, the electricity sector, national and international NGOs and a research institute. In its proposal, IUCN Nepal requested a budget of US$ 25,000, which was financed by GTZ under a grant agreement. The process has defined the following outcomes and outputs:

Outcomes

- Improve understanding of WCD recommendations within Nepal,
- Improve understanding of dam issues, costs and benefits for Nepal,
- Improve participation of all relevant stakeholder groups in developing a national guideline for improved decision making, planning and management of dams and alternatives for Nepal.

Outputs

- Improved public involvement through consultations and workshops on selected strategic priorities under the lead of different individuals/institutions from the Task Force,
- Preparation of theme papers addressing the four selected strategic priorities based on literature review, key informant interviews and possibly focus group discussions,
- Publication and diffusion of the theme papers, both in Nepali and English,
- Preparation of a summary of key recommendations to inform policy makers and mass media,
- Provide background information for future activities,
- Plan future phases of the dialogue.

The initiators of the process as well as GTZ agreed to define the objectives of the dialogue rather generally, even if this posed difficulties at the evaluation stage. This was considered necessary in order to reach a broad stakeholder agreement and promote participation in the process. It was deemed important to guarantee sufficient flexibility and openness of the process in order to increase the likelihood of active participation by a diverse range of stakeholders and to reduce the risk of blocking the process at an early stage.

2.4 Implementation Steps

Accordingly, due to the sensitivity of the topics and the arduous political climate, a flexible implementation approach was chosen. The decision regarding the discussion topics and the selection of individuals/institutions meant to provide input for the discussion, as well as the timeline and selection of participants were left to the Task

Force. Its capacity to properly coordinate the process has been established during the first phase of the dialogue and has been reaffirmed by DDP who had supported this first phase.

Seven Task Force meetings were organized from January to November 2005. They are listed by date in the final report of IUCN Nepal (IUCN 2005). The four thematic papers on the strategic priorities were discussed in four different workshop formats.

A formal Evaluation & Monitoring Program, which would have served as a basis for measuring the degree of achievement of the agreed objectives, was not foreseen when IUCN and GTZ initiated their cooperation on the Nepalese dialogue on dams. However, a reporting procedure was established and IUCN Nepal provided three reports during the project duration. Although the final report mentions the absence of a formal evaluation, during the last Task Force meeting, the participants were asked to provide their reflections on paper cards. Overall, as it will be expanded later, the dialogue provided a platform for discussions between stakeholders with different interests and acknowledged the need to adapt some points of the WCD report for the Nepalese context.

2.5 Results: Outputs and Outcomes

The four reports published under the Nepalese dialogue (Siwakoti and Shrestha 2005; Singh et al. 2005; Dixit and Basnet 2005; Pokharel 2005) differ in the completeness of information provided on the discussion on specific topics, comments of participants and their experience. The WCD strategic priorities have been disseminated together with the reports to a wider audience. However, only the English version of the documents is available to date. In addition, the proposed summaries for key decision-makers and mass media have not been prepared yet; they are scheduled for the end of 2006. A short summary of each report will be prepared and translated into Nepali.

Beyond achieving its intended purposes and apart from the well-organized implementation process, the second phase of the national dialogue accomplished three important outcomes: (i) it provided a platform for stakeholders representing different interests related to hydropower projects; (ii) it allowed a consensus among those stakeholders for the development of future dam projects subject to the avoidance or mitigation of the negative impacts, and (iii) it made clear the need for institutionalizing dialogue mechanisms.

2.6 Lessons Learned

Based on the experiences of the Nepalese WCD dialogue process to date, some key observations for designing national dialogues on dams can be highlighted.

2.6.1 Political Context

The political context was not the most appropriate one for a dialogue process. In fact, the security situation in the country prevented the necessary field visits. This is made explicit by the final report addressed to GTZ (IUCN 2005).

Another issue is the capacity of the public sector to act as a major stakeholder in the process. Nepalese institutions proved to be weak, and their organizational structures often do not fully correspond to their mandates. They suffer weak interagency coordination and lack of skilled professionals, funding and technical and logistic facilities (ADB 2006). Only ministers and deputy ministers are in a decision-making position. This limits the possible impacts of a dialogue process or at least slows it down.

Therefore, the political context and resulting challenges should be briefly examined before deciding to invest resources in national dialogues. At the same time, dialogues have the potential of improving the quality of political confrontation.

2.6.2 Ownership of Dialogue Process and Outcomes

A possible lack of ownership of the dialogue process by the government is discussed in the final report addressed to GTZ (IUCN 2005). In fact, while some consensus was achieved in the discussion among the key stakeholders, a more explicit commitment by the government would have strongly supported the dialogue process and contributed to a greater success.

Consequently, a clear statement of involvement by the government should be required prior to initiating support to dam dialogue processes.

2.6.3 Implementation Approach

The dialogue process suffered to some extent a lack of coordination between the different actors involved and their activities. The steering committee, comprising seven government agencies, two non-government agencies, five international NGOs, two private sector companies and two associations of local communities, was too large to become frequently involved in the process. This was sought to be addressed by establishing a so-called task force, which however ended up having an almost identical membership as the steering committee.

A stronger support by the steering committee would have been helpful. The committee should have defined criteria for selecting participants, representation, duration, issues, methodology, ground-rules, and type of dialogue. The establishment of a more independent advisory group could have been useful. It could have provided feedback during the implementation of the intervention and possibly conducted an external evaluation. Members of this advisory group could have been representatives from other donor agencies, foundations or similar institutions.

Accordingly, an independent advisory body should have been established for supporting the organization of the process and the coordination among stakeholders.

2.6.4 Goal and Outcome Definition

Because of the absence of pre-consultations, participants in the dialogue process worked with a rather superficial definition of objectives (while, as it has been mentioned, the outcomes were described more clearly). Depending on the specific context, and keeping in mind the limitations of each stakeholder, a clear definition of the objectives should have been achieved before the process had begun. The scope and limitations of each objective would have needed to be clarified in order to receive the expected inputs from the participants. Generally, if this does not occur, a superficial treatment of the topics and rather general statements are likely to result from the process.

In the case of the national dialogue in Nepal, the four publications reflect more or less the lack of a clearer definition of the dialogue outcomes. While only one of them (Ensuring Compliance) is more specific on experiences with past projects and provides short case studies, the other three publications tend to be more general. This judgment is confirmed by most of the comments received by the participants of the workshop on the topic "Comprehensive Options Assessment" (Pokharel 2005). The absence of a clear definition of the objectives may in extreme cases even affect the credibility of the entire process.

Therefore, it is important for the proponent as well as for the supporting agency to focus on the definition of the purposes and to specify the outcomes with regard to the impacts of the project. It has to be kept in mind that there are trade-offs between undertaking a process with a clear objective and merely keeping all stakeholders on board. In difficult circumstances in fact, "constructive vagueness" is necessary to ensure the continued participation of key stakeholders.

2.6.5 Appointment and Training of a Facilitator

It was not possible to evaluate the performance of the facilitator of the second phase of the national dialogue process, IUCN Nepal. However, past experiences of other national dialogue processes demonstrated that often facilitators lacked relevant skills like issue identification, reframing, summarizing and recording. Frequently, in such dialogue processes, the participants' discussion centered on debated issues rather than concentrating on the analysis of a topic over a defined period of time. Facilitators can add tremendous value to dialogue processes if they have the required skills.

Training of the facilitator by a professional organization should be a standard component of a dialogue project. The training institution could also serve as an observer at least for the initial meetings and add to the quality of the dialogue meetings by providing professional feedback.

2.7 Comments and Conclusions on the Nepal Case

Dialogue processes in the dam sector will very likely remain an instrument of GTZ in the future. The review of the Nepalese dam dialogue illustrates the major achievements of the process in terms of the remarkable interest in the process of key stakeholders as well as their willingness to provide constructive inputs.

The case of the Nepalese dialogue on dams has demonstrated to be a good example for assessing the impacts of the WCD report and reflects its main strengths and limitations. The main lesson learned concerns the need to contextualize the WCD's recommendations in a given context. Nepalese stakeholders have described this necessity as "Nepalising" the WCD's recommendations. It was possible for Nepalese stakeholders to constructively work on four of the seven WCD strategic priorities, while other areas were seen as too contentious to be covered in the dialogue. One example is the 7th strategic priority (sharing rivers for peace, development and security). These recommendations on dealing with dams in international river basins was seen as highly politically sensitive and the relevant WCD recommendations were considered biased in favor of lower riparian countries by some stakeholders. Similarly, the dialogue participants found that they had to stick to the level of strategic priorities, as there was widespread disagreement about how the 26 guidelines should be applied in Nepal.

This reflects a broader concern in relation to the WCD report, which is shared by many stakeholders worldwide. The Nepalese stakeholders, as many others in other parts of the world, perceived that the WCD guidelines, if followed literally, would make it all but impossible to build large dams. However, it has been clearly stated in the WCD report that the guidelines are meant to offer guidance, not to serve as a regulatory framework. This is in accordance with the need of "Nepalising" the WCD recommendations.

This review and the activities of the Nepalese stakeholders represent the starting point in order to consider the opportunity of continuing the dialogue. In case of further involvement, this experience suggests that it would be desirable to focus on capacity building in local affected communities.

Concerning the role institutions like GTZ can play, it will be necessary for them to prepare comprehensive Terms of References when financing national dialogues. Information such as the timeline of activities and a break down of the budget should be provided in more detail. This method forces the proponent to be more specific on project implementation approaches and hence to recognize constraints and limitations early enough in the planning phase to deal with them.

Processes gain credibility if a time-bound program with clear activities is provided. This helps a broader audience to understand the issue. It is also unlikely that a large number of people will work through rather detailed and abstract publications, which have been for example the main outcomes in the second phase of the Nepalese dialogue. Shorter and more focused versions of these publications would very likely have been of more help for interested groups. Comprehensive case studies would be of interest for politicians and the local population alike. The preparation of teaching materials derived from specific case studies could provide a good introduction to discussions with local people possibly affected by a proposed dam project.

This confirms the necessity to adapt the WCD standards of participation and decision-making to contexts where international support encouraging participation might represent the first national experience in this direction.

3 WWF Initiative "Environment Criteria for Hydropower Development in the Mekong River Basin"

Another activity supported by the GTZ's WCD project is the WWF's initiative "Environment criteria for hydropower development in the Greater Mekong Region."

One of the core demands of the WCD is that environmental and social aspects should have the same weight as economic and financial factors in dams related decision-making processes. This principle should already be applied at the options assessment stage. However, in practice relevant information on environmental and social consequences of dam development is very often not available at the river basin scale. This makes it impossible to fully consider these criteria especially in the early stages of options assessment.

The WWF's initiative addresses this issue by defining relevant criteria and researching necessary knowledge on environmental issues. The aim is to enable dam decision-makers in the Mekong basin to better take environmental aspects into account in dam options assessment.

3.1 The Mekong River Basin Context

Recent economic crises and political upheavals in the Indochina region have "protected" the Mekong from the dramatic changes in landscape and flood patterns that have damaged the ecosystems of many of the world's rivers. However, more rapid development is currently getting under way in the Mekong region, resulting both in threats and opportunities for the environment. The region has a unique opportunity to become a model for sustainable development by improving living standards without destroying the environment.

By 2010, it is expected that international development agencies and country governments will invest an additional US$ 10 billion or more in infrastructure development in the Greater Mekong Subregion. An initial database listed 260 existing and planned large hydropower projects (WWF 2005). Newer information indicates that there is a total of 59 dams already existing or under construction and 151 proposed or potential dam sites. An integral part of development opportunities in the region is represented by the growing demand for electricity. Apart from insufficient generation capacity, access to electricity is still limited, with for example less than 20 percent of Cambodians having access to electricity due to problems in power distribution and limited purchasing power of the urban and rural poor. Energy demand in the region is expected to more than quadruple in the next 15 years. There is large undeveloped hydropower potential, in particular in Laos, Cambodia, Myanmar, Vietnam and in the Yunnan Province of China. It is thus not surprising that governments and the major institutions driving development view hydropower and an extensive regional power grid as the solution to the energy gap. However, hydropower, if developed indiscriminately, is likely to have major impacts on the river and flood plain ecosystems and on the local communities that depend on them for their livelihoods.

The threat to fisheries is particularly serious. Inland fisheries produce 80 percent of the animal protein consumed by the 60 million inhabitants of the Lower Mekong Basin. The annual catch is estimated at 1.5 million tons. These fisheries have been valued US$ 1,700 million per year by the Mekong River Commission (MRC), which represents the four Lower Mekong Governments (MRC 2003a). Experience from the rest of the world (and other parts of the Mekong Basin – such as Pak Moon in Thailand – the last dam funded by the World Bank previous to the WCD) has shown that dams are a major threat to the sustainable management of freshwater fisheries, and that the poorest are usually the direct victims of the decline of fisheries. Biodiversity loss is another potential threat. With around 1,200 to 1,700 fish species, the Mekong River is second in the world (after the Amazon) in terms of freshwater fish biodiversity (MRC 2003b). In addition, the other large river in the region, the Salween, is the last major free-flowing river in Asia, and hydropower development threatens to flood pristine environments and to destroy unexplored tourism potential. There is an urgent need for integrating a strong environmental dimension into the energy plans for the Greater Mekong Subregion.

3.2 *Opportunities for Intervention*

Energy development in the region is currently proceeding mainly on a project-by-project basis, with little attention paid to cumulative impacts or to other options such as natural gas, biomass, wind, and solar power. As the case of Thailand has shown, there is considerable potential for energy efficiency and demand-side management measures, in particular in developing economies (World Bank 2004b). Therefore, hydropower is only one of a number of options for the Mekong countries.

To date, a basin-wide comprehensive options assessment, which is one of the major recommendations of the WCD, does not exist for the Mekong. Furthermore, there is considerable opportunity for reducing the potential negative impacts of hydropower development by looking at cumulative impacts and applying innovative and multidisciplinary basin-wide approaches to site selection. The Mekong River Commission recognizes in its Hydropower Strategy of 2001 the need to assist the Mekong riparian countries in energy development, to promote socio-economic and environmental considerations in project planning by means of environmental impact assessments, and to monitor key parameters to evaluate the cumulative effects of reservoirs, particularly on fisheries (MRC 2001).

3.3 The Role of WWF

The WWF Greater Mekong Programme, spanning Thailand, Cambodia, Laos and Vietnam, and with support from its program in China, is a major environmental initiative with a sufficiently strong institutional presence to engage with all key players. WWF already has signed Memoranda of Understanding with the Asian Development Bank, the Mekong River Commission, and IUCN, and has developed good working relationships with governments in the region. WWF is thus in a good position to work with key stakeholders to ensure that energy development in the region proceeds sustainably. In addition, WWF has created in 2001 the Living Mekong Programme (LMP) under the regional WWF Office, with the aim to marry successful biodiversity conservation with sustainable development, using the tools and approaches of integrated river basin management and eco-region scale planning.

During its initial scoping mission in 2001, LMP identified large-scale infrastructure (and notably hydropower dams) as the single main threat to the aquatic biodiversity of the Mekong basin. This has been confirmed by all assessment and planning exercises LMP has undertaken so far – either individually or in partnership with others.

Experiences since 2003, including the Nam Theun 2 dam project consultation process led by the World Bank, have shown the limits of lobbying on individual projects. The proponent-to-opponent confrontations often lead to detached and sometimes sterile debates that only rarely affect the final decision significantly. They waste time and resources on both sides, and often draw the focus away from other projects that go ahead with very limited consultation processes.

These outcomes led LMP to move away from a case-by-case approach and towards engagement with key partners (governments, basin-scale development organizations like MRC and ADB) in an integrated approach to dam planning on the scale of a single large watershed and/or a region.

3.4 Objective of the Initiative

Under the 1995 Mekong Agreement, the countries of the Lower Mekong Basin agreed to "promote interdependent sub-regional growth and cooperation among the community of Mekong nations, taking into account the regional benefits that could be derived and/or detriments that could be avoided" (MRC 2001, 10). Since both benefits and detriments of hydropower can accrue on a regional scale, options for development should therefore be assessed in a regional context. In 2001, MRC published a hydropower development strategy, which set out policy principles and strategic directions for MRC's involvement in the hydropower sector. It also defined a list of priority activities, including: "Review and update existing studies of potential hydropower development projects in Lower Mekong Basin and establish a tentative ranking of projects in a regional context" (MRC 2001, 10).

Under this political framework, since 2004, WWF LMP is promoting a new basin-scale initiative called "Environmental Criteria for Hydropower Development in the Mekong Basin" to a wide range of partners. The project aims to develop practical tools that result in improved decision-making in site selection as well as construction, operation and decommissioning of hydropower schemes, in particular by promoting basin-wide and cross-basin assessments of options and impacts. These tools are intended to contribute to the implementation of the WCD recommendations in the region, in particular to the strategic priority of options assessment. GTZ's WCD project financially supports the WWF initiative since 2005 with a total contribution of € 30,000 to date.

3.5 Approach, Expected Outcomes and Activities

The initiative is following a two-pronged approach: First, it focuses on the energy supply side by supporting the development of hydropower scenarios and projects which have lower social and environmental impacts without compromising economic objectives. At the same time, it works on reducing the energy demand by supporting the improvement of energy management practices in the countries of the Lower Mekong Basin.

By a decreased and more productively used energy demand and by mitigating the consequences deriving from hydropower projects, the cumulative negative impact of hydropower generation in the Mekong basin can be reduced, while at the same time using the potential of this renewable energy source.

WWF aims to bring together all relevant stakeholders (e.g. multi-lateral financing institutions like ADB, the Mekong River Commission, national and relevant local governments, the private sector – both funding institutions and industry – and the population) in order to develop solutions for the energy sector and hydropower development in a more open and transparent way.

3.6 Preliminary Outcomes of the Initiative

3.6.1 Development of the Habitat Classification Map

In February 2005 the LMP started working with technical support and relevant expertise from the WWF network, mainly from WWF US Conservation Science Unit, WWF Germany and the WWF Global Freshwater Programme, on the development of a habitat classification map. The development process included a training workshop on Geographic Information Systems for staff from the Mekong River Commission, UNEP and WWF Greater Mekong Programme, a workshop to define criteria, methodology, datasets and classification thresholds for the maps. Two sets of maps were produced by digitalizing data for the whole region and including various information layers, presenting a classification of streams and a classification of sub-basins. A draft version of the habitat classification map has been produced at the end of 2005 and subsequently validated by local biodiversity experts, in particular in relation to the classification of the different stretches of the Mekong River and its tributaries. The overlaying with social and economic data has yet to be completed.

The Mekong Habitat Classification map was presented to senior staff of the Mekong Department of the ADB and to the Mekong River Commission. WWF also presented a paper on the map and a concept paper outlining options for hydropower development at the Regional Forum on Integrated River Basin Management (Chiang Rai, Thailand, 28–29 November 2005) organized by the Mekong River Commission. The audience welcomed the paper as a very important decision-support tool for high-level policy makers. The map will be used in the next steps of the initiative by the Task Force for Environmental Criteria for Hydropower Development (see below).

The limits of the map in terms of data gaps have to be recognized. Habitats have been used as a proxy to biodiversity because information on the distribution of freshwater biodiversity is still very fragmented, and despite on-going efforts from different nature conservation organizations, these gaps will not be overcome as quickly as desired.

3.6.2 Task Force and Steering Group for Environmental Criteria for Hydropower Development

Jointly with the Secretariat of the Mekong River Commission and the ADB, WWF identified the establishment of a Technical Working Group as a key element to push forward the concept of options for hydropower development. In a meeting in January 2006, the Secretariat of the Mekong River Commission approved this approach and confirmed its commitment to participate in the now renamed Task Force on Environmental Criteria for Hydropower Development in June 2006.

Subsequently it was agreed that the initiative will move forward through two agreed structures, a task force and a steering group. The task force members include technical specialists from ADB, MRC and WWF, while the steering group is composed of higher level policy and management representatives of the three organizations.

In Phase I, which ended in December 2006, the Task Force has commissioned a group of consultants to conduct an initial assessment on hydropower development in the Greater Mekong Sub-Region. A report has been produced, which includes a literature review of hydropower development plans, safeguard policies and procedures, Environmental Impact Assessments standards and practices, as well as outlines options on the way forward. The first component of the report will be published and distributed as a "stand-alone" report, while the second component will be used in internal discussions within and between WWF, ADB and MRC, in order to plan phase II.

The joint approach has been presented to the hydropower industry at a major conference on water resources and renewable energy in Bangkok in November 2006. In addition, ADB and MRC presented the initiative to the Greater Mekong Subregion Working Group on the Environment as well as the MRC Council.

3.7 Summary of Lesson Learned so Far

The main stakeholders involved in the development of the Mekong region have recognized hydropower development as a very important, but also politically sensitive option. As yet, there is no comprehensive plan for the development and assessment of its cumulative impacts. The main WWF partners have therefore shown support for the initiative from the beginning, but have been very cautious in moving it forward. Therefore, the schedule developed in 2005 took more time to be implemented and more lobbying was needed.

The various partners have shown great interest in the habitat classification map. However, its usefulness will only be proven in the next steps of the Task Force' activities and in the negotiations with the relevant governments on the alternative options for hydropower development. In particular, for the discussion about maintaining free-flowing rivers, e.g., the main stem of the Mekong and valuable tributaries, it will be a very important technical visual and communications tool.

The initial discussion with the members of the task force shows the commitment of the partners for the process. However, the task force is limited to providing technical recommendations. The steering group will make decisions on the way forward on joint work on the environmental criteria, and is responsible for relations with the governments and other external stakeholders. The Governments in the region, the relevant ministries and the multi-lateral and private financing institutions need then to be convinced to integrate the recommendations into their own work. This certainly requires a relatively long process of discussion with those stakeholders.

4 Overall Conclusions

Both the dam dialogue in Nepal and the WWF's initiative in the Mekong basin illustrate the impacts of the WCD. With a global crosscutting policy process like the WCD, some time is needed until its impacts can be properly assessed. Now, after more than 5 years since the publication of the WCD report, it appears that the WCD recommendations have influenced the development of dam policies in many and more subtle ways than it was previously envisaged by both proponents and opponents of the WCD findings. The fact that its recommendations have been discussed – often controversially – in multi-stakeholder processes at the global, national and local level has been valuable for all involved, as demonstrated by the Nepal case. Through exchange of experiences and positions, different stakeholder groups learned from each other and arrived at a broader understanding of the complex tasks of dam planning and building. The actual outcome of dialogue processes consists of consensual recommendations, which however are by nature rather vague and difficult to enforce.

WCD was never intended as a directly applicable blueprint on how dams development should be undertaken in future. GTZ's initiatives are one contribution in interpreting the broad guidance the WCD recommendations provide and making them useful in concrete decision making situations.

However, the authors consider the creation of informal stakeholder networks to be a more valuable, although less visible, outcome. These informal networks are able to contribute in many ways towards better and more inclusive planning processes, and might defuse conflicts to a certain degree, thus indirectly improving dam-related decision-making.

Although the WCD recommendations have not been adopted on a large scale by developers or lenders, they have come to represent a soft benchmark against which new projects are measured. The first post-WCD large dam to be planned and financed by the World Bank (among other public and private banks) was Nam Theun 2 in Laos. The WCD recommendations regarding livelihood restoration, benefit sharing or compliance have strongly influenced project planning and are regarded as state of the art. In other areas, such as equal regard for environmental and social aspects in early stage options assessment, much remains to be done and is tackled by different actors, such as WWF with partners in the Mekong region. It is likely that some WCD recommendations will become mainstream references in International Financial Institutions' projects, while some other recommendations, especially regarding consultation or the decision-making power of multi-stakeholder groups will not be implemented to the extent demanded by the WCD.

Of equal importance is the shift that has been triggered in the behavior of the stakeholders. While confrontation surrounding specific project is still common, increasing multi-stakeholder efforts towards collaboration and the improvement of prevailing practice can also be noted. Efforts like the WWF-led initiative in the Mekong Region are a constructive way of contributing to and shaping future dam development. Dams will continue to be built, especially in developing countries, where legal frameworks are often not adequate to ensure that the negative impacts of dams are minimized, mitigated or compensated. Hence the constructive common efforts of stakeholder groups with diverse experience and skills are key for a sound interdisciplinary planning process.

References

ADB (2006) Environment assessment of Nepal – Emerging issues and challenges, strategy and program assessments, Manila

Bächler G, Spillmann KR (1996a) Environmental degradation as a cause of war, ENCOP-Final Report, Vol. 1, Zurich

Bächler G, Spillmann KR (1996b) Environmental degradation as a cause of war, ENCOP-Final Report, Vol. 2, Zurich

Bächler G, Spillmann KR (1996c) Environmental degradation as a cause of war, ENCOP-Final Report, Vol. 3, Zurich

Baur J (2001) Mehr Nutzen aus Staudamm-Großprojekten? Zum Bericht der World Commission on Dams. In: Aus Politik und Zeitgeschichte B48–49, 23–29

BMZ (2006) Sector concept water, Berlin

Dixit A, Adhikari P, Bisangkhe S (2004) Constructive dialogue on dams and development in Nepal, IUCN Nepal and Nepal water conservation foundation, Kathmandu

Dixit A, Basnet S (2005) Recognising entitlements and sharing benefits: emerging trends in Nepal's hydropower terrain, IUCN Nepal and Nepal water conservation foundation, Kathmandu

GTZ (2004) Promoting a national dialogue on dams and development in Nepal, grant agreement dated 08.11.2004, Kathmandu

GTZ and KfW (2004) Implementation of the WCD recommendations within German development cooperation – Experiences of GTZ and KfW, Eschborn, Frankfurt am Main

IHA (2006) Sustainability assessment protocol, section B – New hydro projects, Sutton

IUCN (2005) Promoting a national dialogue on dams and development in Nepal – Moving forward on strategic priorities, IUCN final report to GTZ, Kathmandu

MRC (2001) Mekong river commission hydropower development strategy 2001, Vientiane

MRC (2003a) State of the basin report, Vientiane

MRC (2003b) Biodiversity and fisheries in the Mekong river basin, Vientiane

Pokharel JC (2005) Comprehensive options assessment for electricity sector in Nepal: dialogue on dams and development in Nepal, IUCN Nepal and Winrock international, Kathmandu

Singh DB, Singh AS, Shrestha S (2005) Gaining public acceptance: dialogue on dams and development in Nepal, IUCN Nepal and Winrock international, Kathmandu

Siwakoti G, Shrestha N (2005) Dams in Nepal: ensuring compliance, IUCN Nepal and WAFED, Kathmandu

WCD (2000) Dams and development – A new framework for decision-making. The Report of the World Commission on Dams, Earthscan, London

World Bank (2001a) Responding to the WCD report: a progress report from the World Bank, Washington, DC

World Bank (2001b) The World bank position on the report of the World Commission on Dams, Washington, DC

World Bank (2001c) The World Bank & The World Commission on Dams report Q&A, Washington, DC

World Bank (2004a) Water resources sector strategy, Washington, DC

World Bank (2004b) Nam Theun 2 hydropower project – Impact of energy conservation, DSM and renewable energy generation on EGAT's power development plan, final report, Washington, DC

WWF (2005) Preliminary data bank on existing and proposed hydropower schemes in the Mekong river basin, Vientiane

How Global Norms for Large Dams Reach Decision-Makers

A Case Study from Turkey

Waltina Scheumann

Abstract Unlike the traditional path of international policy making for e.g. establishing international environmental regimes, with nation states being the decisive actors, the World Commission on Dams (WCD) has followed a different non-state centered approach for developing global norms. The WCD process therefore has been welcomed by many as a prototypical example of how trisectoral networks can help to overcome stalemate in highly conflict-ridden policy arenas. Independent assessments, however, have pointed to the fact that the consensus achieved *within* the World Commission has not translated into a broader stakeholder consensus because nation states were not represented. The following article shows that the WCD's guidelines have gained ground and analyzes the paths and means of policy (norm) diffusion. A case study from Turkey (the Ilisu Dam on the Tigris River) shows that the Turkish Government, who was among those who rejected the WCD's guidelines, has come under severe pressure from diverse actors. However, this can not be attributed to the specific process character employed, but to the WCD's and NGOs' political influence on major finance institutions, governments of industrialized countries and their aid agencies.

1 Introduction

Hildyard[1] and Goldsmith's book *The Social and Environmental Effects of Large Dams*, published in 1984, was an important but not the only indicator that large

W. Scheumann
e-mail: scheumann@die-gdi.de

[1] Hildyard is one of the directors of The Corner House, UK, one of the leading NGOs involved in the Ilisu Dam campaign.

W. Scheumann et al. (eds.), *Water Politics and Development Cooperation*, 55
doi: 10.1007/978-3-540-76707-7_3, © Springer-Verlag Berlin Heidelberg 2008

dams[2] were socially no longer accepted as symbols of progress but blamed for their negative environmental and social effects. Large dams which are actually a technical means for supplying water to cities and agriculture, for flood control and energy production, were criticized as symbols of a westernized style of development, or as Allan (2000) put it, as symbols of the "hydraulic mission". In the perception of their opponents, large dams had failed to achieve their economic and social targets, and it was argued that benefits deriving from large dams would be unequally distributed, favoring some, and disfavoring those being resettled involuntarily. However, the contested issue was not about dams per se but about governance, and perceptions of the appropriate way in which societies should make decisions about water and energy projects. Focusing on large dams included, although less pronounced, critiques on large-scale irrigation systems. But they do not carry this highly symbolic value that large dams have.[3]

The World Bank, in particular, was targeted and made responsible although most of the large dams constructed were not financed by it but by private investors in conjunction with public Export Credit Agencies. However, the World Bank-funded Indian Narmada dam projects, approved in 1985, were perceived by some as the World Bank's Tchernobyl and thus triggered policy reforms inside the World Bank (Fox 2000). Internal reviews of the World Bank's involvement in large dam undertakings and anti-dam campaigns preceded the Gland Workshop, held in 1997, which was organized by the World Bank and IUCN – The World Conservation Union, with broad participation from the governmental and private sector, and civil society groups. As one of its follow-up activities, the World Commission on Dams (WCD) was established.

Unlike the traditional venue of international policy making for e.g. establishing international environmental regimes, with nation states being the decisive actors, the WCD process has applied a different approach: the 12 members to the Commission have been selected on the basis of their personal capacities and were chosen to reflect regional diversity, expertise and stakeholder perspectives, and did not represent states. This process of global norm development has been welcomed by many as a "prototypical example of how trisectoral networks (including the governmental and private sector and civil society groups) can help overcome stalemate in highly conflict-ridden policy arenas," as a "pioneering approach" and a "unique experiment in global public policy making" over an internationally highly contested issue, and as a means for effective governance.[4] On the other hand, independent assessments and analyses have pointed to the fact that the consensus achieved within the

---

[2] ICOLD (2006, 2) defines large dams as having a height of 15 m or more and a reservoir volume of more than 3 million m^3; ICOLD actually proposes two categories: "very large dams" and "large dams".

[3] As Fox (2000, 304) mentions, "mass evictions provided critiques with dramatic photo opportunities, as when Indonesia's Kedung Ombo Dam forced villagers to cling to their homes while flood water rose around them."

[4] E.g. http://www.globalpublicpolicy.net [Cited 15 September 2006]; Asmal (2001); Khagram (2003); WCD Newsletter No 3, June 1999.

Commission has not translated into a broader stakeholder consensus (e.g. Dubash et al. 2001; Streck 2002; Dingwerth 2003).

This article, however, shows that the WCD's non-binding guidelines have in effect gained ground and influenced major stakeholder groups, including states that had expressed serious reservations toward the WCD guidelines. The article is organized as follows: It first analyzes early attempts of developing norms and guidelines by professional organizations which could not accommodate the growing critiques; it then shows the influence anti-dam struggles had on the World Bank's policy which led, in the end, to initiate the multi-stakeholder WCD process. Since the WCD's elaborated guidelines are non-binding, their actual chance of getting implemented relies on whether they are accepted and applied by national governments with major dam-building programs. A case study from Turkey (the Ilisu Dam on the Tigris River) shows that the Turkish Government, who was among those who rejected the WCD's guidelines, has come under severe pressure from diverse actors, a fact that can be attributed to the growing acceptance of the core values of the WCD guidelines. The article concludes by analyzing the actual trajectories and paths, and means of policy diffusion.

2 Norm Development Through International Professional Organizations

The International Commission on Large Dams (ICOLD), an organization of civil engineers and dam designers, started considering environmental and social issues of dam construction after the Year of Environment (1970) and the UN Conference on the Environment (Stockholm 1972) had taken place. In the 1970s, ICOLD was concerned about dam safety, earthquake engineering, sedimentation etc., and still rather critical on taking environmental issues into consideration:

> In recent years the environmental question has influenced decisions on dam design and construction. We are in agreement with this process and welcome discussions on this issue. (...) We feel that in this area many subjective and unrealistic views have been presented, and it appears that governments and politicians, who indeed should offer guidance on this important topic, have not always had a comprehensive view of the consequences of ecologically-related decisions. It is therefore of urgent concern and importance that professional societies contribute to such discussions, for the benefit of fuller and better understanding of the ecology and necessary protection of the environment. (Groner 1976).

In 1978/79, a Committee on the Environment was established under the leadership of E. Hanks from the UK comprising the ICOLD member countries Austria, Brazil, Spain, USA, Ghana, The Netherlands, USSR, Finland, Sweden and Switzerland. However, in 1978 ICOLD still proudly presented "the world's highest and largest dams, man-made lakes, and hydro-power plants" (Mermel 1978). In the beginning of the 1980s, ICOLD discussed how to incorporate environmental and social goals in the decision-making process, and developed guidelines for conducting environmental impact studies for large dam construction and operation. On its XVIth meeting (1988), environmental issues ("Reservoirs and the Environment – Experience in management and monitoring") were put on the agenda. But it needed another decade that ICOLD initiated a world survey on environmental management practice (WPDC 1991) to evaluate 31 country case studies, which led – in 1992 – to ICOLD's statement that "the idea of maximizing the use of river flow should be abandoned." (WPDC 1992) In May 1997, prior to the Gland Workshop, ICOLD released its *Position Paper on Dams and Environment.*

It has not been the intention of this article to analyze ICOLD's efforts in taking social and environmental issues into account nor on its actual influence on dam planning and construction practices. It is, however, sufficient to state here that ICOLD's policy has not been able to accommodate the growing critiques.[5]

3 The World Bank's Policy Under Review

Major policy changes did not come from the professionalists in ICOLD, but from the World Bank's support of the Sardar Sarovar projects on the Narmada River in western India which had sparked worldwide controversy ever since the Bank had approved the projects in 1985. Critiques have focused mainly on the displacement of small farmers and indigenous people, but also on the way environmental issues were addressed. As a response to these critiques, the Bank had commissioned an independent review.[6] The Morse Commission found that "the resettlement and environmental aspects of the project were not being handled in accordance with Bank policies."[7] The Committee on Development Effectiveness of the World Bank's Board of executive directors discussed the completion reports in 1995, concluding:

> (...) the social dimensions of civil works projects need much more attention. (...) Large dams are an important part of economic development... But investments in large dams need to be prepared thoroughly, appraised rigorously, and implemented effectively.... must be sensible to social and environmental considerations... Their efficacy, efficiency, and sustainability depend on participation and institutional development...[8]

Prompted by the Narmada experience, the Bank reviewed the resettlement aspects of all projects active between 1986 and 1993. The Bank's Operations Evaluation Department (OED) concluded amongst other critiques: the Bank's resettlement guidelines were found appropriate but poorly applied. A major impetus for reviewing its own policy was, according to Fox (2000, 311) that the management wanted to know whether there are other "Narmadas" hidden in the portfolio.[9] The most critical

[5] The International Hydropower Association's (IHA) policy is not analyzed in this article since documents on its website (http://www.hydropower.org) are not accessible to non-members.

[6] The first ever independent review of a World Bank-supported project under implementation.

[7] World Bank website, "Learning from Narmada," http://wbln0018.worldbank.org/oed/... [Cited 17 August 2006].

[8] World Bank website, "Learning from Narmada," http://wbln0018.worldbank.org/oed/... [Cited 17 August 2006].

[9] Cf. Fox's 2000 analysis of the effects of external pressures on the World Bank's internal processes.

finding of the review was the understated number of people affected which would not allow even minimal planning and budgeting for resettlement and rehabilitation. Already in mid-1996, when the report was being finalized, OED started negotiations with IUCN for performing a consultative process on the results of the review, and a planned second phase which would look at on-going Bank-funded dams – and at comparative evidence from industrialized countries (IUCN-The World Conservation Union and the World Bank Group, 1997).

According to McCully (2001, 1455), an activist of the International Rivers Network (IRN), anti-dam struggles seemingly influenced the World Bank' dam policy, but "the most important... was the campaign against World Bank-funding of the Sardar Sarovar dam." On a number of occasions, coalitions between affected people and NGOs demanded radical changes from those actors involved in the large dam business, namely:[10]

- The *San Francisco Declaration* (1988) was a position paper of citizens organizations on large dams and water resource management, released on the occasion of an IRN-sponsored international conference which was attended by sixty people from 26 countries.
- The *Manibeli Declaration* (1994) called for a moratorium on World Bank funding for dams. Three hundred twenty-six groups and coalitions coming from 44 countries demanded an independent review of all Bank-funded large dam projects to establish the actual costs, including direct and indirect economic, environmental and social costs, and the actually realized benefits of each project.
- In mid-March 1997, the First International Meeting of People Affected by Dams took place in Curitiba, Brazil. The *Curitiba Declaration* called for an international independent commission that should conduct a comprehensive review.
- In preparation of the Gland workshop, IRN and 44 NGOs and anti-dam movements asked the World Bank President James Wolfensohn to reject the conclusions of the World Bank's OED review and demanded a "comprehensive, unbiased and authoritative review of (its) past lending for large dams" to be done by "a commission of eminent persons independent of the World Bank" (McCully 2001, 1460).

4 The Multi-Stakeholder Approach of the World Commission on Dams

In April 1997, IUCN – The World Conservation Union and the OED of the World Bank jointly organized the workshop in Gland, Switzerland, to tackle the large dams controversy. The workshop's objectives were set as follows:

[10] McCully (2001); http://www.irn.org/basics/ard/index.php?id=sfdeclaration.html [Cited 6 November 2006].

1. "Review the OED desk study of large dams in terms of its data, assumptions, approach, analysis and conclusions and compare the results to documented experience from other sources, including experience of industrialized countries;
2. Develop a methodological framework for the Phase II study that would consider the critical issues that need to be addressed in determining the future development of large dams – including evaluation of alternatives and social, resettlement, environmental, economic, technical and other relevant policy criteria;
3. Propose a rigorous professional and transparent process for defining the scope, objectives, organizations and financing of follow-up work, including a Phase II study incorporating basic guidelines for involvement of governments, the private sector and non-governmental organizations as well as public participation, information disclosure and subsequent dissemination of results; and
4. Identify follow-up actions necessary for the development of generally accepted standards for assessment, planning, building, operation and financing of large dams that would accurately reflect lessons learnt from past experience."[11]

The most important achievement was the agreement to establish a 2-year World Commission on Dams. It was mandated to review the development effectiveness of large dams and establish internationally accepted standards that would improve the assessment, planning, building, operating and financing of dam projects taking due consideration of environmental and social impacts.

The WCD began its work in May 1998 under the chairmanship of Kader Asmal, who was then South Africa's Minister of Water Affairs and Forestry; its 12 members were chosen to reflect regional diversity, expertise and stakeholder perspectives. The WCD was independent, with each member bringing in individual capacity and none representing an institution or a country. However, one third of the commissioners were considered large dam proponents, one third opponents, and the rest to be moderate reformers or supporters.[12] A Forum was established to act as a sounding board for the work of the Commission, the members of whom came from 68 institutions of 36 countries reflecting the diverse range of interests in the dam debate.[13] Public consultation and access to the Commission was a key component of the process. The WCD pioneered a new funding model: 53 public, private and civil society organizations pledged funds to the WCD process. Those contributing were requested to provide untied financial support, and it was informally ruled that

[11] IUCN-The World Conservation Union and the World Bank Group (1997, 5–6).

[12] The Commission's members were Kader Asmal (Ministry of Water Affairs and Forestry of South Africa), Lakshmi Chand Jain (High Commissioner to South Africa, India), Judy Henderson (Oxfam International, Australia), Göran Lindahl (CEO and President of Asea Brown Boveri Ltd., Sweden), Thayer Scudder (California Institute of Technology, USA), Joji Carino (Tebtebba Foundation, Phillippines), Donald Blackmore (Murray-Darling Basin Commission, Australia), Medha Patkar (Struggle to Save the Narmada River, India), José Goldemberg (University of Sao Paolo, Brazil), Deborah Moore (Environmental Defense, USA), Jan Veltrop (Honorary President, ICOLD, USA) and Achim Steiner (WCD Secretary General, Germany). Shen Guoyi (Ministry of Water Resources, China) resigned in early 2000 and was not replaced (WCD 2000, viii–x; Dubash et al. 2001, 6).

[13] Cf. WCD (2000, xix–xxi).

no single donor should contribute more than 10 percent of the total budget. The WCD was equipped with a US$ 10 million budget,[14] which enabled the WCD to commission eight in-depth dam case studies, 2 country studies (India and China), and several thematic reviews and working papers.[15] In November 2000, the WCD Report Dams and Development. A New Framework for Decision-Making was made public.[16]

When the mandate of the WCD ended, a Dams and Development Project (DDP) hosted by UNEP, was created to follow up on the recommendations of the WCD. DDP is financed with contributions from donor countries. Its second phase which ended in April 2007, was dedicated to improve decision-making, planning and management of dams.[17]

5 Critiques on and Reactions Toward the WCD Process and the WCD Guidelines

It has been acknowledged by many that the WCD produced positive results and important lessons both in terms of policy, and civil society networking and strategizing for influencing policy. Since the WCD was an independent commission with an advisory role, implementation would ultimately depend on and involve dam-building governments, and major finance institutions. As its chairman Asmal (2001, 1432) mentioned: "Disagreement arises principally over the intended regulatory force of the guidelines. If intended simply as principles, there is no problem.... The disagreement arises when it appears that they will form the basis for a new sine qua non standard."

A point of major disagreement referred to the formation of the Commission and in particular, to the criteria for selecting its members which have been perceived as crucial, both for the Commission's legitimacy and the acceptability of the recommendation it released.[18] McCully (2001, 1458), for instance, appreciates that "much to the credit of the IUCN staff coordinating the workshop, the IUCN successfully impressed on the OED/World Bank that the Gland workshop would lack credibility and legitimacy unless anti-dam groups were invited" and represented in the constituent structures of the WCD. In this way, dam opponents were able to wield an unusual amount of power, for without their involvement, the process would lose much of its credibility (McCully 2001). On the other hand, the modest representation of national governments in the Commission's formation and their weak formal inclusion in the consultative process has been perceived as weak link to whether its guidelines would be accepted and translated into practices. As Dubash et al. (2001),

[14] Khagram (2003, 13).

[15] Cf. WCD (2000, xxii–xxiv).

[16] For the WCD's strategic priorities, and criteria and guidelines, cf. WCD (2000, Chaps. 8 and 9), or Fink and Cramer in this volume.

[17] Information on the DDP, http://www.unep.org/dams/About%5FDDP [Cited 15 September 2006], and Fink and Cramer in this volume.

[18] For the constituent structures of the WCD, cf. Khagram (2003), http://www.gan-net.net/publications_reports/cases.html#dams [Cited 15 September 2006].

Streck (2002) and ODI (2006) mention, the consensus of the Commission did not translate into a universal stakeholder consensus because major, wary governments could not get on board. In this respect, the strategies of the governments of India and China are instructive: both Indian nationals in the WCD Commission were regarded as anti-dam; an official from the Chinese Ministry of Water Resources served on the Commission in her personal capacity, but resigned halfway, and the Chinese Govt. declined to provide a replacement (Dubash et al. 2001, 8).

However, as will be discussed in the Turkish case study, the WCD's guidelines have made their way and reached decision-makers.

5.1 Professional Organizations and Private Companies

The WCD report was not appreciated by the two major professional dam-building associations, namely the ICOLD and the International Hydropower Association (IHA). Their major reservations were the following:[19]

- The guidelines knowledge base was created by an unrepresentative small sample: out of about 50,000 large dams existing world wide only eight were carefully investigated, most of which were located in developed countries.
- The terms of reference for these eight case studies would have given disproportionate attention to negative impacts.
- Expert organizations such as ICOLD, IHA and the International Commission on Irrigation and Drainage (ICID), and representatives from countries with major dam-building programs were, if ever, only represented in the Forum not in the Commission.

ICOLD expressed "that your (i.e. the WCD's) Report will be viewed as anti-developmental.... The need for structural solutions, including more dams, is undeniable because there are no other practical solutions.... The WCD recommendations are not universally applicable and should not be considered as such by anyone, including funding institutions."[20] Currently, the ICOLD Committee on Governance of Dam Projects is working on a document titled *Role of dams in the 21st Century to achieve a sustainable development target* (ICOLD 2006). The International Hydropower Association too mentioned basic concerns on the report and particularly on the fact that it could only provide comments on the final document but not on the draft. Overall, IHA stated the need to continue dam construction, and claimed that the WCD recommendations were unrealistic and not applicable, but agrees in principle on the core values and strategic priorities.[21] In July 2006, IHA published its *sustainability assessment protocol* as an alternative to the WCD guidelines.[22]

---

[19] Cf. http://www.dams.org/report/reaction [Cited 15 September 2006]; Box 1.

[20] http://www.dams.org/report/reaction/reaction_ICOLD.htm [Cited 13 September 2006].

[21] http://dams.org/commission/forum/f3_iha.htm [Cited 9 November 2006].

[22] http://www.hydropower.org/sustainable_hydropower/sustainability_guidelines.html [Cited 7 November 2006].

A number of international private companies have taken part in the WCD process as Forum members, namely ABB, Alstom, Atlas Copco, Coyne & Bellier, Enron, Harza, Hydro-Quebec, Lahmeyer, Voith Siemens and Skanska. On the day of launch of the WCD report, Skanska (Sweden), for instance, announced in a press release that it "... intends to apply the guidelines;" Harza (USA) too welcomed the report "as a sound approach to the future development of a very old, yet important, water resource technology." *Hydro Review Worldwide*, a leading industry journal, stated that following the WCD guidelines "practical, implementable policies and practices" must be developed.[23] Others who were less enthusiastic adopted the WCD guidelines later on (e.g. Balfour Beatty, UK) due to international protests surrounding the Turkish Ilisu Dam project.

5.2 *National Governments*

Major critiques came from countries with large dam-building programs, e.g. Brazil, China, India, Pakistan, South Africa and Turkey (see Box 1), but also from smaller countries like Ethiopia and Nepal.[24]

Box 1 Reactions from countries with large-dam building programs
Brazilian National Committee on Dams

"... It is quite possible that if not adequately absorbed and treated, the new guidelines and criteria will cause significant cost increases and schedules overruns, and thus lead to the adoptions of less adequate alternate solutions as compared to those that would be the natural choice, considering the prevailing conditions in a given country, as it is the case of Brazil, where there is still an enormous hydropower potential that can and should be developed before other costlier alternatives. ... The WCD Report had the merit of bringing into discussion important points related to the dam business. The results are however unbalanced by what seems to be a prejudice in not properly considering a larger sample of dams, including well succeeded examples." (www.dams.org/report/reactions/icold_brazil.htm) [Cited 6 November 2006]

CHINCOLD (Chinese National Committee on Large Dams)

"In the 21st century and especially its first half, China, a developing country, will continue its step in the river regulation and build dams when necessary to solve the unequal distribution of water resources among different

[23] Friends of the Earth, March 2001. There is no comprehensive review available on which companies involved in the hydropower business meanwhile referred and committed to the WCD guidelines.

[24] Cf. Dubash et al's independent assessment of the WCD (2001, 15-17); http://www.dams.org/report/reaction/reaction_asfaw.htm [Cited 6 November 2006].

areas and periods.... the Chinese government will continue to give priority to the development of water resources in the course of national economic development... to constantly raising the level of water resources development and dam construction.... No force can stop or prevent dam projects, which are so urgently needed in China, include the Three Gorges Project." (www.dams.org/report/reaction/icold_china.htm) [Cited 6 November 2006]

Government of India, Ministry of Water Resources

"Having made impressive strides since independence in developing our water resources, India proposes to continue with its programme of dam construction and create another 200 BCM of storage in the next 25 years or so to ensure continued self-sufficiency in food grain production and to meet the energy and drinking water needs of a growing population. In view of the above, the recommendations and guidelines of the WCD are not acceptable to us."[25] (www.dams.org/report/reaction/icold_india.htm) [Cited 6 November 2006]

SANCOLD (South African Committee on Large Dams)

"... finds the guidelines to be broadly acceptable but with the following reservations: ... In water deficit regions such as South Africa, the role of dams has been an absolutely critical factor in the country's development and will continue to be so. Even though there are sometimes alternatives for small scale or even occasionally major scale supply, dams are inevitably essential as the most viable solution for larger projects in our circumstances." (www.dams.org/report/reaction/icold_south_africa.htm) [Cited 6 November 2006]

TRCOLD (Turkish National Committee on Large Dams)

The General Directorate of State Hydraulic Works "was disappointed with the whole coverage of the WCD report prepared with a partial approach against the water resources development activities all around the world.... Our organization will continue to work to ensure that these dams will be planned and constructed in an environmentally, socially and economically sound way.... In Turkey dam construction is a vital and unavoidable programme for the country. It is supported by all the political parties represented in the parliament." (www.dams.org/report/reaction/icold_turkey.htm) [Cited 6 November 2006]

These countries' reservations were similar to those of ICOLD and IHA, stressing their need for harnessing their hydropower potential and for developing their water resources through dams as viable options.

---

[25] Detailed comments came from India's WCD FORUM member, the Chairman & Managing Director of National Hydroelectric Power Corp. Ltd. (http://www.dams.org/report/reaction/icold_india.htm) [Cited 9 November 2006].

5.3 Major Financial Institutions

Since dam construction is a high-intensity investment requiring considerable financial resources, the financial institutions' reactions toward the WCD guidelines is of particular importance. Often-mentioned sources for dam investments are international loans from multilateral and bilateral sources, and private sector investment, in most cases in conjunction with Export Credit Agencies (ECA) if the projects are perceived as high risk.[26] Public dam developers can also raise funds at the private market or invest their own revenues. Based on diverse statistic sources, Sunman (1999) analyzed in her contributing paper to the WCD, that multilateral and bilateral flows form a very small component of total investments in dams, while the largest source of finances is the public sector in the respective countries.

The response of the various financial institutions to the WCD guidelines was not uniform. The WCD Report was welcomed by some bilateral donors (among them Germany,[27] Switzerland, Sweden), and major multilateral development banks (e.g. the Asian Development Bank, the African Development Bank) made reference to the WCD guidelines.[28]

The World Bank's reactions towards the WDC Report are somehow mixed: In its first response to the WCD's final report (2000), the World Bank supported the core values and strategic priorities, but refers to its own operational policies. Major differences exist on the project preparation and consultation process, on involuntary resettlement, on indigenous peoples, and on projects on international waterways. The Bank mentioned that in "both developed and developing countries the State has the right to make decisions that it regards as being in the best interest of the community as a whole, and to determine the use of natural resources based on national priorities." Referring to international waterways, "the World Bank considers a blanket prohibition on work with an agency that has built a dam in contravention of good faith negotiations to be too broad and to foreclose many opportunities for productive collaboration."[29] Consistent with the WCD recommendations, the World Bank would support strategic planning processes by borrowers for decisions concerning water and energy to enhance the evaluation of options and alternatives, and would support borrowers in financing the priority investments emerging from such processes, and work with borrowers on new projects to map these against strategic priorities articulated in the WCD Report and assess their applicability in the specific

[26] ECAs provide government-backed loans, guarantees and insurance covering commercial and political risks; loan refusal for specific projects by the World Bank means automatically that dam funding is perceived by private banks as "high risk", and they therefore demand support from ECAs before investing money.

[27] Only the German Federal Ministry for Economic Cooperation and Development (BMZ) and the Development Cooperation organizations have committed themselves to the WCD recommendations.

[28] The Asian Development Bank issued an evaluation study on large dams in Asian developing countries as a complementary study to the WCD report (ADB 2002).

[29] http://www.dams.org/report/reaction/reaction_wb2.htm [Cited 9 November 2006].

setting.[30] While supporting the WCD's follow-up as a Steering Committee member of the DDP, the World Bank made clear to its borrowing countries that there would be *no new loan conditionalities stemming from the report* (Dubash et al. 2001, 16, emphasis added). The World Bank's *Water Resources Sector Strategy* which was released in February 2003, particularly has been perceived by major NGOs as walking away from the WCD process, because its strategy includes new investments in high-risk projects such as large dams.[31] In a letter to the then President of the World Bank Wolfensohn, the former WCD Commission members expressed their disappointment that the Bank has dismissed the WCD recommendations in its strategy (12 July 2002).

Different from the World Bank, the European Investment Bank (EIB) subscribed to the principles of the WCD recommendations within its general policy on the environment, and started formulating a specific policy on financing large dams.[32] Relevant for the EIB and the practices of the EU member countries, the European Parliament and the European Council included the WCD recommendations in its Directive 2004/101 (27 October) on the Kyoto Protocol's Clean Development Mechanism, stating that:

> In the case of hydroelectric power production activities with a generating capacity exceeding 20 MW, Member States shall, when approving such project activities, ensure that international criteria and guidelines, including those contained in the World Commission on Dams November 2000 Report..., will be respected during the development of such project activities.[33]

During the 1990s, the ECAs of the G7 (G8) states played a significant role in the boom of private sector financing for infrastructure, with a small part going to hydropower projects. Today, the ECAs are "the single largest public financiers of large-scale infrastructure projects in the developing world, exceeding by far the total annual infrastructure investments of multilateral development banks and bilateral development aid agencies."[34] Compared with recent figures of World Bank-funded dam projects,[35] it can reasonably be assumed that ECAs offer a great potential for future dam financing. Since private sector investment often demands coverage by ECAs, the OECD Working Party on Export Credits and Credit Guarantees stated in December 2003 to strengthen their environmental and social standards. Their Common Approaches for evaluating the environmental impacts of infrastructure projects supported by their governments' export credit agencies ought to ensure that these standards meet established international standards (including those developed

---

30 World Bank July (2001).

31 Hydropower projects are classified as renewable energy projects, cf. Hartje in this volume.

32 Cf. Worm et al. (2003).

33 European Union (2004).

34 Moore in http://www.dams.org/kbase/submission/showsub.php?rec=ECO044 [Cited 30 January 2007].

35 According to World Bank, September (2001) statistics, it financed 3 percent new dam projects between 1970 and 1985, and 2 percent between 1985 and 1995. It is recently involved in about 1 percent of new dam projects in both developed and developing countries.

by the WCD).[36] The 2003 OECD Recommendation on Common Approaches was revised in November 2005 (OECD 2005), and is up for review in 2006.[37] Although the OECD Common Approaches are legally not binding, practice would accord them great moral force.[38]

A GTZ-funded study (2003) investigated the policies of nine Export Credit Agencies[39] and found that "... they (the WCD recommendations) have had only limited visible influence on ECAs' environmental guidelines."[40] The aspects considered in the analysis leading to this statement cover the ECAs' screening processes and exemptions from screening; the policies regarding transparency and information disclosure; the social and environmental standards used; the monitoring mechanisms; the existence of exclusion criteria, and whether special programs exist for environmentally friendly goods and projects.[41]

However, quite a number of ECAs were informed by and/or referred to the WCD guidelines in one or the other way:

- In April 2001, Euler Hermes, Germany, published guidelines entitled *Consideration of ecological, social and developmental aspects*. Applicants are expected to rely on the latest experience and in the light of current knowledge, e.g. the report of the WCD.
- COFACE's, France, sector guidelines have been informed by the WCD report although a number of reservations were made. In the final version of 2003, COFACE stated that best practice criteria are based on the work of ICOLD, the International Energy Agency, the WCD and on the World Bank's operational guidelines.
- In April 2002, the Japan Bank for International Co-operation (JBIC) drew on the WCD Report when working on its *Guidelines for Confirmation of Environmental and Social Considerations*.
- In 2004, Export Risk Guarantee (ERG), Switzerland, explicitly referred to the WCD recommendations.
- In July 2004, the Export-Import Bank, USA, encouraged project participants to address, to the extent practical, relevant principles contained in the Final Report of the WCD.

[36] For the 6th revision of the Draft Recommendation on Common Approaches on Environment and Officially Supported Export Credits, December 2001, cf. Knigge et al. (2003). The US and Turkey had refused to endorse the Draft Recommendations, although for different reasons.

[37] The revised 2005 Recommendation have been criticized by the ECA Watch network. http://www.eca-watch.org/problems/for a/oecd/Common_Approach ... [Cited 28 November 2006].

[38] http://www.adb.org/Water/Topics/Dams/web/common-approaches.htm [Cited 28 September 2006].

[39] Namely the ECAs of Australia, Canada, France, Germany, Japan, Norway, Switzerland, United Kingdom and the United States of America.

[40] Cf. Knigge et al. (2003).

[41] Knigge et al. (2003, 17–29).

- The Export Credits Guarantee Department (ECGD), UK, adopted a set of *Business Principles* in January 2001, which improved environmental and social assessment procedures.[42]

Private financial institutions also referred to the WCD's guidelines:[43]

- The Sustainable Asset Management (SAM)[44] group, for instance, has a joint venture with Dow Jones Indexes to undertake research on top companies for the Dow Jones Sustainability Index. Subsequently, SAM has formulated a questionnaire for companies involved in the dam-building industry based on WCD recommendations.
- In June 2003, the Overseas Private Investment Corporation (OPIC), USA, articulated that policies would have to reflect the findings of the most comprehensive report, i.e. the WCD's.
- In 2003, Swiss Re, one of the largest reinsurance firms, published a *Focus Report on Dams* stating that "it is Swiss Re's conviction that in future, large projects should be handled in accordance with these principles and priorities (the WCD's)."
- In May 2005, the HSBC Group announced its first *Freshwater Lending Guideline*, and declared that HSBC will follow the WCD *Framework for Decision-making*.[45]

Other private banks that referred to the WCD guidelines are Barclays (United Kingdom), Deutsche Bank and Dresdner Bank (Germany), Société Générale (France), and the Union Bank of Switzerland.[46]

Given the non-binding nature of the WCD's recommendations, they seemingly influenced major financial institutions.[47]

6 The Ilisu Dam Case: WCD in Concert with Transnational Movement Shows Impact

The Ilisu Dam case study will show that it is this venue (i.e. finance institutions) that the WCD recommendations had visible impacts on Turkey who is pursuing a major dam-building program. The Turkish State Hydraulic Works (DSI) was among those

---

[42] As ECGD mentions, these changes would refer more to the negotiations at the OECD level than to the WCD process.

[43] Cf. Worm et al. (2003).

[44] SAM Group was established in Zurich in 1995 to focus on the integration of economic, environmental and social criteria into investing. http://www.sam-group.com/htmld/main.cfm [Cited 15 September 2006].

[45] http://www.greenbiz.com/news/printer.cfm?NewsID=28167 [Cited 4 October 2006].

[46] Cf. Worm et al. (2003).

[47] Whether their actual practices have changed, is worth of an in-depth analysis.

who denied the WCD recommendations as universally applicable. Being Turkey's water development organization, DSI has expressed its disappointment with the WCD process and its results, and stated that it would continue its dam-building program because it is "a vital and unavoidable program for the country."[48] It criticized the WCD process as an unbalanced approach. DSI has perceived the WCD recommendations as interfering in under-developed countries' development strategies, and has assumed conspiracy[49] since the WCD's message "comes after completion of the development of water resources of the developed countries and while under-developed countries start to do something."[50]

Almost parallel to the WCD process, the Ilisu Dam project – a key project of the Southeastern Anatolia Project (GAP) – had come under strong international critiques and protests for social, environmental, cultural and international reasons. While some opponents feared an "ethnic cleansing",[51] others claimed negative effects on downstream countries. Meanwhile critiques have centered on the submergence of the archaeological site of Hasankeyf ("It is a holy site!") and on the resettlement issue ("A human-rights disaster").[52]

The Ilisu Dam sited on the Tigris River, 65 km upstream the border to Iraq (Map 1), is expected to create a reservoir with a volume of 10.4 billion cubic metres (BCM) and a surface area of 313 km^2. The Ilisu Dam is designed to produce approximately 3,800 GWh per year with an installed capacity of 1,200 MW, and expected to generate more than US$ 400 million for the Turkish economy (Altinbilek 2000).

From the very beginning of the GAP (the Masterplan was published in 1989), its major hydraulic infrastructure projects had to be financed from the private sector. Due to the absence of trilateral agreements on the use of the Euphrates and Tigris' waters, the World Bank – according to its operational guideline – did not provide credits (which meant, it is a project of high risk).[53] Export Credit Agencies from Switzerland, the United Kingdom, Germany, Italy, Austria, Japan, Portugal, Sweden and the US, coordinated by the Swiss Export Risk Guarantee, considered funding of the Ilisu project (see Box 2). When the project itself and the policies of the ECAs were strongly criticized by environmental, human rights groups and other

[48] http://www.dams.org/report/reaction/icold_turkey.htm, p. 6 [Cited 15 September 2006].

[49] Similarly, a former chief engineer of India's Central Water Commission perceives the WCD's report as a new form of colonialism ("eco-colonialism") with "vested interests who want to dump their surplus foodgrains and other products into these countries in the guise of helping the latter to protect their environment." http://www.hinduonnet.com/2001/08/14/stories/13140411.htm [Cited 6 November 2006].

[50] http://www.dams.org/report/reaction/icold_turkey.htm [Cited 13 September 2006].

[51] http://www.suite101.com/article.cfm/green_home/69623 [Cited 2 February 2006].

[52] http://www.suite101.com/article.cfm/green_home/69623 [Cited 2 February 2006].

[53] This too meant that the World Bank had no influence on Turkey's policy, as it had, for instance, in the case of the Indus River (Pitman 1998).

Map. 1 Location of the Ilisu Dam site
Source: http://www.foe.co.uk/.../images/ilisu_map.gif [Cited 2 July 2007].

globally acting NGOs[54] on social, environmental and cultural grounds, the consortium announced in December 1999 that four conditions would have to be met by the Turkish Government before project funding would be covered by export credit guarantees. These conditions were as follows:

1. Draw up a resettlement program which reflects internationally accepted practice and includes independent monitoring;
2. Make provision for upstream water treatment plants capable of ensuring that water quality is maintained;
3. Give an assurance that adequate downstream flows will be maintained at all times;
4. Produce a detailed plan to preserve as much of the archaeological heritage of Hasankeyf as possible.[55]

In October 2000, 1 month before the WCD released its report, a NGO-organized Fact Finding Mission visited the region to assess the progress made.[56] The Mission concluded that "the conditions have yet to be met, and that the prospect that they will be met in the near future is remote."[57] Shortly before the WCD Report was released, the Swedish dam-building company Skanska which had a 24 percent stake in the consortium, withdrew from the project, followed by Balfour Beatty and most of the other foreign companies in the consortium (see Box 2).

---

[54] Friends of the Earth, the International Rivers Network, the Center for International Environmental Law, the Washington Kurdish Institute (http://www.ilisu.org.uk), the German WEED and the Berne Declaration from Switzerland (http://www.evb.ch).

[55] Quoted from The Corner House, http://www.thecornerhouse.org.uk, p. 3 [30 January 2007]: The Ilisu Dam, the World Commission on Dams and Export Credit Reform, The Final Report of a Fact-Finding Mission to the Ilisu Dam Region.

[56] The Mission was organized by the Kurdish Human Rights Project, The Ilisu Dam Campaign, The Corner House, World Economy, Ecology and Development, Eya on SACE Campaign and Pacific Environment Research Center. http://ilisu.org.uk/news5.html [Cited 28 November 2006].

[57] http://www.ilisu.org.uk/news5.html, p. 2 [Cited 28 November 2006].

Box 2 Development of controversies over the Ilisu Dam

Events	Year
The Masterplan for the Southeastern Anatolia Project includes the Ilisu Dam as a key hydraulic infrastructure on the Tigris river.	1989
NGOs campaign that ECAs should take social and environmental issues into account, and WEED and urgewald (Germany) initiate the Hermes reform campaign. Information was gathered on projects being of environmental relevance, the Ilisu Dam being one of them.	1990s
First consortium established: Sulzer Hydro (later on VA Tech), ABB (later on Alstom), Balfour Beatty (UK), Impregilo (Italy), Skanska (Sweden), Nurol, Kiska and Tekfen (Turkey), with the Union Bank of Switzerland (UBS) taking the lead in forming the consortium and the financing package. The participating firms apply for export credit guarantees.	1997
First draft report of a formal EIA is released.	03/1998
The Swiss ECA ERG declares it would approve the application of the Swiss firms, if the ECAs do so.	12/1998
The UK Export Credits Guarantee Department (ECGD) issues a report on the resettlement issue and an Environmental Impact Assessment study.	1999
The ECAs concerned agree on four conditionalities to be fulfilled prior to any approval of guarantees. ECGD, in particular, demands consultations with the other riparian states.	12/1999
DSI commissions SEMOR (Turkish consulting firm) to conduct a study on affected people and their socio-economic situation.	2000
NGOs organize a number of fact-finding missions and issue a number of studies; results are forwarded to and shared with ECAs, governments, parliamentarians etc.	1999, 2000, 2002, 2005
ECAs demand an EIA report from the consortium; the Swiss ECA commissions a study in order to evaluate the resettlement plan prepared by SEMOR.	08/2000
The EIA is published but does not accommodate the critiques; NGOs complain that World Bank and OECD standards would be violated.	04/2001
Skanska (Sweden) withdraws from the project, followed by Balfour Beatty (UK) and Impregilo (Italy), and the Swiss Union Bank.	09/2000 11/2001 02/2002
Negotiations start between the Turkish Government and VA Tech (former Sulzer Hydro).	Autumn 2004
Second consortium established: VA Tech (Austria), Alstom, Stucky, Colenco and Maggia (Switzerland), Ed. Züblin AG (Germany), Nurol, Cengiz, Celikler and Temelsu (Turkey).	2005

An amended EIA Report and a new Resettlement Action Plan (ENCON, Turkish Consulting) are published, but only in English, not in Turkish.	07/2005
The supply contract is signed between the Turkish Government and the Consortium, but financing is not secured.	12/2005
The EIA Report and the Resettlement Action Plan are published in Turkish, but are only available on the website of the consortium.	1 and 2/2006
Ilisu Dam construction was officially celebrated with the attendance of the Turkish President Erdogan.	08/2006
ECAs agreed in principle to support the project if conditions were fulfilled.	12/2006

Own compilation, e.g. Setton and Drillisch (2006); internet sources

From the Turkish Government's perspective, the NGO-supported Fact Finding Mission was criticized for not having paid enough attention to on-going archaeological rescue activities,[58] and to the Ilisu Dam Lake Area Subregional Development Plan project which were initiated by the GAP Regional Development Administration (GAP RDA) back in the early 1990s. Both projects had to be deferred for almost one decade due to security issues in the region. Thereafter, the salvage project for the documentation and protection of the archaeological heritage of the area started in 1998 with funds provided by GAP RDA. Educational institutes from within Turkey collaborated with international teams from the United States, Germany, Italy and France to devise a comprehensive schedule for the work. Since then archaeological sites in the area have been extensively surveyed and recorded, and excavations and relief works have commenced (GAP 2005). The Ilisu Dam Lake Area Subregional Development Plan could only start in 2002, which caused a delay of implementing the plan because preferable spatial alternatives for resettlement areas had to be developed (Southeastern Anatolia Project Regional Development Administration 2001).

The Turkish Government reacted critically to the international campaign which, it claimed, was led by UK-based activist groups. The Turkish Ministry of Foreign Affairs stated that the dam would neither reduce the flow of the river nor cause pollution (Ministry of Foreign Affairs 2004). The *Turkey Country Report* (2003) to the Third World Water Forum also claimed that the actual facts were somewhat different than those asserted by the Fact Finding Mission. With reference to the transboundary (downstream) issues involved, the report reads:

> The Ilisu Dam is not designed for irrigation, only for power generation: The water passing through the turbines has to flow back into the river bed. River water flowing into Iraq and Syria will not be polluted because the use of water for hydropower is non-polluting. As a result of Ilisu, new sewage treatment facilities will be built in the towns upstream, thus improving water quality. Ilisu will act as regulator holding back water during the winter floods and releasing it during the summer droughts. (Republic of Turkey 2003, 76)

[58] I.e. in Hasankeyf which is the major ancient town on the Ilisu Dam site.

As a matter of fact, there is no international agreement on the Tigris River, while there is at least the Turkish-Syrian Protocol (1987) pertaining to the allocation of the Euphrates river flow between Turkey and Syria. Referring to the Turkish government's resettlement policy, the meanwhile revised Turkish Resettlement Action Plan has been strongly criticized by Cernae (2006, 4), a former World Bank staff member. His key findings are that

> ... the most serious deficiencies of the current RAP (Resettlement Action Plan for the Ilisu Dam and HEPP Project) are: the absence of a full-planning for income restoration and the absence of an adequate plan and outline for creating the organizational set up and capacity for managing the enormous process of displacing, resettling and restructuring the economic basis for over 54,000 people (likely more).

Michael Cernae (2006, 5) believes that "international lenders intent on consistency with accepted international policies and standards cannot regard this RAP version as ready for decision-making on grant export risk guarantees, construction credits and starting actual project implementation."[59]

However, on 5 August 2006, the Ilisu Dam construction site was officially opened and celebrated with the attendance of the Turkish President Erdogan. A new consortium[60] has been established with Austria VA Tech Hydro (Andritz), Ed. Züblin AG (Germany), Alstom Ltd. and Stucky Ltd. (both Switzerland), and a number of Turkish construction firms (see Box 2).[61] But financing is not secured because the ECAs of the respective countries – among them Germany – have yet to decide whether they cover export credits for this risky and internationally contested undertaking. In December 2006, the German Interministerial Committee agreed in principle to support the project if conditions were met (see Box 3).

What Turkish officials first had denounced to be a campaign led by UK-based activist groups, had meanwhile transcended into a forceful transnational social movement, comprising national and international NGOs,[62] joined by Turkish archaeologists and the World Archaeological Congress, which was mirrored in some Turkish newspapers.[63] Local groups have been able to connect with international groups and vice versa. Mayors from the region, local groups, the Chambers of Lawyers and of Engineers, the Diyarbakir Women's Problem Research and Application Centre are supporting the initiative "Save Hasankeyf." Mayors of the municipalities of Batman, Hasankeyf, and Diyarbakir wrote a letter of protest to the ECAs

---

[59] Cf. also eawag aquatic research (2006).

[60] The first consortium pulled out in 2000/01/02 (Box 2).

[61] T.C. Enerji ve Tabii Kaynaklar Bakanlığı, Devlet Su İsleri Genel Müdürlüğü (2006).

[62] Medico international; Friends of the Earth England, Wales and Northern Ireland (FERN); The Berne Declaration, a Swiss public-interest group; WEED (World Economy, Ecology & Development), Germany; International Rivers Network, USA; European Rivers Network; Environmental Defense, USA; Britain's Society for the Protection of Birds; Kurdish Human Rights Project, England; Greenpeace Mediterranean, Energy Campaigner; ECA Watch (i.e. international NGO campaign on Export Credit Agencies, cf. below), and WWF.

[63] On August 8, 2006, after the official ceremony had taken place, the Turkish newspaper Hürriyet wrote "Goodbye Hasankeyf." The Turkish Daily News, 6 August 2006: "The people in the region still have not made up their minds on whether the start of the dam's construction is a blessing or a curse."

of Germany, Austria and Switzerland, and a delegation of representatives from the region visited Brussels to lobby for their interests, and to ask the European Union to keep a close eye on the planned Ilisu Dam project in the context of Turkey's process of accession to the European Union. And they asserted that they would be ready to appeal to the European Court of Human Rights.[64] The delegation also visited Germany, Austria and Switzerland and had discussions with all relevant ECAs and ministries.

Through the formation of the European ECA Reform Campaign, dam opponents capably addressed the ECAs which were to cover funding of this dam project. This group consisting of numerous NGOs[65] gained an unusual amount of influence on national governments not to approve funding because the ECAs' decisions, unlike the private credit market, could be decided by national governments.

German non-governmental actors have been playing a proactive role in the controversies over financing the Ilisu Dam project: *Medico international*, a NGO with a particular focus on ethnical issues, and *WEED* which had devoted many resources in the Ilisu dam case and in organizing the ECA campaign, were especially effective at organizing public pressure.[66] Since the German Government has been among those who committed itself to the WCD guidelines,[67] and is supporting negotiations on Turkey's accession to the European Union, it has great interest to strictly observe in particular any human rights violations and disagreements over transboundary issues. The German Federal Ministry for Economic Cooperation and Development (BMZ) had actively supported the WCD process and its follow-up, and had accepted the WCD report without any reservations. When the German Interministerial Committee containing four ministries (see Box 3) was asked to approve export credit guarantees (Euler Hermes) for the approximately € 100 million for which the German construction firm Züblin AG had applied, it critically reviewed, amongst other issues, environmental impacts, resettlement plans and whether the Turkish Government has notified and consulted the other riparian countries on its planned development projects (Box 3):

Different to this political venue, *amnesty international* Austria directly addressed VA Tech Hydro (i.e. the leader of the second Ilisu Consortium) when it released its report in April 2006, and reminded VA Tech Hydro that being a member of the Global Compact it would have pledged itself to "support and respect the protection of international human rights within its sphere of influence" and to make sure that it is "not complicit in human rights abuses" (amnesty international Austria 2006, 6), a matter for which there would be quite some evidence.

[64] http://www.rivernet.org/prs06_03.htm … [Cited 14 August 2006].

[65] Member-NGOs are: The Berne Declaration, Switzerland; Both ENDS, The Netherlands; The CornerHouse, UK; ECA-Watch Austria; FERN, EU; Finnish ECA Reform Campaign, Iberian ECA Reform Campaign, Spain and Portugal; Les Amis de la Terre, France; Proyecto Gato, Belgium; Reform the World Bank, Italy; Swedish Society for the Conservation of Nature; Finnish Association for Nature Conservation; urgewald, Germany; WEED, Germany.

[66] This statement is selective since NGOs in the UK, Switzerland and the US have too exerted much pressure on their governments. Lately, activists coordinated by WEED built a cardboard dam in front of the German ministry and handed a petition with 35,000 signatures to Mr. Henckel who presided over a decisive meeting on the dam (World Rivers Review, October 2006, 12).

[67] BMZ pressemitteilung Entwicklungspolitik, Nr. 07/2001, 18 January 2001.

Box 3 The procedure for receiving Euler Hermes export credit guarantees

The private construction firm Ed. Züblin AG applied at the German ECA, Euler Hermes, for coverage of approximately € 100 million.

Euler Hermes' examination report was sent to the Interministerial Committee which includes representatives from the Federal Ministry of Economy and Labor, the Federal Ministry for Economic Cooperation and Development, the Federal Ministry of Finance, and the Federal Foreign Office.

The Interministerial Committee assessed the report, and wrote a statement to the private firm mentioning the conditions to be fulfilled prior to any approval for receiving an export credit guarantee: Out of 150 conditionalities, 40 had to be fulfilled before final agreement; among them was the notification of planned measures with downstream states, and resettlement and water quality issues.

Meanwhile, negotiations, coordinated by Ed. Züblin AG, between the ECAs of Switzerland and Austria and the Turkish State Hydraulic Works were ongoing. In December 2006, the Committee agreed in principle to support financing, but demanded conditions to be fulfilled. Among those was the request to secure income for approximately 55,000 people who would be directly or indirectly affected by dam building; if land would be provided as a means of compensation, the value of these lands should be equal to those expropriated; new settlements with modern houses and infrastructure should be built.

Most importantly, a committee of international experts will be established to monitor compliance, the experts of which are selected by the ECAs involved.

http://www.agaportal.de/pages/portal/presse/pms/2006-12-05_ilisu.htm [Cited 31 January 2007].

To conclude, local and international critiques, and the German Government's self-commitment to the WCD's recommendations in development cooperation have created significant pressure on Turkey's dam policy. During this process, which was informed by the WCD's guidelines, Turkey has made considerable changes, and has updated its Resettlement Action Plan in May 2006 (Republic of Turkey, Ministry of Energy and Natural Resources. General Directorate of State Hydraulic Works (2006)). However, negotiations are ongoing and decisions are pending.[68]

7 Conclusions

The World Commission on Dams' process of setting new norms and standards has marked a shift from traditional policy avenues and international professional expert

[68] An in-depth study would be worth in order to assess the actual changes made.

groups to trisectoral networks. They are perceived to be able to overcome stalemate in highly conflict-ridden policy arenas.[69] It is assumed that they could fill the participatory gap of traditional policy making by including "an increasingly large and divers set of nonstate actors in traditional policymaking venues and deliberations."[70] However, the fact that governments with large dam building programs were not directly represented in the process, unlike negotiations over the establishment of international regimes, had important bearings on what is happening on the ground. While the Commission composed of individuals, and not, as is usually the case, of representatives of states, was able to arrive at a consensus, its New Frameworks for Decision-Making has got mixed responses: The major professional organizations have made serious reservations as did the governments with large dam-building programs. Reactions of major multilateral and bilateral financial institution were not uniform, but, as a rule, more positive.[71]

Turkey being one country that expressed strong reservations to the WCD guidelines came under serious pressure by an emerging transnational social movement and Turkish local groups and actors. The Ilisu Dam case study has demonstrated their ability to ally across countries with diverse actors and on diverse aspects. And the Turkish Government has come under pressure from European governments who had committed themselves to the WCD process and guidelines. This self-commitment translated into an effective factor for spreading the WCD norms, when the same governments had to approve coverage of export credits of the respective national firms being part of the Ilisu Dam Consortium (see Box 4).

Box 4 Paths and means of norm diffusion: the Ilisu Dam case

- Self-commitment of governments, of bilateral and multilateral finance institutions, and of private companies.
- Coalitions of international (EU, USA) and Turkish NGOs, and local actors.
- Campaigning of national / international NGOs, influencing and lobbying governments, parliamentarians, multilateral and bilateral donors, ECAs, OECD and EU.
- Conditionalities for approving export credit guarantees.

Almost parallel to the WCD process, non-governmental groups from Europe and the US had joined and effectively organized the ECA Watch Campaign which addressed finance institutions and private companies involved in the dam business. In this transnational alliance ECA Watch has played a central role in monitoring com-

---

[69] http://www.globalpublicpolicy.net [Cited 15 September 2006].

[70] http://www.carnegieendowment.org/events/index.cfm?fa=eventDet... [Cited 15 September 2006].

[71] Cf. Hartje's article in this volume on the World Bank's incentives to re-engage in high risk projects.

pliance with newly international (non-binding) norms, their "dissemination" and implementation. It has been able to heavily influence international finance institutions, national governments and private groups to commit to them. It has been putting pressure on national (OECD) governments not to guarantee export credits for dam projects that seem to offend internationally set standards. It will be seen how these coalitions are able to further influence the World Bank (who re-engages in large dams) and China's export credit agency, the China Exim Bank, who has become the world's third-largest export credit agency financing contested dams (INR 2006, 3).

References

ADB (2002) Study of large dams and recommended practices, Technical Assistance Consultant's Report No 5828 – REG, September. http://www.adb.org/Water/Topics/Dams/web/common-approaches.htm [Cited 28 September 2006]

Allan JA (2000) The water question in the Middle East, Tauris Academic Publications, London

Altinbilek D (2000) The Ilisu Dam Project. In: Water and development in Southeastern Anatolia: essays on the Ilisu Dam and GAP, Proceedings of a seminar held at the Turkish Embassy in London, February, 29–38

Amnesty International Österreich (2006) Human rights aspects of the construction of the Ilisu Dam and hydroelectric power plant project in Turkey, April. http://www.eca-watch.at/downloads/D1.%20Amnesty%20International_IlisuRAP_final.pdf [Cited 9 October 2006]

Asmal K (2001) Introduction: World Commission on Dams Report, Dam and Development. In: American University International Law Review, Vol. 16, No 6, 1411–1433

BMZ (2001) Neue Kriterien für Großstaudämme. Berliner Konferenz einigt sich auf einen Empfehlungskatalog, pressemitteilung Entwicklungspolitik, No 07, 18 January

Cernae MM (2006) Comments on the resettlement action plan for the Ilisu Dam and HEPP Project. Prepared for The Berne Declaration, Switzerland and the Ilisu Dam campaign Europe, 23 February 2006

Dingwerth K (2003) Globale Politiknetzwerke und ihre demokratische Legitimation. Analyse der World Commission on Dams, Global Governance Working Paper No 6, February

Dubash NK, Dupar M, Kothari S, Lissu T (2001) A watershed in Global Governance? An independent assessment of the World Commission on Dams, World Resources Institute: Washington, DC

eawag aquatic research, Switzerland (2006) Independent review of the environmental impacts Assessment Report (EIAR) 2005 on the Future Ilisu Dam (Turkey). Prepared for Erklärung von Bern – The Berne Declaration, 20 February 2006

European Rivers Network (2000) Ilisu Dam fact finding mission: preliminary findings. River-Net NEWS on Turkish Rivers, 16 October 2000. http://www.rivernet.org/turquie/prs00_01.htm [Cited 15 September 2006]

European Union, Official Journal of the European Union, 13.11.2004, L 338/18, amending Directive 2003/87/EC establishing a scheme for greenhouse gas emission allowance trading within the Community, in respect of the Kyoto Protocol's project mechanisms

Fox JA (2000) When does reform policy influence practice? Lessons from the bankwide resettlement review. In: Fox JA, Brown LD (eds) The struggle for accountability. The World Bank, NGOs, and grassroot movements, The MIT Press, Cambridge, Massachusetts and London, England, 303–344

Friends of the earth for the planet for people (2001) Briefing for investors. Balfour Beatty: WCD Resolution, March. http://www.foe.co.uk/campaigns/corporates/resource/past_ briefings.html#2001 [Cited 27 November 2006]

GAP (2005) Devam Etmekte Olan Projeler
Goldsmith E, Hildyard N (1984) The social and environmental effects of large dams, Wadebridge Ecological Centre. A report to The European Ecological Action Group (ECOROPA), Cornwall
Groner CF (1976) The work of ICOLD, Water Power & Dam Construction, April
ICOLD (2006) The Dams Newsletter, No 5, May. http://www.icold-cigb.org [Cited 6 November 2006]
IUCN-The World Conservation Union and the World Bank Group (1997) Large dams: learning from the past, looking at the future. Workshop proceedings. IUCN, Gland, Switzerland and Cambridge, UK and the World Bank Group, Washington, DC
Khagram S (2003) Beyond temples and tombs: towards effective governance for sustainable development through the World Commission on Dams, Center for International Development and Hauser Center for Non-Profit Organizations, The John F. Kennedy School of Government Harvard University. http://www.gan-net.net/publications_reports/cases.html#dams [Cited 15 September 2006]
Knigge M, Goerlach B, Hamada AM, Nuffort C, Kraemer RA (2003) The use of environmental and social criteria in export credit agencies' practices. A study of export credit agencies' environmental guidelines with reference to the World Commission on Dams, Deutsche Gesellschaft für Technische Zusammenarbeit (GTZ) GmbH and Ecologic, on behalf of Bundesministerium für wirtschaftliche Zusammenarbeit und Entwicklung
McCully P (2001) The use of a trilateral network: an activist's perspective on the formation of the World Commission on Dams. In: American University International Law Review, Vol. 16, No 6, 1453–1475
Mermel TW (1978) Major dams of the world, Water Power & Dam Construction, August
Ministry of Foreign Affairs. 2004. Ilisu Dam, Ankara http://www.mfa.gov.tr/MFA/ForeignPolicy/MainIssues/WaterIssues/ILISUDAMsub3.htm [Cited 15 September 2006]
New Collins Concise English Dictionary (1982) The New Collins Concise Dictionary of the English Language, by William T. McLeod (ed), Collins, 30 August
ODI, Research and Policy in Development (2006) The World Commission on Dams: shaping global policy through multi-stakeholder dialogue and evidence-based research, CSO Case Study 18
OECD, Working Party on Export Credits and Credit Guarantees (2005) Updated recommendation on common approaches on environment and officially supported export credits, TD/ECG(2005)3, 24 February
Pitman GTK (1998) The Role of the World Bank in enhancing cooperation and resolving conflicts on international watercourses: The case of the Indus Basin. In: Salman SMA, Boisson de Chazournes L (eds), International watercourses. Enhancing cooperation and managing conflict, Proceedings of a World Bank Seminar, World Bank Technical Paper No 414, Washington, DC, 155–165
Republic of Turkey (2003) Turkey Country Report – Prepared for the 3rd World Water Forum March 2003, Ankara
Republic of Turkey, Ministry of Energy and Natural Resources. General Directorate of State Hydraulic Works (2006) Ilisu Dam and HEPP project amendments on updated resettlement action plan (URAP), May, encon Environmental Consultancy Co., Ankara
Setton D, Drillisch H (2006) Zum Scheitern verurteilt. Der Ilisu Staudamm im Südosten der Türkei, WEED, May
Southeastern Anatolia Project Regional Development Administration (2001) Southeastern Anatolia Project, Status Report 2001. English version on CD-Rom. http://www.gap.gov.tr/Turkish/Kultur/kuldp.html [Cited 15 September 2006]
Stockholm (1972) The International Commission on Large Dams (ICOLD), an organization of civil engineers and dam designers, started considering environmental and social issues of dam construction after the Year of Environment (1970) and the UN Conference on the Environment in Stockholm (1972) had taken place

Streck C (2002) Global public policy networks as coalitions for change. In: Esty D C, Ivanova M H (eds), Global environmental governance. Options & opportunities, Yale Center for Environmental Law & Policy, New Haven, CT

Sunman H (1999) Financing statistics, trends and policies of international finance institutions, Contributing paper for thematic review III.2. http://www.dams.org/kbase/thematic/tr32.htm [Cited 6 November 2006]

T.C. Enerji ve Tabii Kaynaklar Bakanlığı, Devlet Su İsleri Genel Müdürlüğü (2006) Ilisu Barajı ve Hidroelektrik Santrali Temel Atma Töreni, Dicle Üzerinde Altın Gerdanlık

World Bank Fact Sheet, September (2001) Statistics on the World Bank's dam portfolio. http://Inweb18.worldbank.org/ESSD/ardext.nsf/11ByDocName/NewsEventsPressBackgrounders [Cited 28 November 2006]

World Bank Fact Sheet, July (2001) Report of the World Commission on Dams: World Bank position (http://siteresources.worldbank.org/ESSDNETWORK...) [Cited 6 November 2006]

World Commission on Dams (2000) Dams and development: a new framework for decision-making. The Report of the World Commission on Dams, Earthscan, London

WCD Newsletter (1999) No 3, June

Worm J, Dros JM, van Gelder JW (2003) Policies and practices in financing large dams, April, prepared for WWF International – Living Waters Programme, The Netherlands

WPDC, Water Power & Dam Construction, 5/1991

WPDC, Water Power & Dam Construction, 10/1992

Internet Sources

http://wbln0018.worldbank.org/oed/... (Learning from Narmada) [Cited 17 August 2006]

http://www.adb.org/Water/Topics/Dams/web/common-approaches.htm [Cited 28 September 2006]

http://www.agaportal.de/pages/portal/presse/pms2006-12-05_ilisu.htm [Cited 31 Januray 2007]

http://www.carnegieendowment.org/events/index.cfm?fa=eventDet... [Cited 15 September 2006]

http://www.dams.org/report/commission/forum/f3_iha.htm [Cited 9 November 2006]

http://www.dams.org/report/reaction/... [Cited 15 September 2006]

http://www.dams.org/report/reaction/reaction_asfaw.htm [Cited 6 November 2006]

http://www.dams.org/report/reaction/reaction_icold_et_al.htm [Cited 13 September 2006]

http://www.dams.org/report/reaction/reaction_icold.htm [Cited 13 September 2006]

http://www.dams.org/report/reaction/reaction_wb2.htm [Cited 9 November 2006]

http://www.dams.org/report/reaction/report/icold_pakistan.htm [Cited 9 November 2006]

http://www.eca-watch.at/downloads/D1.%20Amnesty%20International_IlisuRAP_final.pdf [Cited 9 October 2006]

http://www.eca-watch.org/problems/for a/oecd/Common_Approach... [Cited 28 November 2006]

http://www.evb.ch/en/p25000551.html [Cited 6 November 2006]

http://www.evb.ch/index.cfm?page_id=557 [Cited 6 November 2006]

http://www.gan-net.net/publications_reports/cases.html#dams [Cited 15 September 2006]

http://www.gap.gov.tr/Turkish/Kultur/kuldp.html [Cited 15 September 2006]

http://www.globalpublicpolicy.net [Cited 15 September 2006]

http://www.greenbiz.com/news/printer.cfm?NewsID=28167 [Cited 4 October 2006]

http://www.hinduonnet.com/2001/08/14/stories/13140411.htm [Cited 6 November 2006]

http://www.hydropower.org/sustainable_hydropower/sustainability_guidelines.html [Cited 7 November 2006]

http://www.ilisu.org.uk [Cited 15 September 2006]

http://www.irn.org/basics/ard/index.php?id=sfdeclaration.html [Cited 6 November 2006]

http://www.rivernet.org/prs06_03.htm [Cited 14 August 2006]

http://www.sam-group.com/htmld/main.cfm [Cited 15 September 2006]
http://www.suite101.com/article.cfm/green_home/69623 [Cited 2 February 2006]
http://www.thecornerhouse.org.uk [Cited 30 January 2007]
http://www.unep.org/dams/About%5FDDP [Cited 15 September 2006]
http://www.weed-online.org/themen/ . . . html [Cited 14 August 2006]

Global Environmental Governance and Its Influence on National Water Policies

Imme Scholz

Abstract Even without a global convention on the right to water, national water policies and politics are already heavily influenced by global environmental governance processes in other areas, e.g. climate change and biodiversity, which offer many opportunities for linking national water policies to global policies aiming at sustainability. These processes are based on simultaneous interventions of multiple actors on the local, national and global levels.

The article asks whether a global regime is crucial for furthering sustainable water policies and analyses the apparent costs and benefits of already existing global environmental regimes in general as well as their interfaces with water policy. Then, an actor-oriented case study from the Brazilian Amazon (conflicts around the dam and hydro-electric plant at Belo Monte) is presented which shows that in Brazil, national water policies and local water-related politics are already permeated by global governance elements and thus contribute to the implementation of global environmental governance as such.

1 Introduction

Even without a global convention on the right to water,[1] national water policies and politics are already heavily influenced by global environmental governance processes. These processes are based on simultaneous interventions of multiple

Imme Scholz
e-mail: Imme.Scholz@die-gdi.de

[1] In 1997, the UN General Assembly adopted the Convention on the Law of the Non-navigational Uses of International Watercourses, which regulates international cooperation on transboundary rivers and lakes. Recently, many NGOs have emphasized the need for an international water convention which would guarantee the individual right to water, protect water as a public good and ensure that government plays a core role in guaranteeing and protecting water rights, in order to contain the trend towards privatization.

W. Scheumann et al. (eds.), *Water Politics and Development Cooperation*,
doi: 10.1007/978-3-540-76707-7_4,

actors on the local, national and global levels. The existence of legally binding global environmental regimes (e.g. the conventions on climate change, biodiversity, and combating desertification), global concepts (e.g. IWRM) and organizations with UN support (e.g. the World Commission on Dams)[2] offer many points of departure for linking national water policies to global policies aiming at sustainability.

The reasoning in this article is based on two arguments: The first relates to the question whether it is necessary to adopt a specific global convention for each environmental problem area or whether the potential linkages of existing conventions to water policies are strong enough to achieve greater sustainability on national and local level. The second argument is that global regimes or conventions, as one manifestation of global governance, need to be set in motion by public and civil actors on local, regional and national level if they are to become relevant regulations. At the same time, global regimes can enhance the agency of local or national actors if the latter succeed in integrating them into their cognitive, administrative, and political structures and strategies.

Consequently, the present article will start out by presenting an analysis of the costs and benefits of global environmental regimes in general and then go on to look at them in relation to their interfaces with water policy. Secondly, the author will analyse the policy processes and conflicts around the construction of a new dam and hydro-electrical power plant in the Brazilian Eastern Amazon (Belo Monte) in order to show how national water policies and local water-related politics are already permeated by global governance elements and thus contribute to the implementation of global environmental governance as such. Here we will make use of an actor-oriented analysis derived from our understanding of the concept of global governance.

The structure of the article is as follows: we start with a short presentation of the analytical concept of global governance as used in this article (part 2). Then we will focus on global environmental conventions and how they relate to water policies on the factual, cognitive, administrative and political levels (part 3). We will then proceed to the case study, the dam and hydro-power plant in Belo Monte, in the Brazilian Eastern Amazon region (part 4). After discussing the main characteristics of the project itself and its expected negative impacts (4.1), we will analyze the legal framework for the Belo Monte dam (4.2) and the multiple actors involved in the conflicts around the dam on the global, national and local levels, their interests and their political strategies (4.3). Here we will put special emphasis on those alliances, strategies and discourses which actively link the local case to global environmental regimes and discourses and in this way contribute to global environmental governance. In the last section we will present our conclusions regarding water policies and global environmental governance (part 5).

[2] For IWRM cf. Neubert et al. (2005).

2 Global Governance as an Analytical Concept

Global governance analysis focuses on rules and norms intended to solve problems which are perceived to be either of a universal or global nature. The analysis considers three dimensions with a view to gaining an understanding of how these rules and norms come to exist and how they are implemented:

- multiplicity of actors (state and non-state),
- multiplicity of policy levels that interact with one another (local, sub-national, national, regional, international and global),
- plurality of governance patterns (public policies vs. private sector governance).

Looked at analytically, political processes of global governance (e.g. the creation of the climate regime or the world trade regime) are characterized by multi-actor configurations in which governments are joined by other actors from science, the private sector and organized civil society in negotiations over problem definition, rule-making, burden-sharing, etc. The analysis of global governance processes can therefore not be state-centric. Another relevant feature is that world politics increasingly take place on and are influenced by processes on the local, national, regional and global political level. As these processes are considered to be inseparably linked, global governance research focuses on the interlinkages between these levels. From these two features – multi-actor and multi-level – we can derive another characteristic of global governance: a plurality of governance patterns. This is the result of the co-existence of a wide variety of forms of governance with no clear hierarchical relation among them and the emergence of new spheres of authority independent of the sovereign nation-state and based on the activities of non-state and supra-state actors. Multi-actor, multi-level and plural perspective together mean that the analytical concept of global governance focuses on the complex interlinkages between different societal actors and governmental institutions (Dingwerth and Pattberg 2006).

The analytical dimensions of this concept reflect the transformations of contemporary world politics in the last decades, which were prompted by technological change, economic globalization and a loss of authority on the part of the nation-state. These transformations have strong implications for the theoretical and political meaning of concepts central to the understanding and working of national and international politics, namely sovereignty, legitimacy and power. Reinicke (1998) distinguishes between a nation-state's internal and external sovereignty. Before economic globalization had intensified, the nation-state had both legal and operational sovereignty with regard to internal issues and acted in the capacity of a central authority. As far as external issues are concerned, relationships among states were characterized by the absence of a central authority and relations were marked by interdependence, in other words, a situation of mutual sensitivity and vulnerability. The process of economic globalization has had two effects: (i) the reorganization of corporate activity in transnational industrial networks under conditions of liberalized international financial markets has reduced internal operational sovereignty; (ii) external sovereignty has become less important, indeed: nation-states see themselves forced to agree globally on domestic policy objectives and to cooperate instrumentally (i.e. by elaborating global rules, norms and standards), in order to preserve their internal problem-solving capacities. The creation of the G7 in the 1970s as a coordination mechanism for industrialized countries and, more recently, the invitation of leaders from emerging economies to G8 summits illustrate this point. The result is that the distinction between domestic and international politics is becoming blurred. In the case of global environmental goods and services, several multilateral agreements elaborated since the 1960s serve to bind sovereign internal decisions to the common interest in safeguarding these goods and services. In a democracy, a government's legitimacy depends both on election procedures and its capacity to deliver. But today the latter is shaped not only by national power relations but increasingly also by external and transnational actors.

The core of global governance consists in the rules, norms and standards that constitute the global regimes designed to shape social action. Global governance research asks where the rules come from, how they are implemented on national and local level, what the relationships are between rule-makers and rule-takers, who loses and who wins, and how all this affects problem-solving capacities. In the following chapter we will briefly present the main global environmental policy regimes and their interfaces with water policies.

3 Global Environmental Policy Regimes, their Effectiveness and their Water-related Interfaces[3]

Global environmental governance relies heavily on a series of legally binding international regimes. In the area of water policy, existing regimes include the UN Convention on the Law of the Non-navigational Uses of International Watercourses (1997), the UNECE Convention on the Protection and Use of Transboundary Watercourses and International Lakes (1992) and the SADC Protocol on Shared Watercourse Systems (2001). But the three Rio conventions (the Convention on Climate Change, the Convention on Biodiversity, the Convention to Combat Desertification) as well as the Ramsar Convention for the Protection of Wetlands also include many principles and provisions with a direct or indirect bearing on water policy.

The Rio conventions have specific characteristics that have led many actors from public administration and civil society organizations to see international law codified in a convention as a panacea for the solution of problems related to the management of global public goods. The climate convention is regarded as the most successful of the three, as it succeeded in creating an instrument which mobilizes considerable resources from the private sector and has generated considerable economic interest in a continuation of the regime. The convention to combat desertification is the weakest of the three, and it is interesting to note that it shows a strong analogy to the case of water: as water scarcity, desertification is not a global phenomenon as such, although it is influenced by global environmental deterioration.

[3] This section is based on Scholz (2004).

In the following section we will present and discuss the present state of our knowledge concerning the benefits and costs and the effectiveness of these conventions as well as the conclusions that may be drawn from experience with these conventions as regards question of whether or not it makes sense to work for a water convention.

3.1 Benefits, Costs and Effectiveness of the Rio Conventions

The experiences made in recent years with the three Rio Conventions – on climate change, biodiversity protection, and combating desertification – show that the following points are among those required for the successful conclusion of an environmental convention (Porter and Brown 1991; Young 1998):

- a consensus among the main actors involved as regards the definition of the problem (causes, consequences, approaches);
- adequate consideration of regional differences (e.g. development levels, ecology) and at the same time agreement on overriding principles and goals;
- for all groups involved, manifest benefits of a cooperative, solution-oriented strategy at the international level;
- involvement of all relevant actors (avoidance of veto coalitions); and
- provision by industrialized countries of additional funds for measures in developing countries and countries in transition.

In all cases of multilateral environmental agreements, a period of between 30 and 40 years has elapsed between the definition of a given environmental problem and the signing of an international agreement designed to address it. The situation was different in the cases of CITES – the convention on international trade in endangered-species – and the Ramsar Convention on the Protection of Wetlands of International Importance, two of the very earliest international environmental conventions, which were concluded in the course of roughly one decade (Pülzl et al. 2004). Two reasons can be cited for this: First, in the 1960s and 1970s the number of competent negotiating partners in the field of international nature conservation was far lower than it is today, and second, nature conservation had not yet been discovered as a global bargaining chip. This meant that at that time the main negotiating parties (individual countries and NGOs or associations of scientists) had far more influence than they have today on the pace of negotiations and the formulation of treaty texts. The increasing complexity of international negotiations due to growth in the number of direct and indirect negotiating partners involved and the need to forge links with adjacent policy fields is one of the most important reasons for the slower pace of negotiations. In other words, today there are tradeoffs between transparency, participation, and efficiency.

One common feature of the Rio Conventions is that they are designed to deal with environmental problems associated with highly complex chains of causes, damage profiles, and impacts. This complexity creates many obstacles when it comes to implementing the conventions and seeking to achieve visible results. By way of contrast, we may point here to the positive experiences made with the Montreal Protocol on protection of the ozone layer, an instrument finalized at the end of the 1980s, and one that has already generated a number of clear-cut successes. This was due to the identification of an unambiguous chain of causes and effects (CFC as the most important causal factor) and a tangible, not all too complex perpetrator structure (producers and users of CFCs) (Parsons 2003).

Now that they have been in existence for a decade or so, however, it is possible to identify some benefits of the Rio Conventions, and they clearly show the relevance of the multi-actor and multi-level perspective mentioned in the second section of this article:

- the political weight of the issues addressed in the conventions has increased at the national level: Civil-society and other interest groups and social movements can point to the obligations assumed by their governments and/or the other signatories and demand that they be complied with; this can serve to more effectively politicize the issue and influence political opinion in the countries concerned;
- the principle of "common but differentiated responsibilities" requires national problems to be placed in a global context and actors to assume responsibility: the Rio Conventions pursue an approach that assigns common responsibility to industrialized and developing countries for finding solutions to global problems. This serves to place the central focus on national and local causes and manifestations of these problems, and thus on the different degrees of responsibility and affectedness of all those involved, and to underline the need for national reforms;
- conventions help to transform environmental goals into actual laws in a number of sectors at the national level: once a convention has been signed, pressure starts to mount at the national level for the adoption of appropriate legislation; even though this may not automatically reduce deficits in implementation or enforcement, it does tend to initiate a gradual processes of institutional learning and change that are needed to bring about altered political constellations conducive to greater reform-mindedness;
- conventions are geared to coordinated international action: the present crisis of multilateralism, brought about by the greater weight attached by the US to national sovereignty than to international cooperation, has served to substantially slow down negotiations, with tangible successes becoming rare. This, however, is no sign that approaches geared to coordinated international action as a means of solving border-crossing problems have become a thing of the past; conventions constitute important learning experiences in global governance and set the stage for the development of formal and informal networks of governmental and nongovernmental actors that may sustain and accelerate learning processes in the fields concerned;
- the political dynamics unfolding immediately after the 1992 Rio Earth Summit made it possible to mobilize additional financial resources for measures in developing countries and countries in transition. These additional financial resources were important in that they demonstrated the willingness of the industrialized countries to act as well as to fund concrete measures. At present these funds are

made available in the framework of the Global Environment Facility (GEF). In the future this responsibility is likely to devolve on special instruments of the climate convention (Clean Development Mechanism and Joint Implementation). Unfortunately, thus far the chance has been missed to harness these dynamics toward the end of sustained increases in Official Development Assistance (ODA). The financing requirements involved continue to overtax the willingness of the industrialized countries to pay, and the budgets of the developing countries and countries in transition have not been restructured accordingly. These issues are on the agenda of the ongoing debate on the financing of global public goods, and they clearly show that while global environmental governance is a difficult process, one still in its infancy, it is nevertheless a process that is absolutely necessary.

Global environmental policy is thus inconceivable without the Rio conventions. Still, we can observe a certain measure of convention fatigue fuelled by the fact that practical changes take considerable time to become visible on the ground. As regards the potential costs of a convention:

- conferences of the parties to a convention are marked by highly costly, formalized negotiation processes: the UN principle that all member states are equal creates substantial space for blockades by veto coalitions. It furthermore reinforces a tendency to assess results in quantitative terms (so and so many countries have prepared a national action plan) instead of focusing on qualitative criteria (Are these countries relevant for the global problem under consideration?). One exception here would be UNFCCC, which links country votes with the share of worldwide CO_2 emissions for which they are responsible;
- the modalities involved favor confrontational negotiating styles: the formation of camps of industrialized and developing countries encourages the parties to adopt negotiating styles motivated more by pursuit of traditional interest policies than by cooperative policy patterns geared to reaching common global goals (global governance). In an environment of this kind position-related gains count far more than progress in changing a given state of affairs (e.g. protecting the climate or biodiversity);
- conventions without any dedicated financial instruments meet with little or no acceptance on the part of developing countries and countries in transition;
- there is a marked gap between agreements on goals and their implementation: reform backlogs in many industrialized countries, which have of course committed themselves to providing input-related funding, tend to undercut the credibility of convention processes. An additional problem, above all as regards developing countries and countries in transition, is that international negotiations require sizable personnel and financial resources which are then no longer available for measures at the national or local level;
- the effects of conventions become visible only over the long term: long-term successes are difficult to explain politically. If it turns out to be impossible to demonstrate such successes in ongoing changes, this is likely to prove bad for the day-to-day business of politics.

However, these costs are not sufficient to warrant declaring environmental conventions obsolete. Indeed, the experiences made thus far clearly indicate the preponderance of the anticipated long-term benefits of a global cooperative approach.

Would this also apply in the case of a water convention focusing on water rights and the sustainable and integrated management of water? Could the short- and medium-term negotiation costs involved in concluding an agreement be justified vis-à-vis the long-term benefits stemming from a water convention? Moreover, is it necessary to have a water convention if we take careful stock of what is already in place?

The hypothesis of this article is that a water convention is not crucial to achieving improvements in global water management. This hypothesis is based on four arguments:

- First, the water community has already developed a guiding concept and leading principles for sustainable water management: IWRM. This concept is internationally accepted and backed by the UN, and it has had a major influence national water policy reforms (Neubert et al. 2005). In this respect (agreement on problem definition, goals and principles), negotiation of a convention could not offer any additional benefits.
- Second, legally binding conventions already in existence have multiple interfaces with water policies which give additional support to the implementation of IWRM and which link local and national water politics with global environmental policies (see the paragraphs below).
- Third, multilateral organizations and bilateral donors are already investing considerable funds in water management. It is unlikely that a convention would succeed in mobilizing additional funds.
- Fourth, there are both global and regional conventions or protocols that deal with the use and management of transboundary water resources and international watercourses, and they include environmental objectives. Moreover, there are ongoing efforts to codify the use and management of transboundary groundwater systems.
- Fifth, the present water crisis is not a global phenomenon, as it is composed of a multiplicity of local scarcities; and it is therefore not likely that a convention would be able to mobilize additional funds and political will for its cause.

What is needed now are efforts to focus on an integrated implementation of both the Rio conventions and water policy reforms. A first step would be to make the interfaces between existing conventions and water policies more explicit and visible. This could be furthered by working out a water work program (in analogy to the forest work program elaborated by the CBD), which should build on the multiplicity of existing water programs related to the implementation of IWRM, the Johannesburg Programme of Action, the Millennium Development Goals and the implementation of Poverty Reduction Strategies (PRSPs). It would in this way be possible to considerably improve the possibilities to actively exploit the synergies between the conventions and water management by developing mutually beneficial activities. Water actors such as administrations, water user associations and NGOs could then point to a work program of this kind, lending more clout to their proposals and demands.

3.2 *Water-Related Interfaces of the Conventions on Climate Change, Biodiversity and Desertification*

In this section we will start out by briefly focusing on the interlinkages between the water cycle and the processes of climate change, biodiversity loss and increasing desertification. Then we will go on to look at the conceptual and operational interfaces between water policy and the Rio conventions.

Climate change will inevitably have strong impacts on the water cycle. Even today, according to the Intergovernmental Panel on Climate Change (IPCC), changes in precipitation levels and patterns, in snow cover and ice cover have been observed, and it is likely that summer continental drying and associated risks of drought have increased. As regards the future, we can note the following:

- "(...) globally averaged annual precipitation is projected to increase during the 21st century, with both increases and decreases in precipitation of typically 5–20 percent projected at the regional scale" (Secretariat of the Convention on Biological Diversity 2003, 33).
- Increases will occur throughout the year across high latitudes, while they will take place over northern mid-latitudes, tropical Africa and Antarctica in winter, and affect southern and eastern Asia in summer. Decreases of winter rainfall are anticipated for Australia, Central America and southern Africa. At the same time, larger variations between annual precipitation are also very likely.
- Extreme precipitation events will be more frequent and stronger, and this will lead to more frequent flooding even in regions where total precipitation decreases.
- Temperature variability will change on a daily, seasonal, inter-annual, and decadal basis. This is likely to lead to more droughts and floods, especially in regions already affected by El Niño. In mid-continental areas, more summer droughts are expected. Nearly all land areas are 90–99 percent likely to experience more hot days and heat waves and fewer cold and frost days.
- Glaciers and ice caps will continue to retreat.
- Lakes and streams will be most affected by temperature-dependent changes in high-latitudes (where the largest changes in temperature are projected), with moderate effects projected at mid-latitudes and the lowest effects predicted for the tropics.

> "Increased temperatures will alter thermal cycles of lakes and solubility of oxygen and other materials, and thus affect ecosystem structure and function. Changes in rainfall frequency and intensity combined with land-use change in watershed areas has led to increased soil erosion and siltation in rivers. (...) Climate change will have most pronounced effects on wetlands through altering the hydrological regime as most inland

> wetland processes are intricately dependent on the hydrology of the river basins or coastal waters." (Secretariat of the Convention on Biological Diversity 2003, 39).

Climate change will obviously also have very pronounced effects on biodiversity. Global warming, changes in precipitation patterns and the intensity and frequency of extreme weather events will lead to changes in the species composition of ecosystems, migration and loss of species. The overall effect will be a weakened capacity of ecosystems to provide their services, be they regulating services (regulation of floods, drought, land degradation), provisioning services (food and water), supporting services (soil formation and nutrient cycling) or cultural services (recreational, spiritual and other nonmaterial benefits). Regulation of the water cycle is one of the fundamental ecosystem services. Changes in the water cycle place additional stress on the capacity of soils for primary production and nutrient cycling, which is already affected by intensive agriculture. The consequences are increasing desertification, and a reduced capacity to sequester carbon in above- and below-ground carbon reserves.

We see that the water cycle is at the centre of the three conventions, be it on the impact side or as a central element of feedback cycles that reinforce negative effects on climate, biodiversity and ecosystem capacities to deliver their services.

Is this pivotal role of the water cycle reflected in the convention texts and the ways in which they are operationalized?

The UN Framework Convention on Climate Change (UNFCCC) establishes two modalities for responding to climate change: mitigation and adaptation. Mitigating climate change by reducing greenhouse gas emissions is crucial to reducing the negative impacts of global warming on the hydrological cycle, and thus also on the conservation of water stocks and flows. Adaptation to the impacts of climate change has a strong focus on coastal management, agriculture and water, as it is in these areas where negative effects are anticipated to be heaviest (IPCC 2001):

- sea level rise due to the melting of the polar ice caps will require the relocation of millions living in coastal areas and in small island states,
- temperature changes, increased drought and flooding as well as shifts in the regional and temporal distribution of precipitation will require profound adjustments in agriculture,
- the same factors will alter water availability and therefore require strong efforts to reduce wasteful use of water and improve water conservation.

The UNFCCC requires the signatory states to develop activities in both areas, mitigation and adaptation, and it has established funds through which the international community will be able to support developing countries' activities in both areas. The main financial mechanism of the UNFCCC is the Global Environmental Facility (GEF), which between 1991 and 2003 invested US$ 1.6 billion in climate change activities, most of it related to mitigation. In addition, it has leveraged US$ 9 billion through co-financing (Greene 2004, 71).

The UN Convention on Biodiversity (UNCBD) has six thematic work programs (agriculture, dry and sub-humid lands, forest, inland waters, mountains, and oceans and coasts) which are based on an ecosystem approach. This approach takes into account the facts that the protection of biodiversity cannot be separated from the functioning of ecosystems and that the linkages between species and ecosystem services as regards soils, water etc. have to be looked at in an integrated way (see Box 1). The ecosystem approach can be criticized for mixing institutional, economic and ecological matters in an ad hoc way, but this apparently unsystematic collection of principles can also be interpreted positively as an attempt to combine normative principles on ecosystem management with scientific insights on how ecosystems work and change. A quick look at these 12 principles reveals some fundamental overlaps with IWRM that could facilitate the design of mutually beneficial activities for water management and biodiversity protection. Operational interfaces with the UNFCCC result from the Marrakech Accords on requirements for the design and implementation of mitigation and adaptation activities. These requirements state that (i) activities in the area of land use, land-use change and forests should contribute to the conservation of biodiversity and sustainable use of natural resources and (ii) the Clean Development Mechanism should assist developing countries in achieving sustainable development.

Box 1 The 12 Principles of the UNCBD Ecosystem Approach

1. The objectives of management of land, water and living resources are a matter on which every society is free to decide.
2. Management should be decentralized to the lowest appropriate level.
3. Ecosystem managers should consider the effects (actual and potential) of their activities on adjacent and other ecosystems.
4. Recognizing potential gains from management, there is usually a need to understand and manage the ecosystem in an economic context. Any such ecosystem-management programs should:
   - reduce those market distortions that adversely affect biological diversity (i.e. eliminate perverse subsidies etc.);
   - align incentives with a view to promoting biodiversity conservation and sustainable use;
   - internalize costs and benefits in the given ecosystem to whatever extent feasible (including full accounting for ecosystem goods and services).
5. One priority target of the ecosystem approach should be conservation of ecosystem structure and functioning with a view to maintaining ecosystem services.
6. Ecosystems must be managed with a view to maintaining their functions.
7. The ecosystem approach should be undertaken at the appropriate spatial and temporal scales.
8. Recognizing the varying temporal scales and lag-effects that characterize ecosystem processes, objectives for ecosystem management should be set for the long term.
9. Management must recognize that change is inevitable.

10. The ecosystem approach should seek the appropriate balance between, and integration of, conservation and use of biological diversity.
11. The ecosystem approach should consider all forms of relevant information, including scientific and indigenous and local knowledge, innovations and practices.
12. The ecosystem approach should involve all relevant sectors of society scientific disciplines.

Source: Secretariat of the Convention on Biological Diversity 2003, 53

The UN Convention to Combat Desertification (UNCCD) declares in its objectives that it will "combat desertification and mitigate the effects of drought..." and for this purpose develop "long-term integrated strategies that focus simultaneously, in affected areas, on improved productivity of land, and the rehabilitation, conservation and sustainable management of land and water resources" (UNCCD Art. 2). In practice, the UNCCD gives more attention to land-related issues than to water issues, which are treated mainly in relation to rain-fed or irrigated agriculture. The UNCCD's principles have a strong focus on participation and policy coordination among different policy levels as well as at the community level. Even though the UNCCD contains no explicit ecosystem approach, the term "integrated strategies" in the convention's objectives as well as Article 4 on general obligations make reference to the need to view land and water resources within their social, economic and bio-physical context. Article 8 explicitly calls for coordinated activities with the other conventions in order to maximize mutual benefits.

In practice, implementation of the three Rio conventions is still being approached in a rather isolated way. Integrated strategies are not being developed as a priority. But activities on the ground often simultaneously meet common objectives of both the three conventions and water policy, due to close interlinkages between these problem areas. There is therefore a great potential for integrated work programs and mutually beneficial and reinforcing activities.

3.3 Potential Cognitive, Administrative and Political Consequences for Water Policy

On the *cognitive level*, we can identify three main consequences:

First: Water policy is increasingly framed as part of global environmental policy and sustainable development. This is not an easy process as water politics is still dominated in large measure by engineers and their technical perspective on water problems and solutions. But increasing water scarcity and shortage are forcing the water community to look more seriously at causes both internal to the sector (e.g. wasteful use in households, industry and agriculture) and external to it (e.g. deforestation, anthropogenic climate change). If there are strong or relevant interlinkages between water scarcity and shortage and environmental processes, the logical next step would be to look for joint strategies and activities with agricultural and environmental actors.

Second: Equity, the sharing of benefits and costs as well as participation are important normative features of sustainable development and global environmental regimes. Through IWRM and interfaces with the Rio conventions, these features are finding their way into water policies and politics as well, which will make them more effective.

Third: The need both to observe climate change and to predict its consequences underlines the need for regular data collection on temperature, precipitation, extreme events etc. These data will improve conditions for policy-making in the water sector as well, especially if they can be complemented by data on water flows and stocks as well as water use.

On the *administrative level*, we can identify two main consequences:

First: Conventions have to be transposed into national law in order to be implemented. Due to the interlinkages between climate change, biodiversity, desertification and water, there is a need for a minimum level of coherence between the objectives, conceptual approaches and instruments established by the respective bodies of law.

Second: Another area where coherence and a clear division of labor are required is administrative competence. In most cases, environmental ministries are in charge of UNFCCC, UNCBD and UNCCD. In many countries, the water administration is a large bureaucracy that enjoys great autonomy on local and national levels. Due to the important interfaces between water and environmental policies, cooperation with environmental departments is a rather new challenge for these administrations. Frictions and conflicts are bound to emerge in this process, reflecting differences in priorities regarding the multiple functions of water resources.

On the *political level*, we can identify two main consequences:

First: State and non-state actors in the water sector can refer to the Rio conventions either positively, in order to add weight to their objectives and demands, or negatively, by rejecting their provisions as illegitimate interference in domestic matters. In any case, the dominant arguments of global environmental governance are likely to make their way into water policies as well, at least in those areas where they overlap with climate change, biodiversity protection and desertification. This means that local and national water-related policies will have to be justified not only with regard to national priorities but also with regard to the protection of global environmental services (e.g. climate stability and biodiversity protection), and this could in the end create additional benefits on the local level.

Second: The biodiversity and desertification conventions set out specific general goals and explicitly support participative procedures with regard to resource management. In practice, this means that the knowledge, good practices and interests of local communities and indigenous peoples regarding natural resource management should be actively integrated into decision-making processes. International NGOs often act as watchdogs in the case of threatened biospheres of regional and global importance, such as tropical rainforests and river systems. Often they find support in the media. When this is the case, recourse to the provisions of conventions regarding participation and parameters for resource use by local actors in defense of their interests can change local dynamics considerably. This external support based on arguments derived from conventions, NGOs and the media is especially important in remote areas where the presence of the state is weak and the rule of law cannot be guaranteed.

The case study presented in the next section is an example of a concrete project, a dam and hydro-power plant in the Eastern Amazon, which has interfaces with three of the four policy areas analyzed in this chapter: climate, biodiversity, and water. The analysis will show which of these interfaces and which elements of global governance were relevant for the political process that developed in the wake of local protests voiced by civil society organizations.

4 The Case of the Belo Monte Dam in the Brazilian Eastern Amazon[4]

The project to construct a dam and a hydro-power plant in the Brazilian Eastern Amazon, at the Xingu river close to Belo Monte, aroused protests, on the local and international level, by peasant organizations, environmental groups and scientists. The Belo Monte project has many faces: it is an important element of Brazil's policy to meet future energy demands and is therefore important for national economic growth. It will have strong impacts on the local social and economic environment, and many fear that these impacts will be mainly negative. It will have strong ecological impacts, as endemic biodiversity at the Xingu is high and the consequences for the reproduction of flora and fauna are unclear. Criticism of the project is based on socio-economic, environmental and water-related issues (4.1). The legal framework offers possibilities to stop the project until some of the open questions are clarified (4.2). The capacity of local actors to voice their protest is supported by their links with international environmental cooperation, national and transnational NGO networks and federal ministries in Brasilia (4.3).

4.1 The Project and its Impacts

At the end of the 1990s, structural power shortages in Brazil revived the idea of increasing electricity generation through the use of hydro-electrical power plants in the Amazon region (see Box 2 for general information on hydro-power in Brazil). In the 1980s these projects had been withdrawn due to large international and local protests and financial bottlenecks created by a debt crisis. In Brazil power generation is still dominated by public enterprises; in the Amazon, Eletronorte is the public monopolist in this field.

[4] This section draws on Scholz et al. (2003).

Box 2 Hydro-power in Brazil

Today Brazil has more than 2,000 dams which flood an area of 34,000 km^2. More than 90 percent of electricity generation in Brazil is based on hydro-power. This amounts to 61,000 MW, that is, 25 percent of the country's overall hydro-power potential. Two thirds of this potential is located in the Amazon region, which means high environmental and transmission costs. Twenty percent of the non-utilized hydro-power potential is located in the south of Brazil, where dams would affect highly populated areas and fertile soils.

In the coming years the government plans to build nearly 500 additional hydro-power plants, which would force 800,000 persons to relocate.

If all plants projected for the Amazon region were built, this would lead to the emission of 231 million tons of CO_2 per year, due to deforestation and the rotting of vegetation in the flooded areas. These emissions would amount to 75 percent of Brazil's 1999 greenhouse gas emissions from the burning of fossil fuels.

Source: www.mabnacional.org.br/modenergetico.html. Cited 23 June 2006

At present Eletronorte runs three hydro-power plants in the Amazon region: Balbina in the state of Amazonas,[5] Samuel in the state of Rondônia and Tucuruí in the state of Pará (see Table 1). The Belo Monte hydro-power complex goes back to a project from the early 1980s, then called Kararaô and Babaquara, which comprised several dams and would have flooded up to 6,000 km^2. As these projected dams generated violent protests in the region, especially by the Kayapó, whose territory would have been massively affected, Eletronorte created a new name for the resuscitated project, and also changed its design in order to improve its quality.

Table 1 Electricity generation and flooded area of different Brazilian hydro-power plants

Hydro-power plant	Flooded area in km^2	Construction time	Electricity generation in megawatts (MW)	MW/km^2
Balbina (Amazonas)	2,380	1975–1987	250	0.11
Samuel (Rondônia)	560	1982–1989	216	0.39
Tucuruí (Pará)	2,430	1976–1984	4,000	1.65
Belo Monte (Pará) (projected)	400		ca. 6,000	15.00
Itaipú (Paraná)	1,350	1975–1991	13,000	9.69

Source: www.mabnacional.org.br. Cited 22 May 2003, www.eln.gov.br. Cited 23 June 2006

[5] The extremely low ratio between MW and flooded area of only 0.11 gives a hint of the enormous environmental destruction which went along with the construction of this dam.

The Belo Monte hydro-power complex would be located on the Xingú River, close to the town of Altamira in the state of Pará and the Transamazônica highway. The power plant itself would be positioned at a bend of the river with a fall of 96 m, which would be used to drive the turbines. Construction of two artificial channels and two dams would be used to redirect the waters of the Xingu, making it possible to reduce the flooded area to 400 km^2. Belo Monte is projected to generate an average of 6,289 MW and a maximum of 11,000 MW per month. Construction will cost about US$ 6.5 billion, which 3.7 billion of which will be needed for the dams, channels and plant and 2.8 billion will go into the transmission lines to the Northeast of Brazil (Pinto 2003).

Due to the high annual fluctuations in the water volume of the Xingu river, the hydroelectric plant could operate fully only during the six-month rainy season, and its capacity would therefore be only 6,000 MW annually instead of the max. Projected figure of 11,000 MW. In order to increase annual production capacity, it would be necessary to build more dams, which would increase the flooded area and the number of affected persons, especially in the indigenous lands of the Xingu basin. Eletronorte maintains that the project is economically viable even with these reduced production levels because electricity prices have risen. Many civil society organizations do not believe Eletronorte's assertions and fear that up to five more dams could be built on the Xingú in the next five decades, especially if energy demand continues to rise in step with economic growth.

Compared with the other dams in the Amazon region, Belo Monte would have the most favorable ratio for generated electricity to flooded area (if Eletronorte keeps the project to its present limits). Even in comparison to Itaipú, which will be the world's largest hydro-power plant until the Three Gorges project in China is completed, Belo Monte would have a good standing.

But the main point of reference for the public debate on the benefits and costs of hydro-power plants in the Eastern Amazon is Tucuruí. Three quarters of the electricity generated in Tucuruí is consumed by the aluminum industry, which does not pay full-cost prices. The flooding led to the displacement of 25,000–30,000 persons. Compensation payments were made on a very irregular basis, many never received any payment, and many had to wait years for compensation. Indigenous groups lost their land and were relocated to a new territory which they can enter and leave only with a permit issued by Eletronorte. In contrast to the neighboring municipalities in the south of Brazil, the municipalities in the Tucuruí area have not received any noteworthy investments in social and economic infrastructure by Eletronorte, as had been promised when the works began. The region did not turn into a growth pole, as had been forecast in the 1980s.

Due to these bad experiences in Tucuruí, Eletronorte has low credibility with(in) large parts of the local population in Altamira. As the hydro-power complex in Tucuruí did not generate sustainable growth effects, civil society organizations in Altamira are very skeptical about the planned project in Belo Monte and very aware of its ecological, economic and social risks.

The ecological risks are massive. The Xingu river basin is the largest continuous forest area remaining in southern Pará. With the dams and the artificial channels, the river basin would be fundamentally disturbed, some parts of it, e.g. the Volta Grande, would dry up and thus be definitively destroyed, preventing the reproduction of several endemic species. A total of 400 km^2 would be flooded, including several parts of the town of Altamira. Flooded vegetation would rot and produce greenhouse gases, carbon dioxide and methane in particular. The Volta Grande, a big river bend located in-between the towns of Altamira and Belo Monte, is a unique ecosystem which depends on the high variation in the water levels between rainy and dry season. A unique population of fish species can be found in the rapids, and this would disappear altogether. The long-term effects on flora and fauna associated with interferences in river systems such as dams (WCD 2000, 73) – which will occur specifically in the Xingu watershed – cannot be estimated because the reproduction cycles of its flora and fauna are yet not sufficiently known. Should more dams be built, the ecological risks would increase accordingly.

From an ecological perspective, the risks of the Belo Monte project should be estimated as high. The benefits of a clean energy source will be associated with considerable impacts on flora and fauna whose scale cannot be measured due to existing, huge information deficits.

Economic and social risks associated with the flooding include the drying up of the Volta Grande and the mismatch between short-lived investment streams and long-term demographic changes. The Juruna, an indigenous people living in the area of the Volta Grande, will suffer under the drastic changes expected in this part of the Xingu ecosystem.[6] The construction would stimulate spontaneous migration to the region and reinforce the rural exodus of peasants in search of urban labor. In total, an additional 50,000–150,000 persons are expected to come to the towns of Altamira and Belo Monte. This would double the present population of Altamira and most likely lead to growth of shantytowns. After construction, a large share of the urban population would be likely to turn to the countryside in order to survive from agriculture, thus increasing deforestation and exacerbating land conflicts.

In order to mitigate these economic and social consequences, Eletronorte worked out a regional development plan which is supposed to be financed with 1 percent of the total investment (Eletronorte 2000, 4). In addition to that, Eletronorte presented a plan for the sustainable development of Belo Monte which includes investments in social and economic infrastructure amounting to US$ 3 billion (Eletronorte 2002, 42). Then there will be compensation and mitigation funds (Rocha 2002). The local population expects these funds to be used to put a permanent tarmac layer on the Transamazônica highway and for the maintenance of the feeder roads. This would be crucial for improving conditions for local economic activities. A substantial part of these investment packages would have to be implemented before construction begins, especially in the rural areas, in order to prevent migration into the town.

[6] In the past the international community has looked closely at processes of relocation and the ways in which indigenous peoples have been affected. In 2003 Brazil ratified ILO Convention 169 on the rights of indigenous peoples. This gives them a much better legal basis for defending their rights vis-à-vis such measures.

4.2 The Legal Framework and the Legal Process

The relevant legal framework for large-scale infrastructure such as dams and hydropower plants comprises the law on the Brazilian environmental administration (see Box 3), legislation on environmental impact assessment and Brazil's water law. In the specific case of Belo Monte, however, the constitutional rights of the indigenous peoples have turned out to be the decisive legal instrument for stopping the project, since the dam will affect indigenous territories and the rights of the people living there.

Box 3 Environmental law and administration in Brazil.[7]

Since the 1980s Brazil has been modernizing its environmental policy, starting out with the establishment of the national environmental system SISNAMA (Law 6938) in 1981. The body of environmental law has been completed since then and administrative units have been created on the three levels of municipalities, states and the federation. Civil society can count on a series of participatory rights regarding environmentally relevant decision-making. SISNAMA has created instruments for the elaboration of transversal and long-term strategies with a view to incorporating environmental considerations and objectives into all policy areas.

The 1988 Federal Constitution (CF) established a complex system for the protection of natural resources. Article 125 CF states that protection of the environment is an objective of the state and a task of all; Article 225 CF states that all citizens have the right to a healthy environment. Each individual citizen as well as the public prosecutor's office (*Ministério Público*) are entitled to take legal action regarding environmental matters.

The *Ministério Público* is an independent institution which acts in the interest of the public and is in charge of defending public law, social and individual basic rights and common goods such as the environment. It has independent parallel bodies on federal and state level, it has special rights to take legal action, and it can issue instructions binding for public administration. The *Ministério Público Federal* has played a very important role in safeguarding the environment in the Amazon. This is due not only to its broad competencies but also to many young, especially committed prosecutors who see their task in supporting civil society and the local population in defending their rights.

In Brazil, licensing procedures for large projects with anticipated environmental impacts are the most important legal instrument in the hands of the environmental administration. In principle, all three administrative levels (municipality, state and federation) are authorized to issue environmental licenses, but the federal environmental authority IBAMA is exclusively in charge of

[7] For Brazilian environmental law and administration, cf. Krell (1993) and Valente (1991)

projects with "regional or national significance." The application of this criterion often creates administrative frictions due to its relative vagueness. In the case of large infrastructure projects, environmental licenses have to be based on the strict parameters of an environmental impact assessment (EIA), which is regulated in several laws as well as by decisions of the national environmental council, CONAMA.[8] The results of the EIA and the license issued are required to be made public.

Brazil has ratified the conventions on climate change, biodiversity and desertification. It has also signed Convention 169 on the rights of indigenous peoples elaborated under the ILO. The provisions of this convention are especially relevant for any intervention in the Amazon r ion, as 20 percent of the region is indigenous territory (Nepstad et al. 2006). Article 231 CF acknowledges the traditional organization and the genuine user rights of indigenous peoples in their territories. The *Ministério Público* is in charge of guaranteeing these rights.

Civil society has broad access to environmental data, guaranteed by the Constitution, the SISNAMA and a special law on environmental information issued in April 2003 and based on the Aarhus Convention.[9] Most relevant laws can be found on the Internet.

Civil society can also fall back on many participation rights and instruments. It participates in the national environmental council, CONAMA, which has legislative powers. It can also take legal action and participate in environmental licensing without having to overcome enormous bureaucratic hurdles.

In September 2000, Eletronorte applied for an environmental license for the Belo Monte project with the environmental authority of Pará state, SECTAM. As the governor of Pará had declared his full support for Belo Monte, Eletronorte assumed that SECTAM would diligently issue the license. At the same time, Eletronorte commissioned the Federal University of Pará to carry out the studies needed the environmental impact assessment.

This licensing process has been stopped by several court decisions. In 2001 the *Ministério Público Federal* had successfully gone to the federal supreme court with three arguments:

1. The environmental authority of Pará state, SECTAM, is not competent: the environmental license can only be issued by IBAMA due to the magnitude of the project and because it is financed by a federal entity.

[8] Federal Law 6938/81, Decree 99.274/90; Federal Law 9784/99, CONAMA Resolution 01/86 and 237/97.

[9] Federal Law 10.650/2003. The Aarhus Convention was elaborated by the EU and refers to public participation in decision-making and access to courts regarding environmental matters.

2. Violation of the legal provisions on awarding contracts without tender: A large contract can only be awarded without tender if the beneficiary has a high scientific reputation. This is not the case with the Federal University of Pará (UFPA). Two environmental impact assessments previously elaborated by the UFPA had been rejected due to scientific deficiencies.
3. Violation of the rights of indigenous peoples: According to Article 231 CF, indigenous people have to be heard by the National Congress for the case that any large-scale water infrastructure projects are planned in their area. Congress has to decide on such projects after having heard the indigenous peoples. Only after Congress has reached a decision can Eletronorte apply for the environmental license.

Reacting to the claims of the *Ministério Público Federal*, Eletronorte withdrew its application from SECTAM and directed it to IBAMA. Also, now no one questioned the obligation to put the studies for the environmental impact assessment out to tender. In July 2005, both Congress and the Senate adopted a decision allowing the federal government to "build the hydro-power plant in Belo Monte at the Xingu river, in a location called Volta Grande in Pará state, after carrying out technical, economic, environmental and other viability studies thought to be necessary" (Etermann 2005). The decision also included a provision on anthropological studies to be made in order to ascertain the opinion of the indigenous peoples affected. Based on these decisions by the legislative power, a federal judge had allowed Eletronorte and IBAMA to go on with the environmental licensing process, alleging that it made no difference whether the indigenous peoples were heard before or during the EIA studies. The *Ministério Público Federal* appealed this decision and won, because Article 231 CF explicitly states that indigenous peoples have to be heard first, which means that Congress' decision is bound by their assessment of the project. When this article was finalized (July 2007), the federal court of justice in Altamira had decided against the *Ministério Público Federal* and allowed feasibility studies and the environmental impact assessment to be conducted, without previously hearing the indigenous peoples concerned by the construction.

Brazilian water law[10] has not been relevant yet in the conflict around Belo Monte. This is not so surprising because its objectives and principles give priority to human and productive uses, although it also acknowledges the multiple functions of water and the need to consider watersheds for water management. Among the relevant features of the law:

- water is understood as a public good;
- water is considered to be a limited natural resource with economic value;
- in case of shortage, human beings and animals have priority for water use;
- water management has the task of ensuring the multiple functions of water;
- water management is based on watersheds; and

[10] Federal Law 9433 from January 8, 1997.

- water management is decentralized and involves public administration, water users and local communities.

Recently, though, the water law has received more attention in the Amazon. In 2005 a severe drought hit the Amazon basin and reduced water levels of the main rivers drastically, thereby effectively isolating many communities and towns. This drought was related to an unusual heating of the Atlantic Ocean, and this was the first time that massive water scarcity was felt in the region. In March 2006 a water week was organized by environmental groups, the church and NGOs in Belém, the capital city of Pará state. The Belo Monte project was at the center of the debate. Civil society organizations called for implementation of the water law in Pará. This would include creation of a water council and of river basin committees in order to facilitate watershed management. Any management plan for the Xingu river basin would have to consider the impacts of the Belo Monte project, which would create additional pressure on the licensing process.

4.3 Actors and Interests

The protest against the Belo Monte project is mainly borne by local civil society organizations, including mainly peasant associations but also women's groups and the church's secretariats for indigenous peoples and for the landless. These 113 organizations have grouped together under the umbrella of the Movement for the Development of the Transamazônica and the Xingu (MDTX). The MDTX is a very well organized and articulate movement. It has engaged in alliances with the *Ministério Público Federal*, the transnational NGO International Rivers Network and the IPAM research organization, which is associated with the US-based Woods Hole Research Center. It has good relations with both the federal environmental ministry and the federal ministry of energy and mines.

The group of the most important defenders of the Belo Monte project includes the majority of the mayors of the municipalities along the Transamazônica, who are organized in the *Consórcio Belo Monte*, the state government of Pará, the Altamira business association ACIAPA, and the federal ministry of energy (MME).

All Brazilian actors are aware of the possibly drastic social and economic consequences of the dam, but there is no consensus about the possibility of compensating them through investments and development plans elaborated by Eletronorte. Some civil society actors regard the environmental consequences as so massive that would be impossible to justify construction. By contrast, most public actors (except for the environmental ministry, MMA) either downplay the environmental consequences or see them as the price that needs to be paid to secure the national energy supply and regional economic development.

All actors along the Transamazônica share the goal of improving living conditions and economic infrastructure, especially by paving the Transamazônica, maintaining the feeder roads and providing better social services. This had been promised to the local population more than 20 years ago, when this part of the Amazon was opened up for colonization. Many actors believe that the construction of Belo Monte offers a unique opportunity to negotiate these investments with Eletronorte and the federal government.

The MDTX believes that economic development is possible without relying on a large-scale project like Belo Monte. They believe in a bottom-up development strategy based on the promotion of smallholder agriculture and the demarcation of conservation units in order to halt (the advance of) illegal timber harvesting, large cattle ranching and soy farming into the region. Their strategy therefore combines economic, social and environmental objectives, the latter having large regional and global importance. The MDTX has both the technical capacity to formulate such a strategy and the political capacity to put it into a broader context that combines both local and global objectives of sustainable development. These capacities have been developed over the course of decades of learning, in the 1960s and 1970s mainly with the support of the church, in the 1980s with the support of foreign NGOs and since the 1990s with support from development cooperation (Scholz 2005). Their involvement in transnational partnerships has taught civil society organizations to see their own struggle within a broader context and to relate it to the causes of external actors. The rise of agroforestry as a new element in their production systems illustrates the point.

At the same time, however, this also requires massive public investment for economic and social infrastructure. This is the reason why there is a faction within the MDTX which would favor negotiations with Eletronorte and the federal government if there is no alternative to Belo Monte to increase electricity generation in Brazil. Its readiness to compromise on the environmental impacts of Belo Monte is not shared by influential members of the MDTX. Another potential conflict within the MDTX is the indigenous peoples question. Many peasants believe that indigenous territories are far too large and that priority should be given to smallholders and their productive potential.

These potential rifts within the MDTX are accentuated by the fact that many of their members are close to the Workers' Party (PT), which has been in power in Brasilia since 2003. President Luiz Inácio Lula da Silva, the environmental minister, Marina Silva, and the first minister of energy, Dilma Roussef, all belong to the PT. Although the PT had formulated a campaign program for the Amazon region based on local development and incorporating environmental considerations, President Lula's assumption of power in January 2003 did not change the basic federal strategies for the region. The Amazon covers more than 50 percent of Brazil's national territory, and most federal actors therefore believe that there is enough space to balance environmental protection with economic development goals. Brazil's economy has been on a growth path since 2003, one that has of course triggered growing demand for electricity. Construction of more hydro-power plants in the Amazon region is the basic response of the government to this challenge.

The federal environmental ministry sought to maintain a balanced position between the opponents and the defenders of Belo Monte, but it sees its credibility increasingly compromised by the rigidity of the ministry of energy. Protection of biodiversity and prevention of greenhouse gas emissions from rotting vegetation and increased deforestation apparently are not shared goals within President Lula's cabinet. It is for this reason that the defense of indigenous rights has gained such importance in the legal and political battle.

5 Conclusions

The main hypothesis of this article was that national water policies and politics are already heavily influenced by global environmental governance processes. These processes are based on simultaneous interventions of multiple state and non-state actors on the local, national and global levels. Global concepts like IWRM and environmental regimes like the conventions on climate change, biodiversity, and combating desertification create linkages between national water policies and global policies.

The analysis of the policy processes and conflicts around the construction of a new dam and hydro-electrical power plant in the Brazilian Eastern Amazon (Belo Monte) illustrates these points:

First, the legitimacy of public policies with heavy impacts on water resources and the environment in the Amazon is questioned by local, national and international civil society organizations, who are seeking to reframe local and national economic development strategies by contrasting or complementing economic and social objectives with environmental objectives (protection of biodiversity, mitigating climate change). This reframing has enabled local civil society organizations to give their alternative local development strategy a transnational dimension and to depict it as a contribution to global environmental policies. The defenders of the Belo Monte project have difficulties presenting arguments of equal global value. Their claim that hydro-power is a clean energy source (as opposed to burning fossil fuels) for sustaining national economic growth is stinted by the methane emissions generated by the lake and its negative impacts on biodiversity.

This does not mean, however, that there is a direct positive link between the ratification of international environmental regimes and an increase in the sustainability of national policies. As far as the Amazon is concerned, there are numerous actors in Brazil, from politics (including left- and right-wing parties), the private sector, civil society and the media, who are fundamentally suspicious of any international effort to support the protection of this ecosystem. International scientific research and development cooperation projects are often alleged to be covering up the interest of foreign powers in exploiting natural resources. This means that concepts generated at the global level need to pass through a process of reformulation from the perspective of Brazilian interests in order to be accepted. As we have seen in the case of Brazilian water policy, this process can be successful.

Second, since the 2005 drought in the Amazon underlined the dramatic effects of water shortage in the region and showed that such a scenario can turn into reality, public attention to the implementation of the water law in the Amazon has increased. With public policies already being exposed to stringent legitimacy tests from the local and global perspective, the application of the water law in the Amazon will face the same challenge.

Third, the alliance of local, federal and transnational actors with the *Ministério Público Federal*[11] has achieved a real contribution to global environmental governance as it effectively managed to stop the construction of the hydro-power complex and drawn additional attention to the risks of subordinating Amazon development to national economic interests.

Fourth, however, thus far it is constitutional requirements that have served as the decisive legal instrument for stopping the project, and not obligations stemming from global environmental agreements ratified by the government of Brazil.

Fifth, the Amazon case study does not support the claim that there is a need for a global water convention. On the contrary, national laws and regulations that would guarantee a sustainable use of the Amazon rivers are already in place. The main bottleneck is the implementation of these laws and regulations, a situation that would not be alleviated by a global convention.

As a result, it can be stated that in the Amazon water-relevant public policies are clearly influenced by the existence of global environmental regimes and transnational alliances of civil society organizations. It may be that the Amazon is a special example since this ecosystem is considered to be of outstanding importance for several global environmental services. Also, conditions for civil society organizations may be better in the Brazilian Amazon than elsewhere because they have received external financial support for quite some time and are backed by two powerful actors on the national level: the media and the *Ministério Público Federal*. To a certain extent, these actors have succeeded in compensating for the weakness or even absence of rule of law in the region itself. In regions where these three conditions are not given, the influence of global governance on national water-relevant policies is likely to be much weaker.

References

Dingwerth K, Pattberg P (2006) Global Governance as a Perspective on World Politics. In Global Governance 12, 185–203

Eletronorte (2000) Complexo hidroelétrico Belo Monte e sua inserção regional. http://www.amazonia.org.br/arquivos/26382.doc. Cited 8 January 2003

Eletronorte (2002) Plano de Desenvolvimento Sustentável de Belo Monte – PDS BM. Versão Executiva Preliminar. Material para discussão

Etermann R (2005) Senado aprova construção de Belo Monte. In: O Liberal, 13 July 2005. http://www.amazonia.org.br/noticias Cited 28 June 2006

Greene W (2004) Aid fragmentation and proliferation: can donors improve the delivery of climate finance? In: IDS Bulletin, Vol. 35, No. 3, 66–75

[11] In 2005 the *Ministério Público Federal* took legal action against the authorisation granted by Congress for the Eletronorte and IBAMA to proceed with environmental licensing without hearing the indigenous peoples first. Greenpeace, the umbrella organisation of the indigenous peoples of the Amazon COIAB, and the influential NGO Instituto Socioambiental (ISA) joined in this legal action.

IPCC (2001) Climate Change 2001: Synthesis Report. A contribution of working groups I, II, and III to the third assessment report of the intergovernmental panel on climate change, Watson RT and the core working team (eds), Cambridge

Krell A (1993) Kommunaler Umweltschutz in Brasilien. Juristische Rahmenbedingungen und praktische Probleme, Frankfurt am Main

Nepstad D, Schwartzman S, Bamberger B, Santilli M, Alencar A, Ray D, Schlesinger P (2006) Inhibition of Amazon deforestation and fire by parks and indigenous reserves, In: Conservation Biology Vol. 20, No. 1, 65–73

Neubert S, Scheumann W, van Edig A, Huppert W (2005) Integriertes Wasserressourcen-Management (IWRM) – Ein Konzept in die Praxis überführen, Nomos Verlagsgesellschaft, Baden–Baden

Parsons EA (2003) Protecting the Ozone Layer. Science and Strategy, Oxford University Press, New York

Pinto LF (2003) A hora de mudar. http://www.estadao.com.br/ciencia/colunas/amazonia/htm/2002/nov/19/11.htm. Cited 6 January 2003

Porter G, Brown JW (1991) Global Environmental Politics, Westview Press, Boulder, San Francisco, Oxford

Pülzl H, Rametsteiner E, Tarasofsky R (2004) Policy advice in options for a legally binding instruments for all forests, BOKU/RIIA, Vienna, London

Reinicke WH (1998) Global public policy. Governing without government? Brookings Institution Press, Washington, DC

Rocha P (2002) Usina de Belo Monte. http://www.amazonia.org,br/noticias/print.cfm?id=14801. Cited 7 December 2002

Scholz I (2004) A forest convention – yes or no? German Development Institute, DIE Discussion Paper No. 6, Bonn

Scholz I (2005) Environmental policy cooperation among organised civil society. National public actors and international actors in the Brazilian Amazon. In: European Journal of Development Research Vol. 17, No. 4, 681–705

Scholz I, Dräger D, Floer I, Neher C, Unger J (2003) Handlungsspielräume zivilgesellschaftlicher Gruppen und Chancen für kooperative Umweltpolitik in Amazonien, German Development Institute, Bonn, also available in Portuguese: Sociedade civil e política ambiental na Amazônia. Os casos da barragem de Belo Monte e da rodovia federal BR-163, DIE, Bonn

Secretariat of the Convention on Biological Diversity (2003) Interlinkages between biological diversity and climate change. Advice on the integration of biodiversity considerations into the implementation of the United Nations Framework Convention on Climate Change and its Kyoto Protocol, CBD Technical Series No. 10, CBD, Montreal

Valente LI (1991) Die Tätigkeit des Ministério Público zum Schutze der Umwelt in Amazonien, in: Paul W, Vieira R (eds) Amazônia und Recht, Umwelt- und arbeitsrechtliche Fragestellungen, Beiträge zur 10. Jahrestagung 1991 der DBJV in Belém do Pará, Frankfurt am Main

WCD (2000) Dams and Development. A new framework for decision-Making, Earthscan, London

Young OR (1998) The effectiveness of international environmental regimes. A mid-term report. In: International Environmental Affairs Vol. 10, No. 4, 267–289

Global Water Governance: Managing Complexity on a Global Scale

Maria Schnurr

Abstract The high diversity of institutions, programs and action plans in international water politics provides a barrier to coherent, efficient action, resulting in increasing implementation gaps in water politics. Applying the principles of governance – cooperation, coordination, common values, integration of decision making levels and subject matters – and establishing a coherent Global Water Governance architecture would guarantee a more efficient use of human and financial resources and close implementation gaps, especially in view of the Millennium Development Goals. The key words describing the necessary transformation process are defragmentation, coordination and commitments and are concerned foremost with the restructuring of UN-Water and the establishment of binding rules or treaties in the water supply and sanitation sector. The proposed path towards a Global Water Governance architecture is not without obstacles. Here, the voice of rational science is needed providing rationales and programmatic support when implementing governance principles on a global level. The design of a global Water Governance architecture would open a new field of interdisciplinary research.

1 Introduction: Governance as a Chance

> The water crisis is essentially about how we as a society and as individuals perceive and govern water resources and services. (WWAP 2003, 383)

Current global water problems are marked by three interrelated dilemmas: (1) high complexity of the subject matter, (2) bad governance, and (3) an implementation gap. That water policy is a field of high ecological, social, and economical complexity does not need to be mentioned. Paired with widespread practices of bad governance this complexity though gives rise to ever widening implementation gaps

Maria Schnurr
e-mail: hmschnurr@web.de

W. Scheumann et al. (eds.), *Water Politics and Development Cooperation*,
doi: 10.1007/978-3-540-76707-7_5,

which jeopardize human and environmental development. The problem is not a lack of technical solutions or programs which could alleviate water problems but a lack of an adequate institutional frame, of commitments and implementation efforts, especially on the international water policy level. This is due to a lack of social resources (*second-order resources*), which are able to cope with fading natural resources (*first-order resources*).

This article argues that, in order to improve performance and close implementation gaps in water politics, the concept of governance, common so far as "Global Governance" and "Water Governance", has to be integrated in global water politics, leading to a structure of Global Water Governance. An accompanying scientific discourse can enhance its conceptualization and support decision-makers when dealing with complex global water issues.[1]

Before introducing elements of a Global Water Governance structure this text will provide a concise overview of present water policy on the international level and its contribution to local water management performance. The elements needed for a transition towards Global Water Governance are then analyzed before at the end of the chapter the challenges on the road to Global Water Governance will be illustrated.

2 International Water Policy and its Structural Deficits

Even though a concept of Global Water Governance will encompass water policy on all levels – international, regional, national, and local – its effectiveness is anchored in the structures and processes on the international level. This level of water policy has been greatly neglected in research and in actual administration.[2] This text argues that international water policy and its integration into Global Water Governance is one key to solving global water problems.

Although the problems of the water crisis have mainly local or national impacts, some of their solutions are found at the international level. Therefore, a close look at international water policy and its structures and processes has to precede a formulation of a concept of Global Water Governance. Water policy in general deals with surface and groundwater and its protection, use, allocation, and management; ideally it also deals with related resources like land or marine aquatic systems and reflects the impacts of and on climate.[3] Given the natural qualities of water, water policy on lower levels deals mainly with the actual management and allocation of the resource, whereas international water policy has the following functions:

- Creating common values and principles of action
- Promoting cooperation between nation states and organizations

---

[1] Compare the objectives of the Global Water System Project www.gwsp.org

[2] For a sample of international water research documents cf. Millennium Project 2004.

[3] The integration of other bioresources is a central element of Integrated Water Resources Management (IWRM).

- Coordinating processes and institutions on different levels
- Knowledge transfer and capacity building
- Funding
- Public relations, awareness building

Despite water scarcity being a universal problem[4] which is often best solved at the local or national level, there are several factors that motivate political actors to deal with water problems at the international level:

- **Large-scale perspective:** Many water related problems elude appropriate solutions on the basin level, and water related processes are often linked to the global hydrological cycle. "Very few problems nowadays are truly local in nature. In the age of globalization, most problems have either direct or indirect causes and/or impacts that have a global nature." (Pahl-Wostl et al. 2006, 4)
- **Transboundary waters:** Many rivers and groundwater systems are shared by two or more states. To effectively prevent conflicts of the allocation and use of shared freshwater resources, internationally valid rules for transboundary waters are essential.
- **Interdependence:** Even though water management in one region does not necessarily affect another region there are other side effects which are enlarged by the interdependence of nation states in a globalizing world. To act single-handedly does not serve any country if they want to prosper and make progress under present circumstances.
- **Mutuality** and a sense of global stewardship obligate countries to help each other.
- **International goals** drive countries to move together towards certain achievements. Water is a means to achieve several mutual international development goals, namely the Millennium Development Goals (MDGs).

Compared to other fields of international policy, e.g. climate change policy or security policy, water policy does not yet enjoy similar structures and powers in order to implement its goals and programs efficiently. What has been achieved in global environmental policy or others may well be achieved in international water policy also (Klaphake and Scheumann 2001, 12).

Water has been high on the international political agenda since 1977 when at the first UN Conference on Water in Mar del Plata/Argentina a Drinking Water and Sanitation Decade from 1980 till 1990 was inaugurated which generated some unexpected progress but still lacked the success needed to avert serious water crises in developing countries. Water was acknowledged as indispensable for all fields of development – from health to economic progress. At the same time, its influence on peace and security issues did receive attention. The social sciences gained ground in water science and policy in the 1990s after the Dublin Conference when more

---

[4] Environmental or development problems are called *global* when their impact is felt globally, e.g. climate change; they are called *universal* when they occur in many places on the planet, but their impact is felt only locally/regionally, e.g. water scarcity or desertification.

and more policy makers recognized the limits of technical solutions and looked towards its management side. The way for the concept of Integrated Water Resources Management (IWRM), promoted by international institutions like the Global Water Partnership (GWP), was opened, requiring a closer integration of actors, interests and water-related resources, mainly on the river basin level. IWRM received international recognition with the Millennium Development Goals (MDGs), which defined IWRM as the solution model for water-related targets. Despite the many efforts and programs on the political level, actual implementation remains behind, leaving 1.2 billion people without access to safe water and 2.4 billion people without adequate sanitation services in 2006 (Millennium Project 2004, 32). This implementation gap is also due to a lack of governance performance on the international level where institutional actors do not seem to cope with the challenge of complexity.

A look at the way international water politics operates and at its structure will help to understand what a lack of governance means here. So far, international water politics has used two so called horizontal coordination modes, namely conferences and policy diffusion (Simonis 2005, 319). Conferences provide a platform for knowledge exchange, establishing an epistemic community and in some cases for decision making. Global policy diffusion is the spreading of institutions and norms into governments and supranational organizations and is part of the new self-organized structures typical for the governance concept.[5] A third mode of coordination in environmental politics is international conventions. While other key problems recognized at the Earth Summit in Rio in 1992 – climate, biodiversity, desertification – have all been covered with international conventions[6] no such convention has been drafted respectively ratified yet for all the uses of freshwater. There are treaties, conventions and action plans on the regional level and for transboundary uses of water like the International Convention on the Non-Navigational Uses of Water Courses (ratification unachieved), but they are missing on the international level for water supply and sanitation issues even though they are crucial for social, economic and ecologic sustainability as proclaimed in Rio 1992. The Millennium Development Goals and their target 10 – reduce by 2015 the number of people without access to safe drinking water and sanitation – is not a binding treaty but a mutual action plan encouraging governments to undertake adequate steps to achieve these goals.

Furthermore, almost no attention has been paid to the structure, i.e. actors and institutions of international water policy. While the – partial – successes of international environment policy are mainly due to institutional rather than mere technical innovations, so far no programmatic action has been taken to develop strong coordinating institutions and processes in international water politics. Instead, the sector-specific approach of local and national water politics has continued on the international level, leading to a high number of international governmental and non-governmental institutions dealing with water resources but lacking the necessary

[5] For examples of policy diffusion cf. Pahl-Wostl et al. 2006, 15.

[6] UN Convention to Combat Desertification, International Convention on Biodiversity, UN Framework Convention on Climate Change.

coordination and cooperation structures in order to efficiently achieve mutual international goals:

> (The) diversity of actors contributes much to the strength of international water and sanitation support and advocacy, but also creates new challenges to coordination to ensure effective coherent action. (Millennium Project 2004, 43).

The results are redundancies and gaps, inefficient use of financial resources, and contradicting values and principles applied in programs, which cause an ever increasing global implementation gap. As this was recognized at the Johannesburg summit where the importance of water for sustainable development was reaffirmed the UN Chief Executives Board for Coordination founded the inter-agency mechanism *UN-Water* in 2003.[7] Its goal is to support nation states in the implementation of the Johannesburg Plan of Implementation (JPOI) water-related provisions and the Millennium Development Goals concerning freshwater. It was designed to coordinate the fragmented activities of the different UN programs and associated organizations dealing with water, e.g. CSD, FAO, UNDP, UNEP, UNESCO, UN-Habitat, WHO and World Bank, the latter being the largest donor in the water sector. UN-Water also cooperates with several non-governmental organizations like the World Water Council, the Global Water Partnership and the Water and Sanitation Supply Collaborative Council. Together with the UNESCO, UN-Water conducts the "Water for Life" Decade 2005–2015.

Despite the ambitious effort of the UN to found such an inter-agency mechanism, UN Water currently "does not have adequate budget or staff to execute the functions at the scale required" (Millennium Project 2004, 43), thus continuing the inefficient use of resources and expanding implementation gaps. The current level of cooperation among UN agencies is insufficient for coping with the challenges associated with international water-related goals, and programs and actions are neither aligned to a common water strategy nor based on common values. The lack of influence on decisions and discourses has been utilized by other organizations, namely the WWC, to replace functions of the UN. This is even more alarming as the values and principles of these and similar organizations do not equal those of the UN-bodies.[8]

Besides UN-Water and the programs dealing, among other assignments, with water named above there are three more water-related institutions within in the UN system:

- The UN Secretary-General's Advisory Board on Water and Sanitation (UNSGAB)[9] is an independent panel of water experts from a wide range of political, economic, civil, or scientific background. Founded on World Water Day March 22nd 2004 and reporting directly to the UN Secretary-General, it assesses progress, raises political visibility of water and sanitation issues, mobilizes more funding and manpower, and works together with monitoring agencies. The UNSGAB is an action-oriented board that drives for implementation and creates

[7] Founding document at: http://www.un.org/esa/sustdev/sdissues/water/CEB_Decisions2003.pdf

[8] Recently though the WWC's principles approached UN principles during the last World Water Forum (2006), especially regarding privatization and human rights issues.

[9] Homepage: www.unsgab.org

awareness for water problems. So far it lacks binding terms of references (ToR) and a timeframe for the accomplishment of its tasks.

- The WHO Joint Monitoring Program (JMP) and World Water Assessment Program (WWAP) are UN freshwater monitoring agencies. While the WWAP assesses the quantity, quality, consumption, and management of freshwater, the JMP completes the assessment of water resources by monitoring water and sanitation services. The two assessment programs complement each other by content but work independently.

In view of ever growing freshwater problems a transformation towards an effective and efficient Global Water Governance architecture that promotes the necessary coordination and cooperation at the international level is long overdue.

3 Global Water Governance: Transformation and Integration

In light of the challenges and shortcomings described above it is suggested that elements and principles of the concept of governance be applied to global water policy and a Global Water Governance structure be established.

The concept of governance emerged in political sciences in the 1990s with the Commission on Global Governance (CGG) convened by the UN which elaborated principles for international cooperation in view of changes precipitated by the end of the Cold War. Coined as a new scientific concept, Global Governance means.

> ...the sum of the many ways individuals and institutions, public and private, manage their common affairs. It is a continuing process through which conflicting or diverse interests may be accommodated and co-operative action may be taken. It includes formal institutions and regimes empowered to enforce compliance, as well as informal arrangements that people and institutions either have agreed to or perceive to be in their interest. (...) Effective global decision-making thus needs to build upon and influence decisions taken locally, nationally, and regionally, and to draw on the skills and resources of a diversity of people and institutions at many levels. It must build partnerships that enable global actors (...) to develop joint policies and practices on issues of common concern. (CGG 1995, 4).

In short, governance favors integral and systemic approaches in problem solving that are necessary for coping with complex interdependencies; its intent is to manage complexity rather than to reduce it. Central elements of a Global Governance architecture are (Messner and Nuscheler 1996, 5–11):

- Shared sovereignty
- Re-empowerment of nation states
- Intensification of international cooperation by binding rules
- Solid basis of values and principles
- Systemic integration of actors, decision making levels, and subject matters (multi-level governance)

For a long period, the international environment and development discourse was dominated by the term "sustainablity" until the term "governance" – matching natural with social resources – took its place at the start of the new millennium. Scientists and politicians alike acknowledged that the more part of environmental and development problems originated in a lack of social resources or poor governance, i.e. in failures to manage resources in an equitable, efficient and sustainable manner. Governance became the panacea for almost every development issue; the term joined forces with global aspects, i.e. global governance, policy fields, e.g. water governance, and normative frameworks, i.e. good governance. Likewise, bad governance was recognized as the main obstacle to most development goals.[10]

The concept of governance mirrors the complexity of the problems of our time – a complexity rising with growing interdependence due to globalization – and the new dynamics of the political processes needed for their solution. The global water crisis with its actors pursuing contradicting goals and with its dramatic challenges especially in developing countries is marked by a high range of complexity. Governance as a structural and normative concept, which so far has proven valid for the local and regional management of water resources (UNDP 2004), is a useful answer to manage this complexity even on a global level. Rather than deepening hierarchic structures and streamlining actors and institutions, the concept of governance allows for a continuous process of balancing diverse interests by promoting cooperative action and effective coordination of different actors on varying levels of political decision-making.

Given the definition and elements of Global Governance and accepting that water governance is a global level issue, a concept of Global Water Governance may be developed accordingly. Water policy on any level will function more effectively and efficiently if a transition to global (i.e. multi-level) governance takes place as experiences show on the local and national levels (UNDP 2004). Rather than intensifying hierarchical structures or setting up one central governing body action should be taken to improve the decentralized coordination of sovereign actors, decision-making levels, values, principles, subject matters, and methods of implementation which allow a comprehensive approach to water problems on the international, national and local agenda. Governments as the single decision making authorities are supplemented by multi-scale, polycentric governance including various stakeholders.

Central elements of a Global Water Governance architecture would be accordingly:

1. Redefinition of sovereignty, along with an empowerment of the nation state in the water sector
2. Intensification of international cooperation with binding rules
3. A common set of values and principles
4. The systemic integration of actors, decision levels, and subject matters
5. The incorporation of water policy in global environmental policy

Naturally, in order to provide legitimacy and efficiency, institutions in a Global Water Governance architecture will be bound to the principles of the normative

[10] e.g. Millennium Project 2004 cites "governance failures" as one of four obstacles to reaching the MDGs.

concept of Good Governance, namely: participation, transparency, coherence, responsiveness, inclusiveness, consensus orientation, accountability, rule of law, and effectiveness (Behrens 2004, 16).

It is suggested to transform the common definition of water governance[11] into the following definition of Global Water Governance:

> The term Global Water Governance encompasses the political, economic and social processes and institutions by which government institutions on all levels – international, national, regional and local –, civil society, and the private sector make decisions about how best to use, develop and manage water resources in order to achieve internationally agreed upon goals, thereby applying the principles of good governance.

In order to achieve this transformation several changes must be effected in the water policy, politics and polity dimension.

3.1 Structural Reforms of International Water Politics

In a sound system of Global Water Governance, international institutions of water policy assist other levels of water governance (local, regional, national) in implementing sustainable management of water resources by providing structural, human and financial resources. Backed by interdisciplinary scientific research, they might initiate discourses about values and principles, foster the transformation of abstract values into concrete, locally adapted goals and targets, and are responsible for the coordination of relevant activities. They integrate the interests and obligations of governments, civil society, and the private sector on all territorial levels in their activities.

Presently, the international system is unable to manage these tasks as it should. As steps were taken by the UN-system to improve the situation by founding, but inadequately funding UN-Water there is room for improvement. Rather than founding a new operational organization – similar to what the FAO performs for the food and agriculture sector or the WHO for the health sector – it has been suggested to upgrade UN-Water to a multi-agency entity, similar to the exemplary UNAIDS, so the UN-system can provide strong and effective support in order to achieve international water-related targets (Millennium Project 2004, 2; Simonis 2006, 2). Instead of stronger hierarchies and the establishment of new programs and organizations, an intensification, acceleration and strategic alignment of existing activities and programs – horizontal institutionalization – would be sufficient (Simonis 2006, 15). This requires equipping UN-Water with adequate staff and budget and providing it with sufficient support and input from and a consistent link-up to UNEP in order to balance the environmental aspects of water management with developmental ones (Rechkemmer and Schmidt 2006, 103). Concerted capacity building efforts of this

[11] A definition of water governance is provided by the UNDP (2004, 10): The term Water Governance encompasses the political, economic and social processes and institutions by which governments, civil society, and the private sector make decisions about how best to use, develop and manage water resources. For others cf. WWAP, Chap. 15 (pp. 370f.).

kind could transform UN-Water into both a global advocate for water and sanitation and a supporting unit for national scaling-up efforts. With such an upgrade, which could be accomplished within a reasonable timeframe, UN-Water would incorporate the following features:[12]

- It combines the competencies of existing programs and institutional experience, thus retaining its character of an inter-agency mechanism.
- It aligns global activities in funding, capacity building, research, knowledge transfer, monitoring, evaluation, and advocacy to the water-related MDGs and the three-fold strategy of MDG interventions for water and sanitation.[13]
- It integrates the interests and needs of non-governmental organizations, the private sector and other relevant stakeholders, especially water users.
- It promotes civil society activities and corporate social responsibility.
- It develops strategic partnerships, especially on sub-national levels.
- It mobilizes additional funding.

These features would make UN-Water an adequate advocate for globally agreed upon water-related development goals and a powerful catalyst for timely, comprehensive and well-coordinated action.

Additionally, water policy would greatly benefit from a revision of the activities of the UNSGAB and the monitoring institutions, namely JMP and WWAP. Advancing implementation efforts is neither possible without a professional strategic focus nor without sound monitoring. The task of UNSGAB needs to be expanded and its activities aligned more closely to the MDGs. It should support UN Water in the design of binding rules and evaluate the inputs and outcomes of water programs. Upon this basis it could derive sustainable strategies, thus strengthening the accountability of institutions, and it could help decision-makers to recognize the integrating character of water for the achievement of all MDGs.

Progress on international goals, namely the MDGs, will require comprehensive monitoring. The two agencies responsible for monitoring, JMP and WWAP, at this time, do not cooperate closely enough. Reasonably, they should produce a joint monitoring report instead of two separate ones; they should also consider cooperation with other (non-governmental) monitoring and assessment programs if At the same time, they will need more funding in order to optimize monitoring and assessment capacities on the country level.

Furthermore, in order to balance environmental and developmental aspects in international water politics, a transformation of the current UNEP into a UN specialized agency, ideally combined with responsibilities for development issues as well, would be of great help.[14] To overcome the still rather splintered approach in

[12] For more detailed suggestions cf. Millennium Project 2004:157f.

[13] MDG interventions for water and sanitation include: (1) Installation and operation of water supply and sanitation services, (2) capacity building, education and empowerment, (3) constructing facilities for storing and transporting water and designing IWRM strategies (Millennium Project 2004, 29).

[14] For details on the suggested reform options of UNEP cf. Rechkemmer 2004, 15f. and Simonis 2006, 13 and 15.

global environmental policy, the rising national autonomies require a strong environmental coordinating institution which aligns strategies and combines financial and political power. Global environment governance as a whole has suffered from a weak institutionalization, but despite highly motivated initiatives by governments and NGOs, solutions to strengthen the relevant UN structures, especially UNEP, are unlikely to be realized soon because the reform models differ strongly.

The suggested changes to UN Water, the UNSGAB and the monitoring programs will greatly enhance efforts to reach water-related development and environment goals because they will promote synergies resulting from coordinated mutual efforts of existing programs and activities. They would form a central element in a Global Water Governance system that fosters cooperation and mutual accountability and are even more important as long as there are no binding rules (treaties or conventions).

3.2 Establishing Binding Rules

One key element of Global Governance is the intensification of international cooperation through binding rules and commitments, i.e. "institutions and regimes empowered to enforce compliance" (CGG 1995, 4). A convention which aggregates rules and principles declaring common rights and duties of users, providers and government concerning freshwater is central to constructing a Global Water Governance architecture.

Presently, there exist only provisions for the management of transboundary waters: the UN Convention on the Law of the Non-navigational Uses of International Watercourses, which guarantees the equitable and reasonable utilization and participation of shared watercourses, as well as several bi- or multilateral transboundary treaties. International provisions for the water supply and sanitation sector and for the general sustainable management of water resources have not yet been designed; the waterrelated MDGs and the Johannesburg Plan of Implementation provisions relating to IWRM are not binding. A complementary version or an annex to the existing UN Convention on the Law of the Non-navigational Uses of International Watercourses comprising the MDGs could be legally based on the General Comment No. 15 of the UN ECOSOC[15] which provides for a human right to water. Since the MDGs are to be achieved by applying IWRM principles, the Johannesburg Plan of Implementation provisions could easily be included. Negotiated fairly and designed in a flexible manner, such a convention would fulfill the provisions of the International Covenant on Economic, Social and Cultural Rights (ICESCR), presenting a cornerstone of a sound Global Water Governance architecture.

[15] Par. 35 of the General Comment No. 15 calls for the design and ratification of international conventions which implement the human right to water. A basis for a draft convention could be the proposal of Friends of the Right to Water 2005.

Designed like similar international conventions, it should consist of coherent principles, norms, rules, and procedures. Ideally, it would (a) provide a common set of values and principles, (b) strengthen the nation states' performance in the water sector, (c) intensify the integration of actors, political decision-making levels and subject matters, and (d) enhance international cooperation and partnerships.[16] According to the principle of the Rio declaration of "common but differentiated responsibilities" for industrialized and development countries,[17] a water convention could not be designed as a one-fits-all solution; but instead, each nation's responsibilities will depend on its capacities and objectives. The design of a water convention should precisely define duties and rights, but allow for the free choice of instruments, thus responding to different needs and capacities while strengthening the self-reliance and accountability of the actors. A model for a global convention with regional/local annexes is the UN Convention to Combat Desertification (UNCCD); the European Water Framework Directive is an example where a framework directive from a supranational body sets the overall frame and targets but where the legal and operational details are decided at national resp. basin level.

Binding international rules for sustainable water management and the water supply and sanitation sector would support a more pro-active international water policy strategy and would join existing initiatives and programs in one powerful mechanism needed to achieve international goals concerning freshwater (Rechkemmer and Schmidt 2006, 102f.; Simonis 2006, 13). They would ensure commitments of nation states and strengthen their role as the main water supply and sanitation provider, thus completing a sound Global Water Governance architecture.

3.3 Global Governance and the Nation State

Governance does not mean more centralization of tasks that are better kept on a local or regional level (Edig and Edig 2005, 150f.). Rather, a core principle of action of Global Water Governance is the empowerment of states because the functioning of Global Governance depends on strong, capable nation states. Similar to Global Environmental Governance, a Global Water Governance system needs to fully respect the sovereignty – and accountability – of the nation states and uses and, if necessary, enhances their problem-solving capacities by providing structural or financial support. The measures taken on the international level need to promote an interactive state which cultivates interaction with society. This means increased attention

[16] For more details cf. Kahlfan 2005.

[17] Full text: "Principle 7: States shall cooperate in a spirit of global partnership to conserve, protect and restore the health and integrity of the Earth's ecosystem. In view of the different contributions to global environmental degradation, States have common but differentiated responsibilities. The developed countries acknowledge the responsibility that they bear in the international pursuit of sustainable development in view of the pressures their societies place on the global environment and of the technologies and financial resources they command." (Rio Declaration on Environment and Development 1992).

to national and local governance as a precondition for mutual accountability in Global Governance.

What applies to the global level also applies to the national: water politics must be included in overall national development strategies, and its integrating character should be considered when designing national water institutions. Here again, binding international rules concerning freshwater would spur nation states' efforts to improve the management of water resources and also help them fully integrate water into general national development plans.

As the latest MDG interim reports have shown clearly, there is a huge gap of progress among countries; despite a few good performers in line with MDG target plans, many governments seem not to be able to cope with the challenges posed by improving water supply and sanitation services, whereas others may have the funds and capacities but are unwilling to allocate more funds and capacities to basic services like water supply. This again suggests the need for a better equipped and more powerful UN environmental governance and development system and/or UN-Water which actively assists and monitors countries in need of help or reform.

4 Challenges on the Road to Global Water Governance

The core question is and will be how to close current implementation gaps. It is not the technologies or programs water policy is lacking – even though in both fields there is still some room for improvement – but rather a sound structural frame with clear rules and binding targets. The key words describing the necessary transformation are *defragmentation, coordination* and *commitments*, which are all core elements of the concept of governance. Applying principles of governance and establishing multi-level, polycentric decision-making structures will foster the much needed inclusion, integration, and alignment of existing water technologies and programs for the benefit of human development and ecological systems. Central elements of Global Water Governance are (1) the establishment of binding rules as suggested above by establishing a water convention based on the General Comment No. 15 of the UN ECOSOC, complementing the UN Convention on the Law of the Non-navigational Uses of International Watercourses and (2) the upscaling of central UN institutions, namely UN-Water and the UNSGAB.

Challenges do always exist when transforming complex social systems. Even if proceeded slowly and carefully there will be unforeseen obstacles. Characteristic of the concept of governance, there is no single leader or decision-making body that can control the transformation process alone. Therefore, strong coalitions of actors willing to change things for the benefit of water and human health need to cooperate.

As in other fields of change – political, social, cultural, and technical – the voice of rational science will reach decision-making levels with more ease than the voice of the actual practitioners or society in general could do. Science as the most modern form of reflection and progress (Hosang 2005, 62) can and must assume its responsibilities by investigating more in detail the chances and requirements of a sound Global Water Governance architecture. Many questions remain, especially regarding the translation of the logic of ecological systems into social systems (Rechkemmer 2006, 105). This would require an even closer integration of social scientists – still a minority compared to ecologists and hydrologists – in the realm of water science. Designing and assessing a Global Water Governance architecture would open a new broad field of research for social scientists with ample opportunities to bring in their views and knowledge and to cooperate with other sciences. Global Water Governance research could be anchored in Global Governance and Water Governance research, thus preventing the reinventing of the wheel, and it should proceed in a descriptive-analytical as well as prescriptive-normative manner. It would also help social sciences to develop a stronger focus on water politics and its different levels from local to global, a topic still broadly neglected even though water resources management is of an inherently political nature.

References

Behrens M (2004) Global Governance. In: Benz A (ed) Governance – Regieren in komplexen Regelsystemen. Eine Einführung, VS Verlag, Wiesbaden

CGG (1995) Our global neighbourhood. Report of the Commission on Global Governance, Oxford

van Edig A, van Edig H (2005) Integriertes Wasserressourcen-Management – Schlüssel zur nachhaltigen Entwicklung. In: Neubert S et al. (eds) Integriertes Wasserressourcen-Management (IWRM). Ein Konzept in die Praxis überführen, Nomos, Baden–Baden

Friends of the Right to Water (eds) (2005) Key principles for an international treaty on the right to water. Draft work in progress for consultation and revision. http://www.alliancesud.ch/english/files/T_WrKr.pdf. Cited 10 May 2006

Hosang M (2005) Die emotionale Matrix, oekom, München

Kahlfan A (2005) Implementing general comment No. 15 on the right to water. In: National and International Law and Policy. Discussion Paper March 2005. In: Bündnis 90/Die Grünen: Wasser ist Zukunft. Nachhaltige Wasserversorgung im 21. Jahrhundert. Dokumentation der Konferenz vom 15. April 2005 im Deutschen Bundestag, Berlin

Klaphake A, Scheumann W (2001) Politische Antworten auf die globale Wasserkrise: Trends und Konflikte. In: Aus Politik und Zeitgeschichte B48-49/2001, 3–12

Messner D, Nuscheler F (1996) Global Governance. Herausforderungen an die deutsche Politik an der Schwelle zum 21. Jahrhundert, SEF Policy Paper 2, SEF, Bonn

Millennium Project (2004) Health, dignity, and development: what will it take? Achieving the Millennium Goals. UN Millennium Project Task Force on Water and Sanitation, Final Report, Abridged Edition, New York, Stockholm

Pahl-Wostl C, Gupta J, Petry D (2006) Governance and the global water system: towards a theoretical exploration, Global Water System Project Paper, Bonn

Rechkemmer A (2004) Globale Umweltpolitik 2005. Perspektiven im Kontext der Reform der Vereinten Nationen, SWP-Studie S45, Stiftung Wissenschaft und Politik/Deutsches Institut für Internationale Politik und Sicherheit, Berlin

Rechkemmer A, Schmidt F (2006) Neue globale Umweltpolitik. Die Bedeutung der UN-Reform für eine nachhaltige Wasser- und Umweltpolitik, Erich Schmidt Verlag (Reihe Initiativen zum Umweltschutz Nr. 64), Berlin

Simonis G (2005) Weltumweltpolitik: Erweiterung von staatlicher Handlungsfähigkeit durch Global Governance? In: Behrens M (ed) Globalisierung als politische Herausforderung. Global Governance zwischen Utopie und Realität, VS Verlag, Wiesbaden, 313–344

Simonis UE (2006) Auf dem Weg zur Weltumweltpolitik – und zur notwendigen Reform der Vereinten Nationen, Wissenschaftszentrum Berlin, WZB discussion papers, Berlin
UNDP (2004) Water governance for poverty reduction. Key issues and the UNDP response to Millennium Development Goals, New York
WWAP (2003) Water for people, water for life. The United Nations World Water Development Report 1, UNESCO/Berghahn Books, Paris
WWAP (2006) Water – a shared responsibility. The United Nations World Water Development Report 2, UNESCO/Berghahn Books, Paris

Part II
Critical Debates Revisited

Strategic Virtual Water Trade – A Critical Analysis of the Debate

Susanne Neubert, DIE in collaboration with Lena Horlemann

Abstract The article analyzes the vigorous debate on strategic virtual water trade in order to save scarce water resources on a global, national or river basin level. It is agreed among experts, that many conditions must be fulfilled if developing countries are to use virtual water trade to their own benefit, while averting or compensating for the likely adverse consequences. Thus, the strategy is rather not suited for least developed countries but better suited to newly industrializing and so-called anchor countries. Regional economic communities with very unequally distributed water resources such as SADC can also benefit from virtual water trade. Before considering the variants of virtual water trade, the article examines also certain fundamental aspects of the strategy. Particular attention is paid to the consequences of increased water use efficiency and high unit water values for different crops on scheme and river basin level.

1 Introduction

The idea of using virtual water trade as a trade policy strategy to offset regional water shortages[1] is currently being debated so vigorously by water experts because it might lead to huge water savings, making it possible – or so it would seem at first sight – to resolve "the water crisis" virtually overnight.

The concept of strategic virtual water trade is based on the idea that, by importing more of their food from water-rich countries, water-poor countries might use their scarce water resources in sectors (e.g. industrial production) in which they achieve a higher value added per volume unit of water, i.e. higher water productivity.

Susanne Neubert
e-mail: Susanne.Neubert@die-gdi.de

[1] In this article strategic virtual water trade is meant whenever a reference is made to virtual water trade. Virtual material flow analyses, which are used solely as an analytical instrument, are included in this article only as facts, not as controversial subject matter.

W. Scheumann et al. (eds.), *Water Politics and Development Cooperation*,
doi: 10.1007/978-3-540-76707-7_6, © Springer-Verlag Berlin Heidelberg 2008

However, protagonists argue that in agriculture, too, productivity increases lead to water savings if the emphasis is placed on the cultivation of crops which are highly efficient in their use of water. Changes in the domestic food supply could then be offset by means of international trade.[2]

As attractive as the concept may appear at first sight, it is a matter of some controversy among water experts. The reason for this is that some of the above pre-assumptions are absolutely questionable, and as successful virtual water trade would be subject to innumerable requirements for the countries concerned, many restrictions relating to groups of countries, areas of application and degrees of implementation would have to be defined.

The debate on virtual water trade has evolved over the past 15 years, having first been postulated by Anthony Allan[3] as an option for Middle Eastern countries. The debate among water experts did not really come alive, however, until 2003, when Hoekstra and Hung analyzed global hydrogen flows caused by the international agricultural trade. This drew greater attention to the considerable potential for saving water by means of virtual water trade, and during the international debate a possible strategy for the Middle East became a possible strategy for water-poor countries in general.

This article describes and analyses the debate on virtual water trade with the aim of identifying controversies and points of agreement and differentiating the subject matter. In addition, the lines of argument advanced by certain disciplines are confronted with those presented by other disciplines, and on this basis, the debate can be carried forward and the possible need for research identified.

2 Approach – the Methodology

The data were collected during a participatory DIE research project carried out in 2005 and 2006 and financed by the German Ministry for Economic Cooperation and Development (BMZ). Ten expert statements were commissioned as part of the research project. The most important questions associated with virtual water trade were raised during a kick-off workshop, and the terms for the statements were formulated on this basis. In addition, the literature on the subject was analyzed, and a number of experts were interviewed. In December 2005 the statements were presented and discussed at an enlarged workshop at the DIE. Particularly controversial issues were considered by additional working groups. While the substantive results

[2] During the debate on virtual water trade, little attention has so far been paid to the possibility of the water released being used to meet household needs. Although irrigation and human consumption most frequently compete for the use of water, this aspect has hitherto played little part in the virtual water debate. The reason for this is that the use of water as drinking water is among the reproductive rather than the productive uses of water and is not therefore accompanied by any directly measurable value added.

[3] Cf. Allan (1996; 1997; 2003a; 2003b) and Allan et al. (2003).

were set out in a monograph (cf. Horlemann and Neubert 2006), they are presented in this article as an analysis of the debate.

3 An Overview of the Controversy

The controversy arising during the debate on virtual water trade can be subdivided into a number of subject areas:

1. controversies on the prospects for strategic virtual water trade,
2. objections to its assumptions and objectives,
3. controversies on the requirements for successful virtual water trade,
4. controversies on certain consequences of strategic virtual water trade.

In the following the various assumptions on links between the effects of strategic virtual water trade are described in "exaggerated" or idealized form, from the viewpoints of protagonists and critics. In this analysis the core of each argument is identified – as required by an analysis of a debate – so that it may then be submitted to a critical examination. Arguments that remain implicit in the actual debate are made explicit here.

3.1 Proclaimed Prospects and Requirements for Virtual Water Trade

The proclaimed prospects for strategic virtual water trade mainly relate to the possible water savings and to favorable effects assumed to result from those savings, as set out in Fig. 1.

The requirements listed here are essential if it is to be understood that virtual water trade is a realistic option for only a few countries or groups of countries, since they must have, for example, sufficient foreign exchange available to pay for food imports if they are to afford virtual water trade. The infrastructure and transport systems of such countries must also be well developed wherever food is to be transported. Countries who apply virtual water trade must, moreover, have a high social absorptive capacity if they are to employ the rural workers who lose their jobs as a result of the reduction of irrigated agriculture. Finally, virtual water trade presupposes good governance, because it must become an institutional fixture and be induced purposefully. Any adverse consequences of virtual water trade must also be compensated for or averted.

The many requirements for the success of virtual water trade cannot be satisfied by any country at a stroke: in principle, their satisfaction is conceivable only in the context of an iterative process over the long term. An ideal-type scenario would be the withdrawal of a newly industrializing country from particularly water-intensive agricultural production chains, accompanied by its gradual entry into virtual water

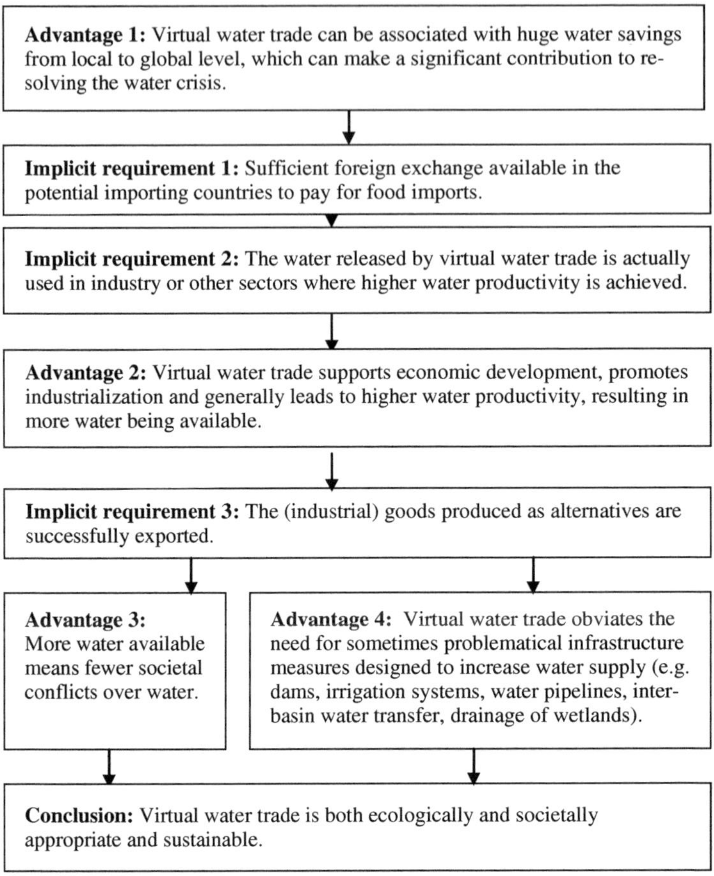

Fig. 1 Proclaimed prospects of virtual water trade from the protagonists' viewpoint (idealized) and implicit requirements
Source: Authors' own presentation

trade that released growing quantities of water. The latter would then be used in an export-oriented sector of industry, and the export revenue would be used in turn to increase virtual water trade further.

3.2 The Critics' Objections to the Premises and Objectives of Virtual Water Trade

In the debate on strategic virtual water trade general and specific criticisms are leveled at the concept. The general criticisms concern (i) the various explicit premises of the concept, (ii) the usually implicit political and economic assumptions on which the concept is based and (iii) the proclaimed cause-and-effect relationships.

The fundamental objections to the concept are independent of one another and can be summarized as follows (see Fig. 2):

These fundamental criticisms are joined by other objections concerning the adverse consequences which critics expect virtual water trade to have. If only because these consequences are so serious and numerous, it is evident at this juncture that, as demanding conditions must be satisfied if virtual water trade is to succeed, not only will restrictions be needed, but implementation will also take a very long time.

The critics argue that pursuing the virtual water trade strategy would widen the gap between poor and rich countries. For the classical and poor developing countries, their economies largely dependent on agricultural production and most having weak institutions, there might be both direct and indirect adverse effects if they themselves or even other countries engaged in virtual water trade for strategic reasons. The indirect adverse effects would occur mainly if trade in food was North–South and so consisted largely of subsidized agricultural products. Classical developing countries that export agricultural products themselves would then become even less competitive. Nor would they then have any supporters in the South standing up for the liberalization of the world market, since, as importers, those opting for virtual water trade would, of course, welcome low world market prices.

1st criticism: market rather than regulation. The idea of politically decreed virtual water trade is based on a planned-economy approach. Instead, water prices should be introduced; the water scarcity problem would then solve itself.

2nd criticism:false assumptions on cause-and-effect relationships. Industrialization is not the consequence of, but a requirement for successful strategic virtual water trade.

3rd criticism: the calculations on global water savings are irrelevant. Global water conservation effects are an arithmetic quantity, which is in fact irrelevant. Nor is there currently any global water scarcity that would make global savings worthwhile.

4th criticism: calculate opportunity costs rather than seeking to increase water productivity within agriculture. Excessive focusing on higher water productivity without considering the opportunity costs of different water uses and sources (in the case of blue and green water resources) leads to incorrect conclusions.

5th criticism: the political risks attached to virtual water trade cannot be predicted. The abandonment of the paradigm of national food sovereignty, which is bound to accompany the strategy of virtual water trade, will make it possible for the potential importing countries to be blackmailed.

Fig. 2 Objections to the premises of the concept of virtual water trade from the critics' viewpoint (idealized)

Apart from the fact that classical developing countries could not themselves engage in virtual water trade because they do not satisfy the requirements referred to above (foreign exchange, organizational capacity, infrastructure, good governance) the critics claim that virtual water trade would have the following negative impacts on such countries if they still tried to adopt this strategy:

- Social distortions in rural areas if, for the sake of virtual water trade, rain-fed and irrigated agriculture was restricted or not expanded further. This would lead to rising underemployment, migration, accelerated urbanization resulting in expanding slums and total impoverishment.
- Drastic changes to ways of life and production methods owing to the urban bias accompanying virtual water trade and the abandonment of irrigated agriculture. This in turn would result in unpredictable societal and cultural changes (cf.Hummel 2005).
- There would be a danger of the reversal of generally valid principles of good water management, such as the IWRM principle of participation, since virtual water trade would tend to be accompanied by a growth in the power of the governments concerned and could therefore be associated with centralization tendencies and corruption (cf. Youkhana and Laube 2006).
- If virtual water trade was engaged in rigorously, food would have to be imported on a grand scale and distributed. Supply bottlenecks would therefore easily occur, accompanied not least by a heightened risk of famine, possibly affecting large sections of the population.

All in all, many classical developing countries would therefore undermine their own (pro-poor) growth prospects by opting for virtual water trade, since the best prospects for these countries lie in the intensification rather than the abandonment of agriculture (cf., e.g., Brandt and Otzen 2004).

The protagonists of the strategy respond to these consequence scenarios by arguing that the virtual water trade strategy is intended primarily for better-off countries and those few countries that suffer from absolute hydrological water scarcity, i.e. those which are de facto dependent on it if they are to feed themselves. From this it can be concluded that for the classical agricultural developing countries virtual water trade is, as a rule, neither a feasible nor a desirable strategy even in the medium term.

As regards the suitability of groups of countries for the virtual water trade strategy, the distinctions are, of course, fluid. Kluge and Liehr (2005) have developed a diagram to show the basic link between successful virtual water trade and a country's level of development and industrialization. This link, which is shown in Fig. 3, can be regarded as a consensus in the international debate.

The arguments for and against considered in the following concern only groups of countries which might, in principle, use virtual water trade to advantage, i.e. newly industrializing and anchor countries[4] and countries suffering from absolute

[4] Anchor countries are large, usually very populous countries characterized, among other things, by the fact that they have dynamic economic sectors (outside agriculture) and considerable growth potential (cf. Stamm 2004).

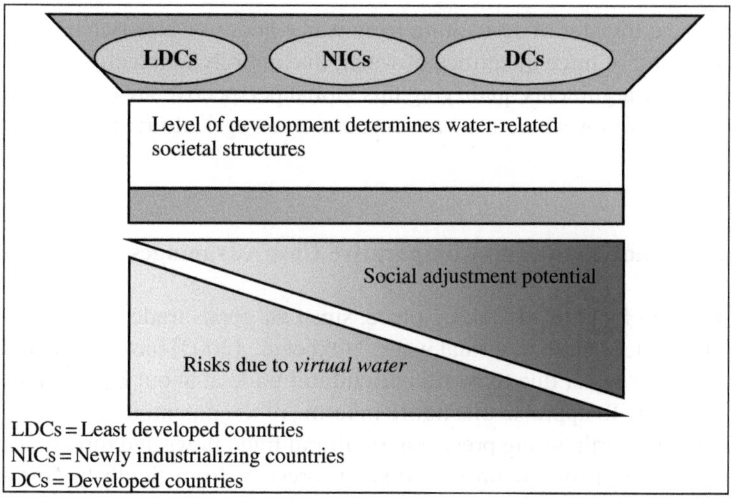

Fig. 3 Societal adjustment potential of water-poor countries
Source: Kluge and Liehr (2005, 10) in Horlemann and Neubert (2006, 102)

water scarcity. The most important controversial issues in the debate are considered first, and variants of virtual water trade are then presented.

4 The Various Criticisms of the Concept of Virtual Water Trade[5]

4.1 Criticism 1: Market Rather than Regulation – or: Virtual Water Trade Cannot be Achieved at the Flick of a Switch

The global material flow analyses by Hoekstra and Hung (2003) and the research by Oki et al. (2003) have led to a debate on the actual and potential scale of global water savings attributable to virtual water trade. The calculations by these authors are based on the fact that the water use efficiency and thus the water requirements of agricultural crops under different climatic conditions vary considerably. Because of higher evapotranspiration the same crop requires, for example, far more water at hot, arid locations than at cooler, more humid locations. These locational differences in water consumption vary from one species and one variety of plant to another by about 30–60 percent. Thus the water needed for the production of one tonne of maize averages 900 m^3 in China, but only 400 m^3 in France. On a global scale, then, 500 m^3 of water per ton of maize would be "saved" if it was produced in France rather than China (cf. Liu 2003). From this it might be inferred, arithmetically, that

[5] The article focuses here on criticisms 1 to 4. Criticism 5, "political dependence", is discussed in greater depth in Horlemann and Neubert (2006).

the increasing strategic transfer of agriculture from water-poor to water-rich regions would lead to the saving of huge quantities of water, due entirely to the effect of locational advantage. The arguments qualifying this global perspective are considered again in Sect. 4.3; what follows here is a brief discussion on agricultural trade flows and their determinants.

4.1.1 Agricultural Trade Flows and Comparative Cost Advantages

Virtual water trade is in fact already taking place, since all goods traded, and especially agricultural products, contain virtual water.[6] Oki et al. (2003) have calculated current actual savings of water due to world agricultural trade at about 8 percent of all water resources used. To appraise this figure in terms of the potential of strategic virtual water trade, it is worth noting present agricultural trade flows and directions. Agricultural trade occurs in two main directions at present, from South to North (from developing to industrialized countries) and from North to South. In contrast, South–South trade is still, on the whole, poorly developed. Whether countries are water-rich or water-poor plays no more than a very secondary role where the direction of the main trade flows is concerned.

According to Brüntrup (2005), these main trade flows are due to various mechanisms, which counteract each other to some extent. For South–North flows they comprise the comparative cost advantages that the two factors of production land and labor give the developing countries over the industrialized countries. On the other hand, trade flows from North to South are due to factors which are inconsistent with the principle of comparative cost advantages. They are:

- The EU and US subsidization on agricultural products, which causes the downward distortion of world market prices.
- Globalization of dietary habits in the South, with the urban population tending to prefer imported wheat to home-grown millet and sorghum.
- Poor infrastructure in many developing countries, making the domestic transport of and trade in foodstuffs from rural to urban areas more expensive.
- Subsidized or free food imports ("food aid") from the USA, etc., which distort competitive conditions.

South–South trade is relatively poorly developed not only because of the agricultural subsidies referred to above but also because of high transport costs due to inadequate infrastructure in and between developing countries — especially in Africa —, the high domestic demand for food in populous countries — as in South(-East) Asia — and the lack of regional economic communities of developing countries that function effectively.

---

[6] This water is designated "virtual" because the quantity contained in the product is very small, but is needed for its production. Seen from the water resource angle, the term also shows that water is not traded here as such, but merely virtually, i.e. indirectly.

Brüntrup (2005) demonstrates in his statement, however, that water shortages have not yet been a decisive factor for agricultural trade. In global terms he estimates the proportion of agricultural trade directly induced by water scarcity today at about 2 percent (cf. Horlemann and Neubert 2006). The most important reason for this small proportion is the general absence of quantity-dependent water rates in countries affected by water scarcity. Water scarcity cannot therefore have an impact as a cost factor or influence, to any major degree, the trade flows which emerge in line with comparative cost advantages. Only where absolute water scarcity prevails and the population's food requirements consequently exceed the domestic capacity to meet them does strategic virtual water trade occur today on a relevant scale (especially Egypt, but also a number of other MENA countries).

If there were scarcity-induced water prices, virtual water trade on a larger scale would have long since evolved. The aforementioned net savings of about 8 percent of total water used as a result of existing virtual water trade must be considered high in these circumstances, because they occur despite opposing economic incentives. If scarcity-oriented water prices were introduced, the comparative cost advantages of agricultural production would shift towards countries better endowed with water, and enormous water savings could thus be achieved.

4.1.2 Calculable Water Savings and Quantity-Dependent Water Charges

The water savings that can be calculated on the basis of these trade flows would, however, have an impact not at global, but at basin level, and they should therefore be appraised from that perspective. If we go back to our example of maize-growing in China or France, the calculated gain from the locational advantage was almost $500\,m^3$ of water per ton of maize. But the actual gain for China at basin level would initially be $900\,m^3$ if the maize was not grown in China, but imported. That quantity could thus be used for alternative productive purposes at the same location.

It can be concluded that, although virtual water trade has a very high water-saving potential at basin level, the proportion of virtual water trade induced by water scarcity is still minimal owing to the absence of appropriate water prices in the irrigation sector. If water prices were introduced, virtual water trade on a larger scale would follow of its own accord.

However, there is as yet no sign of developing countries introducing scarcity-based water prices. Currently under discussion are political prices, which are still too low, as an incentive to engage in virtual water trade. Developing countries have various reasons for rejecting water prices: in the case of agricultural products which can be produced in and exported from both the moderate and the tropical and sub-tropical regions and which therefore compete directly with one another, water prices would further weaken the developing countries' competitiveness and so their export economies. Enforcing water prices in irrigated agriculture is, moreover, a goal that can be achieved only in the longer term because of political, but also technical and administrative constraints.[7]

Without quantity-related water prices and given the state subsidization of electricity in many countries, often making it still worthwhile to raise water even from great depths, virtual water trade on a larger scale is possible only with the aid of direct regulatory measures. Another option for the countries potentially affected is simply to wait until the state of absolute water poverty is reached. Then there will be no alternative but to put virtual water trade into practice rather quickly. Neither scenario appears to be a particular intelligent solution.

4.2 Criticism 2: False Assumptions on Cause-and-Effect Relationships – or: Ways to Achieve Integrated Water Resource Management

Even in countries in which virtual water trade is the ultima ratio because of absolute water scarcity, it must be seen as an iterative process, i.e. a process that continues for lengthy periods. It must also be accompanied by a set of other strategies and measures to cushion the consequences for society. These strategies are outside the water sector, and their implementation would similarly take a relatively long time, examples being:

- the gradual establishment of quantity-related charges for blue water resources,
- the incorporation of the virtual water trade strategy in a national concept for the implementation of integrated water resource management (IWRM),
- the reduction of the EU's and USA's agricultural subsidies (to break the dominance of the North–South track for virtual water trade to the benefit, for example, of South–South trade),
- the internalization of environmental costs in macro- and microeconomic accounting.

If the virtual water trade debate is conducted in isolation, without regard for the socio-economic environment, the consequence scenarios rapidly assume the proportions described in Sect. 3.2. Basic economic and political principles that de facto determine trade and market conditions and cannot simply be turned upside down rarely attract much interest during the debate. Industrialization, for example, cannot be induced by politically decreed higher water productivity, as some protagonists may believe: the fact that water productivity in industrialized countries is on average higher is due rather to the reverse correlation. Industrialization is not primarily caused by individual factors such as higher water productivity: what is decisive is the general economic and investment climate. Nor can farms simply be deprived

[7] The introduction of quantity-related water prices also presupposes expensive infrastructure and entails a high administrative workload. Such requirements can generally be satisfied only with difficulty.

of their land use and water use rights. This is, however, what some protagonists of virtual water trade implicitly suggest when they refer to a reduction in irrigated agriculture. Farmers are usually private landowners, or they have long-term rights of use which cannot be simply withdrawn by the state. Agriculture cannot therefore be cut back at the flick of a switch. What is possible in the short term, if water prices cannot be charged, is, for example, the imposition of specific bans on the growing of particularly water-intensive crops, optional legislation on virtual water trade (as in Botswana and South Africa) or the conclusion of trade agreements.

Virtual water trade can also be indirectly promoted if countries potentially concerned stop implementing infrastructure projects that increase the water supply or implement only selected projects of this kind. The money saved could be used to import food. Refraining from implementing costly projects to increase the water supply, many of which are also socially and ecologically problematical, would thus be totally consistent with the Integrated Water Resource Management (IWRM) approach.

Critics object here that IWRM essentially entails managing water resources in such a way that countries and regions do not find themselves in a situation of absolute water scarcity in the first place. As long as a country lacks a flourishing non-agricultural economy, the critics argue, it is therefore better in every respect for their economic growth and for poverty reduction to keep the option of irrigated agriculture open. This, they claim, is possible if water is used sustainably, i.e. without thoughtless exploitation. Exploiting a country's water resources until absolute scarcity is reached is, on the other hand, a cynical solution. Virtual water trade thus tends to be a silent strategy for avoiding good water management at home by resorting to the water resources of other countries.

A conclusive assessment of the compatibility of virtual water trade with IWRM cannot, in principle, be undertaken here and probably needs to be related to specific cases. The arguments for and against in the debate have so far been unable to cancel each other out because they concern different aspects of the IWRM concept. The critics also emphasize that virtual water trade would tend to promote processes of political concentration by which such principles of IWRM as subsidiarity, participation and good water management might be counteracted (cf. Youkhana and Laube 2006). The counterargument is that virtual water trade, too, can be decided at river basin level and would not in any way need to be controlled from the centre. It cannot be decided here which is the more realistic scenario. Whether compatibility can be achieved will tend to depend on the political shaping of the actual virtual water trade policy.

4.3 Criticism 3: Calculations of Global Water Savings are Irrelevant – or: Water is a Spatial-Temporal Resource

How appropriate it is to seek to save water resources at global level, as some advocates of virtual water trade suggest, depends on whether the resource is actually scarce at global level. As, however, it is estimated that there is sufficient fresh water on Earth for some 20 billion people, there cannot yet be said to be a global scarcity of water. Realistically speaking, then, humankind is heading not for a global water crisis, but rather for a crisis at basin level. Scarcity is apparent or manifests itself in practical form at this level, and water must consequently be saved there (alone).

It is not usually made clear in the debate on virtual water trade that all of Earth's water reserves are subject to the hydrological cycle and are therefore, in principle, renewable, although they change their aggregate state. The total quantity of water is completely exchanged over specific periods, which vary with its location. The global quantity of water therefore always remains the same, and global "water savings" and absolute water losses are utterly impossible. Figure 4 shows that, after entering the atmosphere, water is unavailable for other human uses only during the circulation periods.

The availability of water for human consumption could be increased if periods of use were lengthened in a sustainable manner. In most countries, however, the opposite occurs de facto: as a result of mismanagement and climate change, water can no longer be stored in sufficient quantities in natural or near-natural reservoirs but, with increasing speed, reaches the seas through evaporation or as river water and is there much quicker than in former times made virtually unusable by salinity.

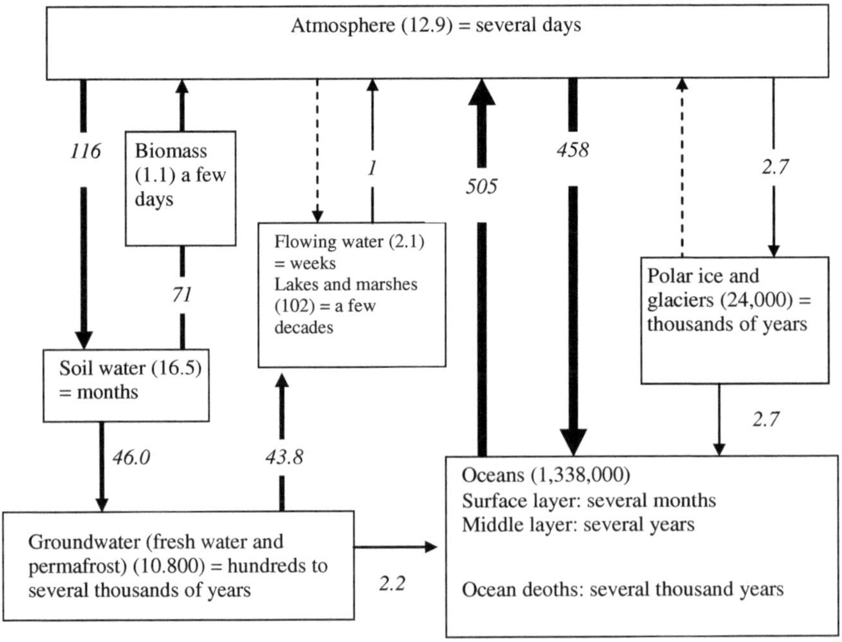

Fig. 4 Global water cycle (reservoirs in 1,000 km^3, rivers in 1,000 km^3 year-1, (italic) and typical residence periods)
Source: WBGU 1997, 59

Artificial reservoirs, such as dams, are able to compensate for the storage functions of near-natural reservoirs to only a disappearingly small degree. They also entail high investment and operating costs, need to be maintained, unlike ecosystems, and do not, moreover, perform any comparable ecofunctions. Taking account of the temporal nature of water resources is therefore of the utmost importance for an appropriate water policy, the protection of ecosystems and decisions on use. A major goal in this context should be to lengthen the residence periods of water in natural reservoirs or buffers and, at the same time, to make optimum use of it when it is available. At river basin level rival claims on water use, for example, might thus be defused through improved coordination and regulation of the times when water is used.

The conclusion to be drawn is that the advantages of virtual water trade in terms of possible water savings would not occur at global level and that they could have a physical impact only at the level of the basin.

4.4 Criticism 4: Calculate Opportunity Costs Rather than Seeking to Increase Water Productivity – or: The Efficiency Trap

Water productivity can be improved not only through increased industrial production,[8] but also through adjustments within agriculture, e.g. new crop-growing patterns. An argument that recurs in the debate on virtual water trade is therefore that crops using water more productively should be grown in water-poor regions.

Water is the most important productivity factor in agriculture. The natural yields of the same crop can be increased by 30–100 percent if it is grown under irrigated rather than rain-fed conditions. Although most typical cash crops, such as soya beans, sugar cane, cotton, vegetables and fruit require far more water per unit area than such typical subsistence crops as cereals, far higher water productivity can usually be achieved with cash crops, since their market value is many times higher.[9] This being the case, it might be inferred that water-poor regions should be recommended primarily to grow such cash crops as bananas, cotton and vegetables. And indeed Liu et al. (2007a) propose such a scenario when they say:

> ... The results suggest that agricultural structure adjustment towards high water use value crops in water scarce countries is a way to optimize agricultural water use ... (Liu et al. 2007a)

However, many of the crops that demonstrate high water productivity, such as bananas, cotton and some other cash crops, are "water-intensive" in relation to their

[8] Water use efficiency is meant here in the biochemical sense, whereas water productivity describes the monetary value of water as a result of its use by the plant ($\$/m^3$ of water used). This virtual water value of a crop plant varies both with the plant's degree of water use efficiency **and** – even more importantly – with its market value.

[9] The division into "subsistence crops" and "cash crops" is not unambiguous, since in principle almost any crop can be both used for subsistence and sold. The distinction made here is therefore no more than roughly accurate and may vary in individual cases.

water needs per hectare, per unit of time or -per ton. Bananas, for instance, require, at an average of 12,900 m^3 of water per hectare, three times as much water as millet (4,400 m^3 per hectare) and more than twice as much as cotton (5,750 m^3) (cf. Figs. 5 and 6) but they can still be regarded as water-productive because of their high production rate per ton and their high market value.[10] The question now is which figures are decisive, those related to yield or those related to units of area or time, when it comes to recommending crops for water-poor regions.

To make it easier to find an answer to this question, we consider the practical situation at farm level and determine on what basis decisions are taken at that level. To make his farm profitable, the farmer must endeavor to cultivate the whole area of the farm while achieving the closest possible crop rotations, since only then can he maximize his profit.[11] Let us now compare two scenarios on this basis: a rain-fed

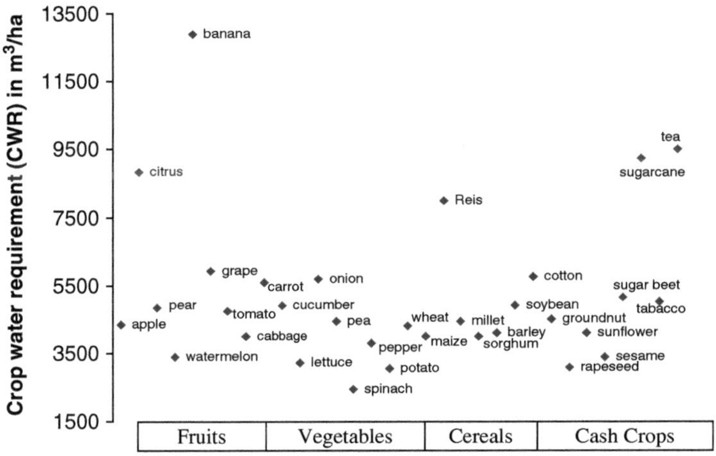

Fig. 5 Crop water requirement (CWR) per unit area
Source: Authors' own calculation based on Table 1A in the annex, as compiled by Liu et al. (2007a) Comment on the water requirement of rice: according to Liu et al. (2007a), rice requires 8,000 m^3 of water and is thus one of the highly water-intensive subsistence crops. However, their figures on this aspect are not conclusive: they maintain that rice consumes some 4,550 m^3 of water per hectare in its growth and needs a further 3,450 m^3/ha or so of water so that optimal anaerobic conditions obtain for its growth ("seepage water"). Liu et al. add the two quantities together and so arrive at the enormous total of 8,000 m^3/ha. What they do not consider is that much of the seepage water is passed on directly and is therefore usually available again to the lower reaches of the river, to other users or to the same area

[10] All data relate to comparable locations in China. Cf. Table A1 in the annex.

[11] This fundamental objective applies even though, in fact, the farmer can only partly succeed in making this effort over the long term, since his decision on what to grow is also influenced by other factors: seasonally different growing periods and locational requirements of different crops, the possible overlapping of schedules, limited capacity of the farm, including the availability of seasonal labour, limited liquidity, difficulties with the marketing of perishable cash crops and, possibly, unfavourable marketing conditions. It is therefore important for the farmer always to have several crop options so that he can adjust to changing conditions.

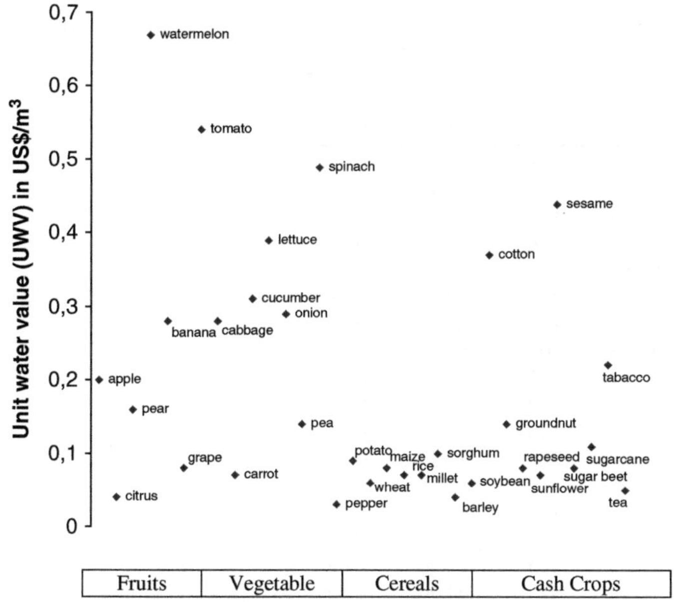

Fig. 6 Virtual water value (US$/m^3 water)
Source: Calculated on the basis of Table 1A in the annex, as compiled by Liu et al. (2007a)

farm (scenario A) is equipped with irrigation infrastructure (scenario B). While the farmer in scenario A plants equal areas with wheat, maize, potatoes and onions, he is now able in scenario B to guarantee a constant water supply and so to grow crops with a higher market value. In scenario B he therefore opts for bananas, water melons, tomatoes and cotton. If we assume that here, too, equal areas are given over to the various crops and if we remember that, because of their short growing period, tomatoes can produce two harvests in one season, the basic data in Table 1A reveal the following differences between the two scenarios:

In scenario A the farm income amounts to about US$ 2,534.80 and total annual water requirements to 17,030 m^3. In scenario B, on the other hand, the farm income is about US$ 13,258, i.e. roughly five times higher,[12] and total annual water requirements are 31,550 m^3, i.e. two times higher. From this it can now be concluded for the typically real situation is that, while water productivity is far higher in scenario B because of the huge increase in the market value of the crops produced, total water requirements per unit area in scenario B have almost doubled.

[12] The different operating costs for scenarios A and B do not include the location because of the data situation; gross figures must be used instead. It goes without saying, however, that the operating costs in scenario B are far higher than in scenario A, because not only is the maintenance of the irrigation system expensive, but the cost of water pumps, pesticides, fertilizers and seed is similarly far higher in scenario B than in scenario A. Despite this, scenario B is usually much more profitable for farmers. For a comparison of water productivity levels and the total quantities of water required, however, the method of description chosen is adequate. In the usual calculations of virtual water value (US$ per m^3 of water) farm costs are again not included, only the market values of the agricultural products.

Table 1A Crop water requirement (CWR), virtual water content (VWC) and unit water value (UWV) for various crops (1999–2001 averages) and net yield values

Crop	CWR (m^3/ha)	Yield (ton/ha)	VWC (m^3/ton)	Net value (US$/ton)	UWV (US$/$m^3$)
Apples	4500	9.09	495	97.80	0.2
Citrus fruit	8850	8.00	1106	47.45	0.04
Pears	4850	8.30	584	91.25	0.16
Water melons	3400	32.00	106	71.19	0.67
Bananas	12900	20.60	626	178.00	0.28
Grapes	5940	12.19	487	40.00	0.08
Tomatoes	4750	25.13	189	102.60	0.54
White cabbage	4010	19.25	208	59.24	0.28
Carrots	5600	17.79	315	22.00	0.07
Cucumbers	4900	16.97	289	90.56	0.31
Lettuces	3220	21.47	150	58.00	0.39
Onions	5690	23.47	242	70.00	0.29
Peas	4430	8.05	550	75.07	0.14
Spinach	2450	13.63	180	87.43	0.49
Paprika	3800	1.43	2651	70.00	0.03
Potatoes	3040	13.53	225	21.06	0.09
Wheat	4300	4.41	975	64.30	0.06
Maize	4000	4.74	844	68.25	0.08
Rice	8000	6.72	1190	91.91	0.07
Millet	4440	1.72	2586	190.37	0.07
Sorghum	4000	3.23	1240	121.72	0.1
Barley	4100	2.33	1760	65.72	0.04
Soya beans	4900	1.53	3203	184.65	0.06
Cotton	5750	3.23	1782	667.76	0.37
Groundnuts	4500	2.94	1532	221.70	0.14
Rape	3090	1.53	2020	157.23	0.08
Sunflowers	4090	1.70	2401	161.74	0.07
Sesame	3400	1.05	3238	1412.10	0.44
Sugar beet	5150	25.54	202	16.14	0.08
Sugar cane	9230	74.40	124	13.96	0.11
Tobacco	5000	1.78	2809	604.97	0.22
Tea	9500	0.78	12227	581.88	0.05

Source: Rice, wheat, maize and soya bean yields are derived from NBSC (2001), average yields in the main growing regions of China being shown. The yields of the other crops originate from FAOSTAT (2003) and represent average national yields. The crop water requirement (CWR) for rice consists of the sum of CWR and seepage water

If, then, it is argued that achieving higher water productivity is decisive for the sparing use of water, the actual situation shows that this is often wrong, since total and time-related water requirements are the core factors, determining the residual water flow in the river.

The conclusion is that in agricultural practice water resources cannot be saved simply by growing crops that use water more productively. There is an equal need to consider the amount per hectare and per time unit if the use of water is to be sustainable.

In our example, it would therefore be in the interests of water conservation to grow water melons, spinach or sesame, all crops that require little water per unit area, **but also** achieve high levels of water productivity. It is obvious at this point that not many kinds of crop fulfill both requirements, and we also have to take into account that what the farmer decides must always be appropriate to the farm environment and to marketing opportunities. It is therefore rather difficult to save water by optimizing cropping patterns at farm level.

4.4.1 Green and Blue Water Resources give Rise to Different Opportunity Costs

When it is also considered that in scenario B there has been a change from rain-fed farming to irrigated farming, it becomes even more difficult in the vast majority of conceivable cases to assess scenario B as a means of saving water.

To enable this to be understood, the characteristics of the various water resources will first be explained: as "green" water resources cannot be transported, they can be used only for plant growth, i.e. in agriculture and ecosystems. In terms of water productivity, agriculture is, as a rule, superior to the other uses referred to, and the opportunity costs of agricultural use are therefore normally low. As "blue" water resources can be transported, on the other hand, they can be used in industry and to generate energy. Consequently, the opportunity costs of using blue water resources in irrigated agriculture are, as a rule, far higher than those of green water resources.

Let us now return to the scenarios. While the opportunity costs for water use in scenario A were very low, they are far higher in the irrigation scenario B. Taking these opportunity costs into consideration, it now has to be asked what form an appropriate scenario at farm level can take if water is to be saved? According to the above arguments, it would be appropriate only for efforts to be made "within rain-fed" or "within irrigated farming systems", but not when the new cropping system is accompanied by a shift from a rain-fed to an irrigated system.

For a comprehensive assessment of the virtual water trade strategy, however, it is essential to include the opportunity costs. For agricultural use these considerations mean that, as a rule, green water resources should be used before demands are made on blue water resources for agricultural production. Charging prices for blue water alone can act as an incentive to adopt this approach. At the same time, however, it must be ensured that rain-fed farming does not advance into areas that are essential for nature conservation or for other important sources of income generation (e.g. tourism). This balance could probably be achieved only by imposing restrictions.[13]

All in all, it also seems important for the decision on the use of water to be adjusted to human needs, even if these uses are of a reproductive nature. This means

[13] In a comprehensive assessment not only must a distinction be made between green and blue water resources, but possible losses of water quality must also be considered. Such losses may be very high both in agriculture and in any alternative industrial use. Depending on what industries are meant and what environmental protection requirements there are, industrial uses may lead to pollution that more than cancels out the advantage of higher use efficiency.

that, besides the productive use of water, other uses, such as its use as drinking water or in ecosystems, should be taken into account. The value of these uses is not easy to calculate, but they are nonetheless essential for the survival of the planet. How this balance between the various uses is to be struck and controlled as intelligently and fairly as possible has yet to be adequately described.

5 The Variants of Virtual Water Trade

During the international debate on virtual water trade a total of five variants have emerged in recent years, each focusing on different aspects or trade relations:

- **Variant 1** sees virtual water trade as a way of offsetting water shortages in countries with absolute water scarcity by means of increased agricultural trade from North to South. Proponents of this variant usually fail to undertake an economic analysis.
- **Variant 2** focuses on virtual water trade as an area of global governance and aims at the greatest possible global savings of water through the exploitation of locational advantages. What is missing here is a geographical analysis (where does it make sense to save water?).
- **Variant 3** singles out water productivity as the most important premise, in that the aim is to achieve the highest possible levels of water productivity in agriculture through changes of farming systems or through shifts to industrial uses. This variant overlooks the determinants of decision-making in practical agriculture and does not include the opportunity costs of the various uses of water.
- **Variant 4** relates virtual water trade to trade within regional economic communities of developing and newly industrializing countries and focuses on South–South trade, particularly in the SADC region. Member countries with more abundant supplies of water, such as Zambia, might thus export agricultural products to countries with less water, such as South Africa. This variant sounds very attractive, although the current direction of trade is in fact the reverse.
- **Variant 5** emphasizes the advantages of domestic virtual water trade, an approach that appears particularly interesting for countries comprising several climatic regions as e.g. most anchor countries do. If this approach was to be adopted in practice, prices would have to be introduced for blue water resources in particular.

The first three variants have already been discussed in this article; variants 4 and 5 will be considered in the following, before conclusions are drawn.

5.1 Re Variant 4: Regional Virtual Water Trade

Malzbender (2005) and Meissner (2005)[14] have put forward the idea of using virtual water trade to exploit water shortages within regional economic communities – in

[14] Cf. Horlemann and Neubert (2006).

this case, the SADC. The SADC member countries vary widely in terms of available water resources. While Zambia, Angola, Mozambique and Congo, for example, are well endowed, South Africa, Namibia and Botswana can be regarded as soon to be water-poor countries. The water crisis in this economic area could thus be alleviated if the water-poor countries imported more of their food from the water-rich member countries. This option seems very attractive, because at least some of the requirements for successful virtual water trade are already satisfied in the potential importing countries: South Africa is industrialized and also has good governance to some extent. At the same time, the potential exporters of virtual water could benefit from the strategy as it would stimulate their growth (pro-poor growth).

A win–win situation of this kind could be created if, for example, Zambia did more to develop its agriculture and exported the products to neighboring countries, whereas South Africa could increase its food imports, while using its scarce water resources for industrial production, as drinking water or for ecosystems. Such virtual South–South water trade might stimulate growth in both groups of countries and so be appropriate in both ecological and trade terms.

When, however, the actual situation today is considered, the main trade flows are found to pass in the other direction: despite its enormous agricultural potential, Zambia is unable to feed its population, and despite water scarcity, South Africa exports more agricultural products than it imports.

> At the current level of development in the potential net food exporting SADC states Angola, DRC, Mozambique and Zambia, the regional implementation of the virtual water strategy remains theoretical. All four countries are currently net food importers and some of them have in fact in the recent past been recipient of large amounts of food aid [. . .]. Ironically, the largest cereal exporter in the SADC region is the water scarce South Africa, which exports large amounts of maize into neighbouring countries. (Malzbender 2005, 5).

Both groups of countries should therefore begin by investing. The water-rich countries should invest in the development of their rural regions, they should expand their infrastructure, and they should combat urban bias. To cope with these cost-intensive tasks, a poverty-oriented policy is needed and more financial support. The water-poor neighboring countries, on the other hand, should charge quantity-related water prices and, for instance, prohibit water-intensive cropping patterns, so as to provide an incentive to import more food from neighboring countries that have more water.

5.2 Re Variant 5: Domestic Virtual Water Trade

In the past three years the debate has also turned to the possibility of solving the problem of water scarcity in some countries through increased domestic virtual water trade. Obuobie et al. (2005) have shown in their study how differently the various regions of China use their green and blue water resources (cf. Obuobie et al. 2005). The use of green water, for example, ranges between 83 percent in the province of Henan and 32 percent in the province of Shandong (cf. Fig. 7). To protect blue water resources in the water-poor North, the potential of the provinces that grow

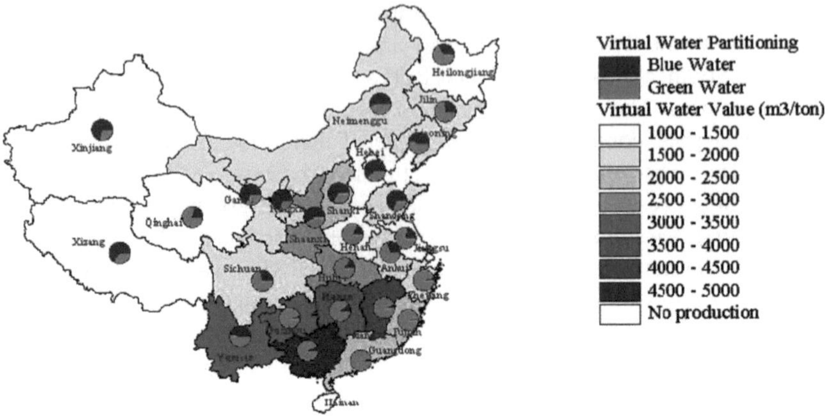

Fig. 7 Shares of green and blue water resources in the virtual water content of wheat in China, 1999
Source: http://www.nideco.ethz.ch/news/past_events/colloquium_2003/Liu_Presentation.pdf.Cited 20 Feb 2007

food in rain-fed farming could be tapped to a greater extent, with the agricultural products increasingly traded within China. Domestic virtual water trade might be an appropriate solution in China to the problem of saving national water resources and counteracting the looming water crisis.

Conclusion

The analysis of the debate on virtual water trade has shown that, although agreement has been reached on certain aspects in the debate, some immature ideas on the feasibility and on the advantages and disadvantages of the concept still persist. Water experts agree that virtual water trade can generate highly relevant water savings, and this at basin level. There is broad consensus that virtual water trade can be successfully engaged in only by countries that are already partly industrialized and also satisfy the other conditions for virtual water trade (infrastructure, good governance, organizational capacity, high social absorptive capacity). Consequently, only newly industrializing and anchor countries, i.e. countries that have dynamic economic centers and so sufficient economic strength to import food, can derive any benefit from virtual water trade. An exception is formed by countries already suffering from absolute water scarcity. For them engagement in virtual water trade is essential – whatever the cost – because they are no longer able to feed themselves.

The debate also makes it clear that virtual water trade cannot be achieved at the flick of a switch, but – if at all – only as an iterative process. Long time-horizons must therefore be allowed for when the discussion turns to virtual water trade.

Although virtual water trade cannot be decreed, it can be promoted directly or indirectly with a number of political regulatory instruments. The most appropriate solution consists in eventually introducing quantity-dependent water prices, so that comparative production advantages can evolve in the various countries of their own accord. This step is very complicated and momentous for the countries concerned and therefore requires lengthy adjustment periods. It should also be accompanied by such other measures as the reduction of agricultural subsidies, with the aim of improving the competitiveness of the southern countries and so giving the water-rich countries of the South a chance to engage in virtual water trade.

Water experts are also likely to agree among themselves that a positive assessment of virtual water trade can be given only if it forms part of integrated water resource management. Advantages consisting in the possibility, because of food imports, of avoiding projects geared to increasing the water supply must not be offset by disadvantages consisting in otherwise doing without good water management.

On the whole, virtual water trade is only one of several approaches to solving the problem of regional water shortages through the reallocation of the resource. However, water savings should be assessed realistically, since no water as such is saved as a result of higher water productivity in or outside agriculture. In the final analysis, it is essential to limit the total quantity of water used and to optimize the temporal coordination of the various uses, so that management is sustainable. In an assessment of the virtual water trade strategy it is also important for opportunity costs, which, as a rule, vary widely in the case of blue and green water resources, to be included as a decision-making criterion. An attempt should always be made to exploit green water resources before blue water resources are used in agricultural production. As there are more rivals for the use of blue water resources and the opportunity costs must therefore be set higher, such water should, where possible, be reserved for other purposes (especially for use as drinking water). But this decision, too, must ultimately be taken at basin level.

Generally speaking, the debate on virtual water trade is overly focused on ways of using water as productive as possible, and what is usually overlooked is that this objective on its own is not enough to protect water resources against overuse. For that, water needs to be used as intelligently as possible, with such aspects as water productivity, essential human needs and ecological sustainability all taken into account. If this goal can be achieved, virtual water trade can be included in the toolbox of Integrated Water Resource Management as an option in certain, suitable situations and contexts.

References

Allan JA (1996) Water, peace and the middle East: Negotiating resources in the Jordan Basin, Tauris Academic Publication, London

Allan JA (1997) 'Virtual Water': A long term solution for water short middle eastern economies, Paper presented at the 1997 british association festival of science, University of Leeds, Leeds

Allan JA (2003a) Virtual water eliminates water wars? A case study from the middle East. In: Hoekstra AY (ed) Virtual water trade – Proceedings of the International Expert Meeting on Virtual Water Trade, IHE, Research Report Series 12, Delft, 137–145

Allan JA (2003b) Virtual Water – the water, food, and trade nexus – useful concept or misleading metaphor? IWRA. In: Water International 28 (1), 4–11

Allan JA et al. (2003) Policy options for water-stressed states (POWSS) – building lessons from the Middle East and Southern Africa into decision support for policy makers, African Water Issues Research Unit, Overseas Development Institute, London, Pretoria (Final Report/pre-publication)

Brandt H, Otzen U (2004) Armutsorientierte landwirtschaftliche und ländliche Entwicklung, 1st edn, Nomos, Baden–Baden

Brüntrup M (2005) Ökonomische Überlegungen zum virtuellen Wasserhandel: Expert statement for the BMZ project "Virtual Water Trade – a realistic concept for resolving the water crises?" German Development Institute, Bonn

Hoekstra AY, Hung PQ (2003) Virtual Water Trade: A quantification of virtual water flows between nations in relation to international crop trade. In: Hoekstra AY (ed) Virtual Water Trade – Proceedings of The International Expert Meeting on Virtual Water Trade, IHE, Research Report Series 12, Delft, 25–47

Horlemann L, Neubert S (2006) Virtueller Wasserhandel – Ein realistisches Konzept zur Lösung der Wasserkrise? German Development Institute, DIE-Studies 22, Bonn

Hummel D (2005) Wasser und Ernährungssicherheit: Was sind die demographischen, institutionellen und sozio-kulturellen Bedingungen für virtuellen Wasserhandel? Expertenstatement im Rahmen des BMZ-Projekts "Virtueller Wasserhandel – Ein realistisches Konzept zum Umgang mit Wasserarmut in Entwicklungsländern?" German Development Institute, Bonn

Kluge T, Liehr S (2005) Anpassungsmodalitäten, Regionalisierung und Skalenübergänge im Konzept des virtuellen Wasserhandels: Expertenstatement im Rahmen des BMZ-Projekts "Virtueller Wasserhandel – Ein realistisches Konzept zum Umgang mit Wasserarmut in Entwicklungsländern?" German Development Institute, Bonn

Liu J (2003) Virtual water, green water and blue water in China. http://www.nideco.ethz.ch/news/past_events/colloquium_2003/Liu_Presentation.pdf.Cited 15 March 2006

Liu J, Zehnder AJB, Yang H (2007a) Historical trends in China's virtual water trade. Water International. In press

Liu J, Williams JR, Zehnder AJB, Yang H (2007b) GEPIC – modelling wheat yield and crop water productivity with high resolution on a global scale. Agricultural Systems, Vol. 94, Issue 2, May 2007, 478–493

Malzbender D (2005) The political implementation of the "Virtual Water Trade" concept: Expertenstatement im Rahmen des BMZ-Projekts "Virtueller Wasserhandel – Ein realistisches Konzept zum Umgang mit Wasserarmut in Entwicklungsländern?" German Development Institute, Bonn

Meissner R (2005) Virtual water trade strategies for an industrialised and a developing country: The case of South Africa and Zambia: Expertenstatement im Rahmen des BMZ-Projekts "Virtueller Wasserhandel – Ein realistisches Konzept zum Umgang mit Wasserarmut in Entwicklungsländern?" German Development Institute, Bonn

Obuobie E, Gachanja PM, Dörr AC (2005) The role of green water in food trade, Zentrum für Entwicklungsforschung (ZEF), Bonn (Term paper for the interdisciplinary course, International Doctoral Studies)

Oki T et al. (2003) Virtual Water Trade to Japan and in the world. In: Hoekstra AY (ed), Virtual Water Trade – Proceedings of The International Expert Meeting on Virtual Water Trade, IHE Research Report Series 12, Delft, 221–235

Stamm, A. (2004) Schwellen- und Ankerländer als Akteure einer globalen Partnerschaft. Überlegungen zu einer Positionsbestimmung aus deutscher entwicklungspolitischer Sicht, German Development Institute, DIE Discussion Papers 1, Bonn

WBGU (1997) Welt im Wandel. Wege zu einem nachhaltigen Umgang mit Süßwasser. Jahresgutachten 1997, Springer, Berlin, Heidelberg

Youkhana E, Laube W (2006) Virtueller Wasserhandel und Konflikte um Wasser: Expertenstatement im Rahmen des BMZ-Projekts "Virtueller Wasserhandel – Ein realistisches Konzept zum Umgang mit Wasserarmut in Entwicklungsländern?" German Development Institute, Bonn

The Debate on "Water as a Human Right" and its Implications for Development Assistance

Danuta Sacher and Michael Windfuhr

Abstract The article describes the basic concept and content of the human right to water and its implications for development policy on water delivery, as well as some current initiatives towards the implementation of the rights approach in development and national water politics. It summarizes the critical steps in the development of the debate on "water as a human right". The first two chapters outline the current debate on the human right to water; the following chapter discusses measures and policies for the implementation of the right to water. The fourth chapter examines the added value associated with a rights based approach to water. The article ends with a short presentation of the important challenges for the future debate on the right to water. It is important to see that for ESC-rights in general a common standard of interpretation is emerging and that similar categories and standards are used to describe the content and the State obligations of all these rights.

1 Introduction

"The world has the technology, the finance and the human capacity to remove the blight of water insecurity from millions of lives. Lacking are the political will and vision needed to apply these resources for the public good". This statement from the latest Human Development Report (UNDP 2006, 28) confirms and repeats the main message of the first United Nations World Water Report (UNESCO 2003), which also stated that the solution to existing water problems requires political will. The human rights approach is increasingly gaining attention as a key concept, putting the needs and rights of poor and excluded people at the heart of the issue. The rights-based approach provides an ethical and legal framework for mobilizing support for the prioritization of water and sanitation and also a set of tools for policy makers and civil society groups.

Danuta Sacher and Michael Windfuhr
e-mail: d.sacher@brot-fuer-die-welt.de

W. Scheumann et al. (eds.), *Water Politics and Development Cooperation*,
doi: 10.1007/978-3-540-76707-7_7,

This article describes the basic concept and content of the human right to water and its implications for development policy on water delivery, as well as some current initiatives towards the implementation of the rights approach in development and national water politics.

Until the 1980s, the right of every person to adequate drinking water was an underlying assumption of the international debate on water delivery. Similarly the principle that the state was mainly responsible to ensure the realization of this right was another assumption of the international debate. This changed during the 1990s, after the first UN Water Decade (1981–1990), which unfortunately missed one of its central goals to create access to sufficient water to all then excluded persons.

The decade after the end of Cold War brought with it stormy globalization and a strong faith in market-driven forces. Within this context, the official debate shifted towards the concept of water as an economic good, which found its first prominent expression as one of the Dublin principles at the International Water Conference in Dublin 1992. The World Bank adopted this as their new guiding principle in their sector policies (World Bank 1993) and established the controversial conditionality for loans in the water sector in southern countries, making the reduction of state subsidies a condition of new credits, introducing the cost recovery principle within tariff systems and opening up the local water markets for privatization and foreign investment. Huge privatization projects in many metropolitan areas were initiated, from Argentina to Indonesia, from the Philippines to Uganda. At the same time, the World Bank promoted and supported international platforms like the World Water Council (WWC) and the Global Water Partnership (GWP), and created jointly with them the World Water Forum, held every 3 years. These bodies became important agenda setters for global water policy discussions. Water issues also play an increasingly important role in other intergovernmental organizations in the UN system, such as the WHO, UNESCO, UNEP. A debate has emerged as to which are the more successful approaches to deal with the water problems.

In 2001, the Freshwater Conference in Bonn tried to balance the rights-based approach and the perspective on water as an economic good. As part of the run-up to the Johannesburg Summit, the Bonn Conference closed with a clear vote to keep the responsibility for the water sector in public hands without excluding private participation via public private partnerships (International Freshwater Conference 2001). In 2002, the World Summit for Sustainable Development in Johannesburg clearly backed the latter by promoting Public Private Partnership (PPP) as key for water policy in developing countries (WSSD 2002).

At the end of 2002, The United Nations human rights system took some initiative in the debates, contributing substantially with the adoption of General Comment No. 15 to the International Covenant on Economic, Social and Cultural Human Rights (ICESCR) concerning the right to water, by the UN-Committee for Economic, Social and Cultural Rights (CESCR). The General Comment confirmed the fundamental human right to water and demonstrated its importance for the realization of any other human right (CESCR 2002).

This article summarizes the critical steps in the development of the debate on "water as a human right". The first two chapters outline the current debate; the following chapter discusses measures and policies for the implementation of the right to water. The fourth chapter examines the added value associated with a rights based approach to water. The article will end with a short presentation of the important challenges for the future debate on the right to water.

2 Water as a Human Right – its Legal Status and the International Discussion

The right to water is part of international law as one of the rights implied in the International Covenant on Economic, Social and Cultural Rights (ICESCR), which, together with the International Covenant on Civil and Political Rights (ICCPR) and the Universal Declaration of Human Rights (UDHR) constitutes the basis of the human rights protection system. Although the ICESCR does not explicitly mention the right to water, it is protected by Articles 11 and 12, according to the UN CESCR. In 2002, the Committee on Economic, Social and Cultural Rights, as the UN body[1] responsible for the surveillance of the ICESCR, drafted e General Comment No. 15, an authoritative legal interpretation of the human right to water, deriving the right to water from the right to an adequate standard of living (Art. 11) and the right to the highest attainable standard of health (Art. 12) of the ICESCR.

Both of these rights are also mentioned in the Universal Declaration of Human Rights. Several other sources of international law can be found, with a reference to a right to water, or relevant for the interpretation of it. Some of the sources are in legally binding instruments, such as the human right covenants while others are so-called "soft law instruments", or non-.binding instruments, such as declarations, voluntary guidelines etc. These latter instruments cannot be discussed here in detail.[2]

Because the right to water is not mentioned explicitly in one of the basic human rights covenants, many commentators demand the development of a precise legal definition of the right to water. But even without the word "water" being mentioned directly, few question the existence of the right to water in principle. The CESCR overcomes this problem in GC 15 by stating: "The human right to water is indispensable for leading a life in human dignity. It is a prerequisite for the realization of other human rights". Other commentators conclude: "The human right to water does exist, as water is the most essential element of life." What is missing is a clearly defined formulation in an international law source (Scanlon et al. 2004, 12).

---

[1] All major Human Rights Treaties have a monitoring body, which is normally set up by the treaty itself. E.g. the International Covenant on Civil and Political rights is monitored by the Human Rights Committee; the Convention on the Rights of the Child is monitored by the Child Rights Committee. These committees are called treaty bodies in the human rights system The ICESCR does not have its own monitoring body. In order to allow a regular monitoring of the Covenant, in a resolution in 1987, the Economic and Social Council of the UN established a separate monitoring body for the ICESCR, the Committee on ESC-Rights. While it has a different legal origin than the other treaty bodies it is de facto working and accepted by State Parties as a treaty body and has also started to develop General Comments.

[2] A good overview of the relevant instruments can be found in Appendix 1 to Scanlon et al. (2004, 35ff.).

That a human right to water can be derived from the existing legally binding sources is therefore not questioned any more. The recent discussion and resolution of the Human Rights Council on the right to water also demonstrate substantive State support of the right to water. It is more difficult to find a binding or qualified interpretation of the content and the related State obligations of that right. Some commentators demand that a more detailed guide to implementation in legally binding language is needed, before applying the right to water. However, hardly any of the civil and political human rights or the economic, social and cultural human rights are described in detail in the human rights treaties. They all needed interpretative guidance. Such guidance can normally be given either by national or regional court decisions or legal opinions given in quasi judicial complaint procedures of the UN human rights system or is provided by authoritative interpretations given by the treaty bodies in the form of general comments. These interpretations are not legally binding, but their relevance comes from their recognition by States party to the respective treaty or by courts and international lawyers referring to them. The general comments from all treaty bodies are compiled regularly by the Office of the High Commission of Human Rights (OHCHR), the latest compilation of which can be found on the OHCHR web-site.

The example of the right to food shows the relevance of General Comment No. 12 of the CESCR (1999). The GC 12 was the background reference for the development of the Voluntary Guidelines on the right to food, developed by the Committee on World Food Security of the FAO in 2003 and 2004.[3] The text was finally approved by all 187 member states of the FAO in November 2004 in the FAO Council (FAO 2005), indicating the support of states to the definition of the right to food and description of state obligations given in the GC 12. GC 15 on the right to water follows the same structure as GC 12, using this to develop the content of the right and the description of State obligations. Since its adoption in 2002, GC 15 has also received broad support by many States, in the discussion in the Human Rights Commission and in other forums (cf. Chap. 3). While interpretative in their nature, GCs must be seen as important steps towards the development of a full and comprehensive understanding of each of the internationally recognized human rights. Their role is normally characterized in the literature as "authoritative interpretations".[4]

Within the UN human rights system, the right to water is a relatively new issue. While the right to an adequate standard of living is contained in the UDHR, the specific formulation only refers to food, clothing, and housing – although it is formulated in such a way to indicate that the list is not exhaustive and that other aspects are not excluded in principle. Article 11 of the ICESCR is formulated in a similar way. For this reason, it is necessary to explain why the issue of water or the right to water is not explicitly mentioned in older texts and why it has become important now. Three trends can be identified that have contributed to intensify the debate on water and water-related problems:

[3] The formal title of the voluntary guidelines is: "Voluntary Guidelines to support the progressive implementation of the right to adequate food in the context of national food security" (FAO 2005).

[4] Cf. among many different articles: Craven 2001; Riedel 2006; Windfuhr 2006.

1. The demand for water for uses other than drinking water is increasing fast. Water use for energy, irrigation, industry etc. is competing with to the need for water for domestic use. Developing an understanding of the human right to water will help to describe more precisely the state obligations vis-à-vis water for domestic use. The initiation and implementation of large infrastructure projects have been the cause of some serious conflicts in recent years.
2. The public discourse about the privatization of water supply systems or parts of them, both in the North and the South, has highlighted the issue of the right to water in recent years. Supporters of privatization claim that in developing countries, only by working in co-operation with the private sector will enough investments be raised to improve or reconstruct water supplies. On the other hand, critics fear that privatization will increase prices and lead to a deterioration in accessibility to this scarce resource, especially for poor groups in society.
3. The extent of the global water problems: Water is becoming an increasingly scarce resource and access to water is becoming a cause of conflict – especially with regard to distribution and usage. The issue of access to water is therefore increasingly relevant for the enjoyment of many other related human rights, such as the right to food or the right to health. Current and future conflicts over use will become more difficult to resolve. The legal understanding of water as a human right can help to deal with such conflicts.

In recent years, the right to water has been taken up also be several institutions in the human rights system of the United Nations, e.g. in the work of the two Special Rapporteurs on economic, social land cultural rights, the Special Rapporteur for the right to housing (Miloon Kothari) and for the right to food (Jean Ziegler).[5] The debate surrounding the right to water and its recognition is influenced by the general debate on economic, social and cultural rights. Many of the prejudices against ESC-rights are gradually being overturned thanks on the one hand to the work of the Committee on ESC-rights and on the other hand to the practical work of human rights organizations which have documented numerous cases of ESC-rights violations over the years. While it will still take several years to overcome the historic neglect of ESC-rights, it is becoming evident that all five groups of human rights – civil, political, economic, social and cultural human rights are interrelated and are comparable with regard to their legal character.

The UN CESCR derives the right to water from the right to food and the right to health. The **normative content** of the right is described by several elements: the right to water primarily encompasses the right to access of every person to an adequate system of water supply and protection against interference through for example arbitrary disconnection. Available water should not be contaminated. Water systems and facilities should be organized and managed to ensure that access to water is guaranteed to be continuous and provides equality of opportunity for all.

---

[5] Both special rapporteurs have written a report to the UN Human Rights commission every year. The reports from 2001 and 2002 of both discuss issues related to the right to water in their respective mandates. A short overview on Ziegler's work on the right to water can be found in Ruspekhofer 2006.

According to the CESCR, access to water refers to water that each person needs for personal and domestic use. Originally, the General Comment was supposed to refer to the "right to drinking water". During the committee's consultations and discussions it became clear that the category "drinking water" is too narrow and that the right to water must encompass personal use and domestic use in its entirety since health and sanitary aspects could be excluded otherwise. At the same time, the use of water for irrigation and other non-domestic uses is excluded from the definition. The issue of access to irrigation in order to produce sufficient food is evaluated by the CESCR as being part of the right to food. The 2006 UNDP Human Development Report uses two terms to differentiate between these two functions of water supply, using the term "water for life" for domestic use of water and the access to basic sanitation to differentiate this from "water for livelihood" which refers to the use of water for agriculture and industry (UNDP 2006, 3).

The CESCR further defines what can be understood by "access to water". Based on comparable definitions found in the general comments on the right to food and the right to health, the committee adheres to the policy that access primarily not only refers to the physical accessibility of water. Water must also be available and of adequate quality and each person must have the economic means to obtain water. Access should not be limited by discriminating practices, for example minority groups or women may not be prevented from using water. Accessibility also encompasses the right to sufficient information, both regarding his or her rights as well as information concerning water issues, thus enabling democratic monitoring.

The essence of all ESC-rights is that States are obliged to guarantee full implementation of these rights for all persons. General Comment No. 15 includes almost 20 paragraphs in which the CESCR describes States' obligations for the implementation of the right to water. Twelve paragraphs deal with national obligations. When characterizing States' national obligations in more detail, the CESCR has used the established difference between three levels of state obligation: respect, protect and fulfill. This differentiation has guided the work of the Committee over the last few years and is today broadly accepted among international lawyers.

1. The obligation to respect obliges States to refrain from engaging in any practice or activity that "interferes directly or indirectly with the enjoyment of the right to water". This obligation includes the protection of water sources in crisis situations.
2. The obligation to protect the right to water requires States to prevent third parties, such as individuals, groups or corporations, from interfering in any way with the enjoyment of the right to water. Protection includes adopting effective legislative and other measures to protect disadvantaged groups in society and to avoid discrimination. According to the committee, this includes the obligation to adopt effective legislative and regulative measures in the case of privatization of water systems.
3. Under the obligation to fulfill the right to water, States are required to take positive measures to assist all individuals and communities without access to water to enjoy the right to water. The State should ensure that water is available for each person at an affordable rate. The State is obliged in the ICESCR to use the

maximum of its available resources to progressively realize the right to water.[6] Moreover it has to show that the resources used are directed to the most vulnerable groups in the country.

For many years, the wording "maximum of resources available" and the obligation to implement progressively has created resistance to all ESC-rights among some international law experts who believe that this definition is not precise enough to allow a clear legal understanding. Some experts believe that rights should only be negative rights granting freedoms and that only civil and some political rights would match such a recommendation. This old dichotomy between negative civil and political rights and costly positive economic, social and cultural rights has been overcome in the last decade to a large degree among international legal scholars, but it is still an argument sometimes used by government representatives. The obligations presented above cover both negative and positive ones. It often takes time to explain such a differentiation. ESC rights have a strong negative component. They require under the obligation to respect and protect state to refrain from destroying existing enjoyment of the right.

One important step in the clarification was the adoption of the Limburg principles in 1987 by a group of international lawyers, which met at the Maastricht University (United Nations 1987). This states that each government has to show that it has taken steps toward realizing each right with respect to the most negatively affected groups. These steps taken should be as expeditiously as possible. Such provisions might still seem to be very open for interpretation, but whether a state has used the maximum of available resources in a given situation cannot be analyzed according to principle but requires a judgment by court analyzing each specific case. In a recent court decision in South Africa on housing rights, the judges introduced the category of reasonableness in order to discuss the use of available resources. The government was obliged to show that it had used the available resources reasonably well (Liebenberg 2006).

It is particularly important that the obligation to fulfill does not demand the impossible from states. It is not expected that the State provide all citizens with water, food, housing, health, and work immediately. Rather, the State should draw up policy measures in such a way that the "maximum of its available resources" (Article 2 CESCR) is used. Measures that put people in a position to secure their own water supply belong to this category. When such measures are not sufficient, the State is obliged to provide direct access to people affected by violations of the right to water. The CESCR divides the GC's obligation to fulfill into three different subcategories:[7]

- **Facilitation**, requiring the state to take positive measures to assist individuals and communities to gain access to water supplies themselves. The government should develop national strategies for the progressive implementation focusing on appropriate low cost techniques and technologies (para. 27f.).
- **Promotion**, requiring the state to ensure that there is adequate education on the hygienic use of water.

[6] For further interpretation of "maximum of available resources" see below.

[7] Cf. para. 25–29 of the General Comment No. 15 (CESCR 2002).

- **Provision**, if facilitation does not have sufficient effect. Provision requires the state to directly provide water when individuals or groups are unable to do so. The obligation to provide finds its limit in the clause, "the maximum of resources available". States must prove, however, that the maximum of resources available have actually been exhausted, and in doing so is not discriminating and has already identified who is in need so that the individuals and groups that are especially disadvantaged are supplied. Therefore it is not an obligation to respond to unlimited demands on governments, but rather to focus the resources on the most vulnerable groups. The General Comment outlines States' national obligations in much more detail, which cannot be presented here, because of limited space.

Interestingly General Comment No. 15 also describes **States' international obligations** with regard to the right to water. By dividing States' obligations into national and international obligations the CESCR continues to use the differentiation which was first established in General Comment No. 12 on the right to food from 1999. States' international obligations refer to potential violations of each States' activities within international organizations and the potential impacts of a States' policy measures on people in other countries. Many international law experts limit States' obligations to the national level since human rights regulate the relationship (the rights) of the individual before the state. More recently, non-governmental organizations and an increasing number of international law experts have called for the need to consider the international impacts of State policy measures. An analysis of the frameworks for State policy making in terms of globalization clearly demonstrates that there is good reason to treat the international obligations as part of the human rights protection system. Today State polices can have far-reaching impacts on citizens in other States. European agricultural policy regulations, for example, can have significant and well-documented impacts on the right to food of farmers in Africa, e.g. when farmers lose their income through export subsidies that ruin prices for staple food on local African markets. African governments cannot easily influence these and other macroeconomic policies because conditions for their own agricultural policies are often fixed by the World Bank or by regulations of multilateral or bilateral trade agreements. Many countries are not allowed to use protective measures against imports that come highly subsidized on the national market.

International Obligations[8] taken up in GC 15 include (1) regulations on development co-operation that can be divided into two areas: "Positive" obligations and "negative" obligations in development co-operation. Development co-operation should first of all ensure that it does not contribute to violations of the right to water in other countries (negative) (para. 31). It can assist countries in their efforts to implement the right and to fulfill its obligations (para. 34). The General Comment stipulates that (2) in international relations no embargoes shall be imposed affecting water (para. 32). (3) Each State is required to adequately control private companies and persons that invest in other countries in order to ensure their activities do not contribute to a violation of the right to water (para 33). Furthermore, states should always be aware of their human rights obligations (4) when drawing up international

[8] International Obligations are taken up in para. 30–36 of the GC 15.

agreements and should not enter into any agreement that contradicts human rights agreements (para. 35). For further commitments of intergovernmental organizations (IGOs), States should ensure (5) that these IGOs, within the scope of their own programs and projects, are not partially responsible for violations of the right to water (para. 36). According to the Committee, this is especially the case for interventions by the World Bank and the International Monetary Fund.[9] It is important to note here that in General Comment No. 15, the Committee on ESC Rights clearly differentiates between the role of states in the international arena (international obligations) and the obligations that international organizations themselves have.

3 Social Movements and Civil Society Campaigns for a Human Right to Water

The human right to water is relatively new in the work of civil society organizations. Generally, civil society has only recently started working with economic, social and cultural rights and the right to water has been a latecomer, only being taken up in a more consistent way since the mid 1990s. The first to work with the right to water were those human rights organizations that had already started to work with other economic, social and cultural rights.[10]

However, during the 1990s several civil society organizations had already started to address the global water crisis and its related problems in a more visible way. Water problems were addressed for along time particularly by environmental organizations or through health and gender related activities. Over the years more and more groups in many countries started to work on water issue. Although there is no formalized structure or representative system, the impressive multitude of local and regional civic organization and activities around water is already being called by some an emerging "global water movement".[11] It includes groups in resistance against privatization and private sector participation in urban water supply systems, groups of affected people by the impacts of dams and industrial water contami-

[9] Para. 60 of General Comment No. 15 deals particularly with the obligations of other actors than states, such as Intergovernmental agencies. Special mention is given to the International financial Institutions: "The international financial institutions, notably the International Monetary Fund (IMF) and the World Bank, should take into account the right to water in their lending policies, credit agreements, structural adjustment programmes and other development projects (see further General Comment No. 2), so that the enjoyment of the right to water is promoted. . . ."

[10] The UN-Committee for Economic, Social and Cultural Rights was established in 1987. At the beginning only very few civil society organisations regularly attended the session. Habitat International Coalition on the right to housing, FIAN-International on the right to food and the ICJ were among the first international human rights organisations that took up the issues. Some regional and national organisations followed, such as the American Association of Jurists. HIC and FIAN were also among the first organisations to document specific cases of violations of the right to water in the 1990s.

[11] The term "global water movement" has been used since 2006 by Tony Clarke, one of the authors of the "Blue Planet". (Barlow and Clarke 2002).

nation, traditional environmental groups working on water quality issues as well as rural defenders of traditional water rights against industrial demands. (Also a number of civil society organizations implementing their own water and sanitation solutions.)

The starting point for the water issue to move citizens in a more concerted way can be set in 2000. In this year, the so called "Water War" in Cochabamba against the impacts of privatization of the urban water supplies gave birth to a water movement in Bolivia and the world, this almost coinciding with the declaration of the commoditization-promoting "Water Vision" of the 2nd World Water Forum in Den Haag, both events together generating a first global water justice discussion at the World Social Forum in Porto Alegre 2001. It was however, still a long way to travel before the wide and inclusive range of involved social actors, from poor urban dwellers to indigenous groups, from urban women's to ecologist groups, from Asia, Africa, North- and Latin America and Europe, built step by step a bridge between their different issues and constructed a common cause, an inherent part of which today is the human rights approach.

Before that, a range of prejudices and reserve had to be addressed. Ecologists rejected the human rights approach because they feared excessive use of water by humans and damage for nature; indigenous groups were afraid that their traditional rights could be subordinate to anyone else's right; also, there were strong reservations against the United Nation system as a whole. Actually, the General Comment No. 15 on the Right to Water was a good platform to have detailed discussion on these issues, and it proved to be wide and precise enough to become part of a common understanding of the upcoming movement, and to become the conceptual reference for the paradigmatic discussions on the character and strategies on water.

Even with all this progress in the recognition of the human right to water, there is still a long way to go to get a common understanding regarding water as a human right, the content and the State obligations linked to it. The process amongst civil society groups around the concept of human rights and its implications for political action is still at the early stages. At the same time, tools of how to use the right to water in local, national and international cases are under development. Some civil society organizations have already prepared manuals on how to use the right to water in local court cases. More and more civil society organizations are becoming interested in using a rights based approach.[12] The human right to water was firstly included in an international water justice conference declaration in New Delhi 2004. Today the human rights approach in the interpretation of the UN-Committee for Social, Economic and Cultural Rights is becoming part of the basic principles of the emerging global water movement.

[12] The Ecumenical Water Network – a global faith-based network on water problems is currently setting up a website with background information and tool-books on how to use the human right to water in specific conflicts. During the World Social Forum in 2007 in Nairobi, an African Water Network was created as a civil society platform to work with water problems on a continental level. One of the important issues in the network will be using the human right to water as a tool in local conflicts.

Promising discussions and international networking are underway regarding necessary keystones and possible strategies for strengthening the right to water further. One of the options discussed is the proposal to promote an international legally binding instrument in the form of treaty or convention.

4 Perceptions and Interpretations of the Human Rights Based Approach

As with other economic, social and cultural rights, the discourse on the right to water is still a new one. It is therefore not astonishing that still a number of misconceptions exist on ESC-rights in general and specific rights such as the right to water in particular.

Beside these normal misconceptions one can observe that the debate on the right to water is particular rich of new forms of interpretations. Many actors involved in international water issues are currently trying to influence this emerging debate on the right to water. Most of them have started using the reference to the right to water, but some of them are far from using an adequate understanding of water as a the human right. In this chapter we present recent strategies or discussion papers that refer to the right to water.[13] Some of them are close to the emerging mainstream in understanding the right to water, other are based on misconceptions. The overview is given to show the extent of the current debate and the need to be precise in order to respect the specific nature of the human right to water.

A core misconception is that the human right to water is often discussed in the context of other legal forms of guaranteeing secure access rights to water sources. The Overseas Development Institute was among the first that tried to provide clarity in the different approaches to water rights and the right to water. In a briefing paper on the "Right to Water: Legal Forms, Political Channels" the ODI differentiated three legal forms of water rights: as a human right, as a contractual right of access and as a property right (see ODI 2004). As we will see by examining the different position or discussion papers presented below, it is this differentiation that is causing the biggest problems or misconceptions regarding the human right to water. In the ODI definition, water rights (as contractual access rights and as property rights) are legally enforceable property rights on the use of a certain amount of water (private–private contracts as well as public–private contracts). Water rights are promoted by some actors in the discussion as one of the tools that might help people to have their access to water guaranteed. In this respect water rights are one of the tools available in national water policies to broadening and securing access to water for as many users as possible.

[13] Several other interesting position or discussion papers were developed e.g. by specialized organizations such as WaterAid, by intergovernmental organizations such as the contribution of the WHO to the debate (WHO 2003) or by scholars and experts (Riedel and Rothen 2006). They cannot all be presented here and this article is not intended to be comprehensive. This chapter focuses on extremes to show the current broad span of the debate.

Such a discussion on water rights must be separated from the human rights debate on water, which is grounded on obligation of governments vis-à-vis the most vulnerable groups in society. Potential conflicts between the three legal forms of a right to water can exist. A contractual right for water services or the normal form property rights in the water arena in a country can both be in conformity or in breach with human rights norms and standards. Human rights are claims of individuals against its government. They create State obligations. Therefore human rights set the normative framework under which other forms of contractual security should operate.[14] States have to make sure that all forms of private–private or private–public contracts that impact on water rights are done in a way respecting the essential standards set under the human right to water. For example, States have to ensure that certain individuals or groups are not discriminated against in accessing water. States have to make sure that the most marginalized and vulnerable groups get priority attention and that water supply systems under their jurisdiction, whether private or public water companies do not breach these human rights standards.

4.1 The World Bank Water Policy and the Perception of the Right to Water

Until recently, the World Bank declared that it was not mandated to deal with human rights and that it was even prohibited to do so. It was argued that human rights were essentially political and therefore the World Bank was prohibited to raise human rights issues with its member countries (World Bank 2006). However, a recent Legal Opinion[15] by the World Bank's General Council has softened this argument, providing the legal ground for the World Bank to "consider human rights explicitly as it engages with its member countries".[16] In short, the Legal Opinion gives the following guidance for the World Bank's engagement in human rights:

- The World Bank may support member countries in implementing their international human rights obligations.
- The World Bank should take human rights violations into consideration where they have an economic impact.

---

[14] This is a basic understanding of all human rights texts that the have to be transferred into national law and that they are setting the framework for the overall body of positive law developed in a countries. The positive law shall not be in violating human rights standards. While this is a principle, concrete conflicts often needs to be decided by national constitutional or supreme courts of if available with human rights courts in regional human rights protection systems.

[15] Senior Vice President and General Council "Legal Opinion on Human Rights and the work of the World Bank", January 27, 2006.

[16] The change in interpretation of the Articles of Agreement is justified in the following way in the same document: "The manner in which the Bank's purposes and mission are now understood makes consideration of human rights essential. Human rights relate substantively to many of the activities of the World Bank. They are deeply interconnected with the purposes outlined in Article I, in large measure because they are directly relevant to the Bank's mission of poverty alleviation."

- In some areas, human rights are directly relevant to the World Bank's work; an example is public participation in Bank supported projects.
- The World Bank has a responsibility to stay engaged with a member state even if human rights are violated by that state, provided the Bank can continue to achieve its purposes.
- The World Bank is not an enforcer of human rights. The legal opinion essentially deals with the question of how human rights influence the relationship between the World Bank and its member states. While the legal opinion carries the spirit of liberating the World Bank to engage in human rights, it ignores the restrictions that human rights impose on its own activities.

The environment department of the World Bank additionally made a detailed analysis of General Comment No. 15 (World Bank 2003). The analysis is based on an affirmative perception that the right to water can be a helpful framework for the development of national water policies. However, the World Bank has not yet started to use such a framework and its potential in its own sector policy and in its support to Country Water Assistance Strategy.

The ongoing criticism of the World Bank policy in the water sector is focused on the practice of privatization-biased credit conditionalities and market-centered policy advice reflected in Country Water Assistance Strategies (Public Citizen 2004; WDM et al. 2007). These strategic pointers are often decisive for the policy choices that countries make. Until now the World Bank has not started to analyze and to adjust its strategic framework to a human rights perspective. Such an analysis would have to be based on the framework for national implementation strategies given in the General Comment No. 15.[17] Countries should identify first the most vulnerable groups in the society. They should review the impact of the existing legislation, strategies and policies and should then modify or change the legislation, strategies and policies in order to achieve compliance with the central human rights norms. Countries are encouraged to develop a rights based national water strategy and action plan.

The critics of the World Bank policy advice argue that its current approach in the water sector all too often contributes to a deterioration in the effective access of poor and vulnerable groups to water for domestic use or to sanitation, which is documented even in reports of the World Bank's own evaluation department (World Bank 2003). Adopting a human rights based approach would definitely be an important contribution to the urgent revision of its water sector policies.

A critical discourse on policy choices from a human rights perspective has not yet been initiated, either between the World Bank and its critics or in academic circles. It is this type of policy discourse which is needed if a human rights based approach should gain momentum.

During the 1970s and 1980s, the World Bank supported publicly administered and financed water sector policies in order to improve the supply with drinking water and sanitation. A major change was initiated in the policy paper from 1993 for

[17] The recommendation for national implementation can be found in GC 15 in Chap. V "Implementation at the national level" in paras. 45–52.

the management of water resources. As one of the first development institutions, the World Bank started to promote here an integrated water resource management approach, which today is also supported by many bilateral donors. Such an approach tries to link all potential water user forms, from dams, irrigation, drinking water, sanitation and water for industrial use. One of the central elements of that strategy is that the Bank insists on the development of an integrated political and institutional framework at the national level, based on decentralization and privatization of management and infrastructure. It is this strategic orientation that is challenged by critics of the World Bank, particularly because the World Bank is one of the most powerful donors, both in the conceptual and the financial sense. They claim that the World Bank is reducing the role of governments to a role of creating an enabling legal and institutional regulatory framework. As lead financial donor for many developing countries, the World Bank can also implement such strategies. Eighteen countries have adopted in between a "Country Water Resources Assistance Strategy" (World Bank 2007). The new sector policy papers so far are not built on human rights based criteria and do not mention the human right to water explicitly.

4.2 The Right to Water as Defined by the World Water Council

The World Water Council published a booklet on the right to water during the World Water Forum in Mexico in March 2006. The report is based on the program: "Right to water. What does it mean and how to implement?" initiated by the World Water Council and financed by the Swiss and the French development agencies (Dubreuil 2006). Many authors from intergovernmental agencies, private companies and NGOs contributed to the report, which in the main introduces the interpretation of the right to water as it was developed in the General Comment No. 15. It also contains references to other legal sources such as the Convention on the Rights of the Child from 1989 or the Convention on the Elimination of Discrimination against Women from 1979.

The text differentiates between the human right to water and other form of water rights. Water rights generally refer to accessing or using water for specific purposes ("abstraction rights"). The law concerning water rights may define who can use water and under which circumstances. The human right to water contains minimum standards to any water law in order to avoid discrimination in access, or other forms of violations of that right. The text is also clear in arguing that water cannot be treated only as an economic good because important functions of water cannot be monetized. Regulation of access to water based only on market mechanisms will not work.

Besides background information given on the content and on state obligations concerning the human right to water, the text again mixes the human right to water with other forms of water rights, when it starts talking about the implementation of the right to water. Page 11 has a chart that presents the rights and duties of users and the right and duties of authorities. The duties of the user contain elements such as "to pay a fair price for the service, including fees and taxes; to pay the charges for extending services. Such a duty cannot be seen as an element of the human right to water, due to the fact that human rights define State obligations. The distinction between the human right to water and other water rights is ignored. While contractual water rights will have duties and rights for all involved stakeholders, the human right to water requires that governments control all actors involved in private contracts to respect human rights principles. In mixing the conventional discourse on contractual water user rights and duties with a human rights-based approach the text negates any meaningful understanding of the human right to water and becomes highly problematic. It destroys the essence of the human rights based approach as a minimum standard for government laws and regulations and replaces it with rights and duties of private contract law.

One gets the impression that particularly the transnational water industry and private companies get nervous when the human rights based approach is mentioned. They are concerned that they have to provide water for free to the very poor of their society (in the UK water companies no longer have the right to disconnect non-paying clients whose duty to pay remains).[18] Therefore it is useful for private suppliers to use the rights and duty analogy to legitimize the pricing of water supply and the expansion of guaranteed prices.

4.3 The Human Right to Water in the Second United Nations World Water Development Report

A similar substantial argument or misconception can be also be found in the second United Nations World Water Development Report (WWDR). This report, published before the second World Water Forum in Mexico in March 2006, is a valuable compendium on information related to the current water crisis. It was published with the contribution of many specialized organizations within the United Nations family. In terms of content, it is a very wide-reaching book, providing the reader with a richness of information and analysis. Many case studies discuss reasons for the current problems related to access to water and sanitation. The report contains at the same time lessons on how to effectively improve national water policies. Its central message is that it is political will that is needed to speed up action. While the text is detailed and differentiated in many parts, the reception of the human right to water approach is marginal in the text and based on a substantial misunderstanding. By not recognizing the potential and opportunities attached to a rights based approach the report underestimates the role the human right to water can play in generating the

[18] Concerning a possible price for drinking water, GC 15 demands that the price has to be 'affordable"' for all users. Affordable means that a person can afford it, without compromising its access to any other essential human rights. This requires government policies being based on human rights to check whether water is affordable. For particularly poor users it often requires to provide at least a minimum essential level of water for free.

missing political will. The right to water is described purely as a moral obligation: Key recommendation No. 1 of the WWDR reads as follows:

> We need to recognize that access to clean water is a fundamental right. In 2002, the UN Committee on Economic, Social and Cultural Rights affirmed that 'sufficient, affordable, physically accessible, safe and acceptable water for personal and domestic uses' is a fundamental human right of all people and a pre-requisite to the realization of all other human rights. Although not legally binding for the more than 140 countries ratifying the International Covenant on Economic, Social and Cultural Rights, this decision carries the weight of a moral obligation on the signatories to progressively ensure that all the world has access to sate and secure drinking water and sanitation facilities, equitable and without discrimination. (WWAP 2006, 520).

Characterizing the right to water only as a moral obligation eliminates the substance of any meaningful human rights-based approach, which is that individuals can sue violators and go to court based on internationally recognized minimum standards set by the human right to water. A purely moral obligation does have the quality of the MDGs or many other declarations or action plans from international UN-Conferences. While such texts are of huge importance because they contain benchmarks for action that governments have voluntarily accepted, it is the binding legal nature of human rights that represents the value added of a human rights based approach. Human rights contain binding obligations for the States that are party to the respective human rights treaties. The International Covenant on ESC Rights has currently 153 State parties, which are bound by norms contained in the treaty.[19] State parties must develop national laws that guarantee effectively the enjoyment of all human rights. The provisions of General Comments are interpretative guidance for legislators and for the judiciary for the design of public water policies and the appliance of the human rights norms to specific case situations. They guide the action of governments that are willing to implement the right to water.

4.4 The Right to Water in the 2006 Human Development Report

The latest of the substantive reports of the United Nations with a particular focus on water is the Human Development Report 2006 from the UN Development Programme. Because it was published close to the other mentioned reports, it cannot present much new empiric results or new strategies, but the report is quite unique in focusing on justice issues in access to water. It also places the right to water in a central role.

The report has two main parts. The first three chapters deal with the issue of "water for life", the missing or often problematic access to drinking water and to

---

[19] The text mentioned that the International Covenant on Economic, Social and Cultural Rights (ICESCR) has 140 State parties, but this is an old figure. The ICESCR today has 153 State parties and more States are in the process of ratifying it. It has close to the same number of State parties as the International Covenant on Civil and Political Rights. Both central human rights treaties are on the way to universal ratification.

sanitation for millions of people world wide. The following three chapters deal with "water for livelihood". Here the report focus on water as a productive resource shared within countries and across borders. Both parts are written from a judicial perspective. The report starts with a clear recognition that water scarcity is not at the heart of the global water crisis, but that it is rooted in power, poverty and inequality rather than in physical availability. In order to address the current scarcity of water for so many persons what needs to be developed is water security as an integral part of overall human security.

> In broad terms water security is about ensuring that every person has reliable access to enough safe water at an affordable price to lead a healthy, dignified and productive like, while maintaining the ecological system that provide water and also depend on water (UNDP 2006, 3).

The concept of justice used in this report is a broad one: Water insecurity is violating several human rights principles such as: (1) equal citizenship, (2) social minimum, (3) equality of opportunity and (4) fair distribution. In that broader concept of justice, the right to adequate water is first taken up as one of the central concepts and used in proposals for solutions of many of the current problems related to water.[20] The right to water is mentioned under "(2) social minimum" as a way to define the essential minimum on access to drinking water, in order to avoid death or unnecessary illnesses. The provision in the constitution of the Republic of South Africa, that the first 20 liters per day should be free for every person, is taken up as the bare minimum of the human right to water. "Ensuring that every person has access to at least 20 liters of clean water each day to meet basic needs is a minimum requirement for respecting the right to water – and a minimum target for governments." (UNDP 2006) The quote highlights that the implementation of human rights should be in the long run more then just guaranteeing the bare minimum – as important as that minimum is. Human rights norms are universal norms, which allow the monitoring of governments and allow to be held them accountable. The other elements of justice mentioned are basic human rights principles, central for the realization of all human rights including economic, social and cultural rights.

Therefore the text is built upon the recognition that upholding the right to water is an end in itself and a means for giving substance to the wider rights in the Universal Declaration and for giving substance to human dignity. In comparison to the UN World Water Development Report, the Human Development Report takes up the current understanding of economic, social and cultural human rights:

> Human rights are not an optional extra. Nor are they voluntary legal provisions to be embraced or abandoned on the whim of individual governments. They are binding obligations that reflect basic values and entail responsibilities on the part of governments. Yet the human right to water is violated with impunity on a widespread and systematic basis – and it is the human rights of the poor that are subject to the gravest abuse. (UNDP 2006, 4)

---

[20] The report contains two special contributions from Kofi Annan, then Secretary General of the United Nations and from Luiz Inácio Lula da Silva the President of Brazil. In both contributions the two authors highlight the importance of a rights based approach to water issues (UNDP 2006, 78–79).

This broad accountability perspective allows it to put government action under the scrutiny of internationally agreed minimum standards. If a government violates the right to water of an individual with its given set of policy choices, it has to change the policies. The minimum standards have always to be guaranteed for all persons living in the country. If this cannot be met, the State has to prove that it has used the maximum of the available resources and that the existing resources are not sufficient. The report does not mix up the different concepts of the human right to water and water user rights. It understands contractual water rights as one of the instruments of water policies available to governments. Their use has to be consistent with human rights standards.

5 From Global Norms to Implementation

Designing the best implementation process for the human right to water is a difficult task, because the availability and quality of water is influenced by decisions in different policy areas. They all need to be recognized and modified when necessary in order to realize the human right to water. The realization of those human rights that require standards and policy measures in many policy areas is therefore in principle a difficult task.[21] The voluntary guidelines to support the implementation of the right to food have a structure that recognizes these problems. Each of the guidelines is directed to one of the policy areas at the national level. Similarly implementation measures for the human right to water require a discussion of policy measures and policy outcomes in different fields of politics.

An initial examination of the different levels of State obligations is helpful to further specify implementation policies for the human right to water. The obligation to respect requires that government policies should not harm the legal norm of the human right to water. Government policies should for example not destroy existing access to water; they should not arbitrarily disconnect individuals or groups from access to water or should not discriminate de jure individuals or groups in their access to water. Under the obligation to protect, governments are under the obligation to control all third parties involved in water politics. Private actors as well as public actors or government institutions must be controlled in order to guarantee that no violation of the human right to water happens. This is the challenge in most cases of large dam constructions. The discussion of state obligations to fulfill is even more complex. Every government has a number of policy choices possible to implement the right to water and to set up water sector policies or national water strategies. It is often difficult to know in advance which of the choices will be more effective. Effectiveness in creating access for poor people to drinking water or to sanitation would be a human rights criterion for assessing certain policy choices.

[21] Some economic, social and cultural rights are more restricted, such as the right to education. Education is in most countries the task of one ministry. Even if forms of discrimination in access to education and other potential violations or not only cause in the education system, many of the necessary policy responses are to be taken in education policies.

Therefore the approach chosen by the UN-Committee on Economic, Social and Cultural Rights is correct. General Comment No. 15 does not prescribe a certain policy choice as the only possible one. It also does not derogate certain policy choices as being the best or only policy option to implement water security. But it is clear that all policy choices should contribute to the full implementation of the right to water, that they should not violate the right and that they should focus government attention on the needs of the particular poor groups and individuals. Governments should assess the impact of policies and adopt or change policy choices if these objectives are not achieved in coherence with the human rights obligations. While effective human rights implementation policies will vary according to the local situation, nevertheless a common learning and exchange of experience is required. If certain policy choices do cause severe forms of discrimination in access these should only be used transitionally by governments where local circumstances do not allow for other options. The implementation of economic, social and cultural rights requires that policy choices are discussed by using human rights norms and standards. Progress as well as steps backwards need to be monitored and should lead to policy learning.

Implementation measures are possible at all different levels of governance. The following examples are chosen to illustrate successful implementation measures.[22] The first example refers to changes at the constitutional level. The second shows the importance of national legislation. The following one discusses options of how to document violations of the human right to water. The implementation work on the human right to water is still at the early stages. More pilot experiences are needed to find out best policy options. The lessons learned need to be documented and shared widely in order to promote the necessary national and international discussions on future water policies. The limitation set for policy choices by the human rights norms is that policies should not contribute to violations and that all policies should contribute to the full realization of that right.

5.1 National Experiences from Uruguay and India

In October 2004, a plebiscite on water policy took place in Uruguay, the result of which is of importance for the international rights approach discussion. After 2 years of intense sensitization and campaigning work by a broad civil society alliance made up of a diverse range of environmental and social groups, trade unions and academic and religious groups, a proposal for constitutional reform concerning water was submitted to the general popular vote. With a surprising support of 64.7 percent of all registered voters, Article 47 of the Uruguayan constitution was modified by

[22] More examples could be given for successful implementation measures, but the brevity of this article does not allow further examination. It would be desirable for the design of future implementation measures for the human right to water to set up a common learning process to explore successful measures for implementation. A common pool with information on lessons learned should be available to governmental and non-governmental actors involved in water politics.

introducing explicitly the concept of the human right to water. Besides the general affirmation that access to water and sanitation is a human right, it stated:

- that water must be managed in a sustainable way showing solidarity with future generations and the needs of the hydrological cycle,
- that users and civil society participation must be guaranteed at all levels of planning, management and control of water resources,
- that water management should be organized along regions and watershed areas, prioritizing human needs,
- that water supply management systems must be governed by social welfare principles before economic ones (Achkar 2006).

The constitutional reform confirms also that water has to be managed in public trust and excludes explicitly the privatization of water supply services. The new Uruguayan constitutional paragraphs, by deducing more concrete and unequivocal policy principles for public water policy from the right to water in the interpretation of General Comment No. 15, manifest the potential of the rights approach as a policy guiding tool, and thus contribute considerably to the international discussion on rights based legal frameworks (Achkar 2006).

Concerning the human right to water discussion, it is also interesting to have a look at the jurisdiction around the conflict between the Coca Cola Company and the Plachimada community in the Palakkad district in the state of Kerala, southern India. The nucleus of the conflict is, that since 1999 Coca Cola has a soft drink factory in this rural area, overexploiting the groundwater by extracting 500000 liters of water daily, and according to the local community council, causing damage to the local population whose wells are drying up, constraining the water available for human and agricultural needs. In 2004, the Kerala High Court passed a judgment on the case which was clearly deduced from the right to life, which constitutes one of the normative pillars for the recognition of ESC-rights and the right to water in the view of the UN-Committee on Economic, Social and Cultural Rights. The Kerala High Court judged that the underground water belongs to the public, with the state as trustee. The state has the duty to prevent overuse. It was ruled that, "The inaction of the state in this regard will be tantamount to infringement of the right to life of the people guaranteed under Article 21 of the Constitution of India". Therefore, the state was duty-bound to ward off excessive exploitation. The judgment follows the human rights perspective, that access to water for private consumption is the first obligation of governments. The judgment came under revision through upper courts and the case is still not concluded legally.

5.2 Categorizing Violations of the Human Right to Water

Among the first international organizations working on economic, social and cultural rights, FIAN-International and Bread for the World have taken up and investigated several cases[23] of violations of the human right to water. The types of violations observed can be divided into three broad categories.

The first group of violations is situations in which individuals' access or accessibility rights are destroyed or interrupted when water sources are directly privatized or water resources are overused by a competing user. An example of this is the loss of access rights of nomad groups to traditional water places. There are a number of cases in which accessibility rights are violated due to overuse of scarce water resources by other users. Some accessibility conflicts arise when property/user rights to water resources are changed. Others occur when direct forms of privatization of water resources happen, e.g. wells or rivers streams that were previously accessible to many users are sold to a private user.[24] This group of cases also includes cases taken up by FIAN concerning the loss of accessibility to irrigation water or the destruction of irrigation systems. Here, the destruction or interruption of water supplies for personal and domestic use was also identified as a parallel problem.

The second group covers cases in which access to water was destroyed as a result of water pollution. This was the case in Ecuador's lowlands where oil production caused widespread river and water resource contamination. There are also cases of cyanide accidents in surface gold mining sites in Ghana, which destroy water and food supplies for whole villages. Surface mining sites lead sometimes to the deviation of rivers for pits or waste dumps. Entire villages can lose their access to water for domestic use as well as for irrigation. Cases involving the contamination of water resources due to intensive chemical application in agriculture – in flower or banana plantations, for example – also belong to this group.

A third group of cases included those in which changes in water supply systems occur. These cases are those discussed most at the international level in the emerging global water movement, which has concentrated its work on cases of privatization of urban water supply systems. In these cases the change of ownership of water supply systems can impact particularly on poor peoples' access to drinking water, e.g. through a rapid price increase after water privatization.

In 2004, FIAN, together with Bread for the World, undertook an international Fact Finding Mission to India and investigated several cases of violation of the right to water. The report of the Fact Finding Mission contains examples related to all three categories mentioned[25] and comes to several conclusions: First, it is methodologically possible to document violations of the human right to water based on solid proof of facts. The human rights standards which are breached must be clearly spelled out, also detailing the responsibility of the government for the context for the given violation.[26] Secondly, the examples show particular problems that might

---

[23] The publication presents cases of violations of the right to water from Bolivia, India, Ecuador and Peru (Gorsboth 2005).

[24] See for example the above-mentioned Plachimada case from India (Chap. 4.1).

[25] The report summarizes the findings of the four groups of the Fact Finding Mission in four regions of India (Bread 2004).

[26] Some initial court cases with a direct or indirect reference to the human right to water has been filed and decided upon. The example of the Plachimada case mentioned above is one. Another one, which is well documented, is a lawsuit against the Municipality and Province of Córdoba in

occur with the introduction of contractual commercial water user rights. Such an introduction can lead to access problems for particular poor persons or groups. When introducing contractual water rights it is necessary to take explicit account of the particular access problems of poor and vulnerable groups. It is important in this context to regulate private investors from the first moment. If the contracts are already signed it will become difficult to do so. Thirdly, several cases indicate that pollution problems and conflicts among different groups of water users might become one of the core problems in the future, if legal and production patterns do not change.

Case based analysis concerning violations of the human right to water is still at the beginning, even if several cases are already well documented. The further analysis of cases will be the essential tool to improve the understanding of the human right to water in the coming years. The more cases that are documented and assessed,[27] the richer will be the understanding of what can reasonably expected from governments when implementing their human rights obligations under the right to water.

5.3 The Discussion on Monitoring and Indicators

Part of any national strategy for the implementation of the right to water is the monitoring of government policies. Monitoring should identify acts of omission in government policies that lead to violations of the human right to water. Moreover, it should help to identify any loopholes in the legislation or in the policy choices of the government, where needs of certain vulnerable groups are overseen. The monitoring should be set up to assess also the effectiveness of the given set of policies implemented by a government.

General Comment No. 15 calls for States to use indicators to monitor the right to water:

> To assist the monitoring process, right to water indicators should be identified in the national water strategies or plans of action. The indicators should be designed to monitor, at the national and international levels, the State party's obligations (...). Indicators should address the different components of adequate water (such as sufficiency, safety and acceptability, affordability and physical accessibility), be disaggregated by the prohibited grounds of discrimination, and cover all persons residing in the State party's territorial jurisdiction or under their control (...). (CESCR 2002, para. 53)

Argentina. The case was filed by a non-governmental organisation because the lack of maintenance and capacity of a public sewer-water treatment facility led to a contamination of local water sources for years of several poor neighbourhoods (Gorsboth 2005; COHRE 2004).

[27] An international complaint mechanism for Economic, Social and Cultural Rights is currently under discussion and will be elaborated in a working group of the Human Rights Council. It shall have the format of an optional protocol to the International Covenant on Economic, Social and Cultural Rights. Similar complaint mechanisms already exist for most of the other international human rights treaties. Such a complaint mechanism would allow – under certain restrictions for the admissibility of cases – the analysis of specific cases of violations of ESC-rights and the right to water.

In 2004, Bread for the World together with the Heinrich Böll Foundation and the international human rights organization Centre on Housing Rights and Evictions (COHRE), organized an international workshop which for the first time worked on a framework for indicators responding to the human rights framework of General Comment No. 15 (Bread for the World et al. 2005). Participants included experts and practitioners in human rights, statistics and water and sanitation policy from a range of organizations and countries, including UN organizations, government and civil society representatives. The focus of the workshop was to consider the particular requirements for indicators – may it be based on quantitative or qualitative information – to make them human rights sensitive. The indicators should capture the three dimensions of State party's human rights obligations. Also, they should be designed for use by official statistics systems, but also by civil society groups. One of the most important requirements is whether the indicator can be disaggregated for gender, ethnicity, race, religion, nationality and social origin. Human rights based monitoring has to focus on the changes for the most excluded groups. Current international indicators and survey systems – the WHO-UNICEF Joint Monitoring Programme probably being the most authoritative source on water – focus mainly on access aspects, but less on affordability. Another observation is that besides indicators for the national and international level, there is a role for indicators at the local level with the participation of communities and groups, whose right to water is threatened.

This first international exercise on indicators for the right to water confirmed that General Comment No. 15 is not only a valuable source for policy principles, but also instrumental for the definition of quantifiable indicators and qualitative criteria regarding the implementation of the right to water. A draft matrix was worked out with potential quantitative and qualitative indicators and guiding questions following the structure of General Comment No. 15, which might be of immediate use even before further elaboration (Bread for the World et al. 2005).

6 The Value Added of a Rights Based Approach to Water for Development Assistance?

The implementation of the right to water and the adoption of a rights based approach on water for development assistance will be an important step towards overcoming the current world wide crisis of missing access to water and sanitation.

6.1 General Considerations of "Value Added"

The value added of a rights based approach becomes obvious, when the problems underlying the many violations of the right to water are analyzed with a human rights lens. They are essentially caused by deprivation of individuals or groups in access to water. While water availability is or will become a problem of majority population groups in some countries in the near future, the current forms of scarcity at the heart of the global water crisis are rooted in power, poverty and inequality. Even in poor and water stressed countries, the rich members of the society consume huge amounts of water. In high income areas of cities in Sub-Saharan Africa people enjoy several hundred liters a day, while in poor suburbs people often have difficulties to have access to only five liters.[28] The 2006 Human Development Report is of the opinion that deprivation in access to water is at the core of the current access problems. While household water requirements cover only a tiny fraction of the water use,[29] it is characterized by a tremendous inequality in access to clean water and to sanitation. The situation is particularly difficult in rural areas and for women and girls, who spend a huge amount of time collecting water and girls who are often not sent to school, either because they have to help collecting water at home or because schools do not offer separate sanitation for girls.

Even when considering the productive use of water in agriculture and industry, which is not part of the human right to water as described by the UN-CESCR in its General Comment No. 15, but are seen as part of the right to adequate food, there are justice issues linked to user patterns that are important to tackle. The 2006 UNDP Human Development Report contends that this is less a hydrological constraint than a man made problem. Those users that can afford irrigation techniques are often pumping and overusing existing aquifers, while particularly smallholder farmers often do not have access to such water sources. In many cases the political and institutional constraints determine the deficiency of the supply and not the physical availability. In many countries scarcity is created by pubic policies that have encouraged overuse of water through subsidies or under-pricing. Key to the current access problems is the fact that particularly poor people are systematically excluded from access to water often by their limited legal rights or by public policies that limit their access to the infrastructure that provides water both for domestic and for the productive use, or as the HDR puts it: "... scarcity is manufactured through political processes and institutions that disadvantage the poor" (UNDP 2006, 4).[30]

Because issues such as deprivation in access or ineffective public policies are at the core of the problem, a rights based approach is crucial for finding solutions to the problems. Human rights create entitlements of persons vis-à-vis their government. These entitlements can be legally claimed and are therefore a good tool to hold governments accountable. Moreover state obligations under the human right to water become clearer when checked in recourse procedures (court decision, investigations etc.) A rights based framework, therefore, clarifies government obligations

---

[28] The UNDP Human Development Report 2006 shows in detail the different rates of consumption in countries which do not have a sufficient overall supply of water.

[29] According to the same HDR-Report it covers less than 5 percent.

[30] The problems related to water for irrigation cannot be discussed here in detail, because it would go beyond the scope of the article.

and assists in developing criteria for designing and evaluating policy policies and institutional processes.

Human rights are individual entitlements. They therefore set limits on the restrictions and deprivations that individuals can permissibly be allowed to bear even in the promotion of other social goals or overall development objectives, such as economic growth. Human rights standards are therefore minimum standards that do not allow tradeoffs with other policy goals. Moreover general achievements in human development are not always accompanied by achievements in human rights fulfillment and cannot always be equated with steps forward. On the contrary, certain development processes may go hand in hand with a deterioration of the livelihood and the human rights fulfillment of particular vulnerable groups. Moreover, the development way of thinking seldom asks "how" results were achieved, while within a human rights framework the quality of the process of policy development and implementation is of importance. Issues of transparency, participation and due and fair procedures are not only an add-on but a basic condition for the conduct of government institutions, for the design of public policies and the use of public budgets. A rights based assessment and framework must not only look into the obligations and responsibilities of national governments, it should also assess the potential impact of policy measures or effects of one country on persons living in another country, the extraterritorial obligations.

While the water crisis will require policy changes in a lot of policy areas the right based approach set standards for the design and implementation of such policies and regulations. Moreover it creates the space for recourse procedures, which are essential to correct policy mistakes. Recourse mechanisms allow faster learning for policy makers.

6.2 The Position of the German Government with Respect to the Right to Water

The German government has become internationally one of the most visible supporters of the right to water over the last few years. Nevertheless, the different ministries and departments support that trend not all in the same way, but differ in nuances and pro-activity.

The relevant resolutions or decisions regarding the right to water in the UN-Human Rights Council until now have been prepared by Germany together with Spain. In October 2005, he German Ministry for Foreign Affairs held an expert consultation on the right to water. The human right to water is seen as an important instrument to deal with the global water problems, which will be aggravated in the coming years and decades. Particular conflicts over the use of such a scarce resource will become issues which might be better solved involving human rights norms and standards (Riedel and Rothen 2006).

While the Ministry for Foreign Affairs was supportive of the human right to water, it was some time before the Ministry for Development Cooperation (BMZ) started to include the right to water in their water sector policies. After Japan and the USA, the German government is the third largest donor for water-related development aid, donating approximately € 350 million per year. This makes German sector policy on water internationally important. With the background of disputes regarding the paradigms and strategies for reaching the Millennium Development Goal on water and sanitation of the last few years, an overall evaluation of the German sector policies and ODA sponsored water programs and projects was undertaken during 2004. The evaluation ended in a process of developing a new water sector strategy, which was then adopted in 2006.

The evaluation confirmed what has been feared by German civil society organizations working together in the water working group of the German NGO Forum Environment and Development: One of the observations was that water-related ODA of Germany does not really focus on the poorest countries and does often not really reach the poor, which holds true for the overall contributions of OECD countries (Brugger 2004; GKKE 2004, 37ff.). A similar observation was made when the newly created sector team for human rights of the GTZ analyzed the water sector policy of the BMZ in Kenya. While the projects are running efficiently it was observed that the focus was the connection of private households to water infrastructure and that the targets groups were not among the poorest in Kenya.

The newly developed water sector concept has taken up the human right to water explicitly in the water policies related to human settlement (BMZ 2006, 8). Progress towards a common understanding of the rights approach can be seen, when the German water sector concept for the first time recognized poorest peoples right to water being superior to the cost recovery principle (BMZ 2006, 12). Interestingly, instead of the human rights approach, the authors choose the concept of Integrated Water Resource Management (IWRM) as their overarching concept as it allows the consideration of water from an overall perspective of ecosystem management and protection. Nevertheless the full integration of a human rights approach to water within water sector policy should not be limited to the aspect of individual access to a daily minimum amount of water, but must necessarily shape all aspects of long term water management, such as watershed management, water rights, distribution of water among different uses or rural water policy (i.e. should guide the whole IWRM).

The sector concept also considers a range of important principles such as protection of the ecosystems, human rights approach, poverty reduction, preference to rain-fed agriculture before irrigation, participation of civil society, among others. The proof will be the implementation in form of a concrete action plan. Its focus and priorities needs to be monitored within the broad framework of the concept in the coming years and it has to be observed if the implementation of the human right to water becomes an integral part of the water sector policies (BBU et al. 2006).

Additionally, some years ago, the BMZ included in their policies a broader recognition of a value of a rights based approach to development. The Ministry agreed on an overall human rights policy and adopted a human rights action plan in June 2004 (BMZ 2004b). The action plan can only be seen as a starting point as it is poorly financed and institutionally integrated.

7 Future Plans and Challenges

This article has shown examples of how the human rights to water can be interpreted and implemented through legislative and other policy measures. The different examples show that we still have to go a long way to achieve a secure use and understanding of the norms and standard of the human right to water. Still some misconceptions exist. This is partially due to the fact that the overall perception of the new developments in the interpretation of the application of economic, social and cultural rights, which has emerged among experts and international lawyers during the last decade, is still quite new and it will take time to spread and become common knowledge. On the other hand, we should recognize that at the core of every water policy and its guiding principles, there is an often unequal and difficult negotiation of contradictory interests and control of power. In this context, every new case helps to improve the interpretation and application of the right to water. It is important to see that for ESC-rights in general a common standard of interpretation is emerging and that similar categories and standards are used to describe the content and the State obligations of all these rights. The most relevant interpretation of the human right to water is therefore the GC 15.

There are continued misconceptions and uncertainties in the application of the right to water, which influences the understanding and debate in two ways:

- There are only a few countries that do not accept that the right to water is a human right.[31] However, while the General Comment No. 15 is welcomed by many countries, further clarification regarding the content and the respective state obligations is necessary. To fulfill this need, in its second session in November 2006, the Human Rights Council requested the Office of the High Commissioner for Human Rights to study the scope and content of the relevant human rights obligations related to equitable access to safe drinking water and sanitation under international human rights instruments.[32] This study shall be submitted prior to the sixth session of the Council in September 2007.
- A key area for misconception is the role different actors can and should play in the implementation of the right to water. The misconception is further aggravated due to the fact that the different legal forms a right to water can have, as a human right, as contractual or property rights are often confused. It is a strength of the UNDP Human Development Report 2006 that it differentiates clearly between the human right to water and water user rights. Other reports mentioned earlier in this article confuse these different legal forms. Private water user rights go hand in hand with rights and duties for all stakeholders. This is different from the human rights perspective, which describes the State obligations vis-à-vis individuals living in the country. Particular water user rights can be one

[31] Among them is the United States of America, which has problems with all ESC rights, but also some other countries are sceptical for the time being, such as Poland.

[32] The decision was taken on 27 November 2006 (UN-Doc: A/HRC/2/L.3/Rev.3).

of the possible tools for building security of access rights. If done badly it can also cement the exclusion of certain groups or individuals from access to water sources.

The human right to water puts a focus for governments on the poor and vulnerable of society. Without measures against the injustice in access to water and sanitation there is a high risk that the interpretation of water user rights will exclude particularly marginalized groups such as women, smallholder farmers and pastoralists. The HDR highlights the responsibilities of governments to act properly and that access to water and sanitation will seldom be achieved through private sector involvement. Key to understanding why some countries do much better is that they have public policies in place that help to convert general income increases of the country into human development results.

Much of the resistance against ESC-rights still comes from countries that are afraid that they might ratify or might have ratified a blank check, although the conclusive interpretation of theses rights is still at the initial stages. Such fears will only be overcome by a development of a jurisprudence of specific cases, which proves that court rulings do approach ESC-rights in a reasonable framework, without putting undue expectations on governments. The development of the interpretation of ESC-rights so far, as for example in the cases of the right to food and to housing, is very fair and these experiences should help to overcome unjustified fears of governments. On the other hand it is obvious to note that many governments will have to improve their right to water related policy outcomes, some substantially. They might also fear a meaningful recognition of water as a human right, because they have to make such changes.

The strength of the General Comment No. 15 is that it is not policy prescriptive. Applying the right to water does not require determined policy instruments per se. It allows each country to select from a variety of policy measures those required or best suited by the specific country situation. The human right to water describes minimum requirements for the policies chosen and sets limits to the tradeoffs possible. Moreover the human rights approach to water will help to focus government action on the most vulnerable groups. It also allows the analysis of government policy choices and scrutiny of the impact specific policies choices have on these groups. The experiences with water policies already identify lessons that can be learned such as that markets alone will not work adequately to protect and guarantee access to water or access to sanitation. The regulation of market forces as well as public investment and regulation is essential.

More work is needed with specific cases. The perspectives mentioned are an encouragement for civil society organizations to work with water cases from a human rights perspective. Cases can and should be taken up by national human rights institutions as well as by courts and other institutions of the national legal systems. Responsible governments are the key to solve most of the water problems. The human right to water helps to describe the essential elements of responsible governance.

References

Achkar M (2006) Una Expresión de Democracia Directa: La Comision Nacional en Defensa del Agua y la Vida (CNDAV) y la Lucha por el Agua en Uruguay, Bread for the World Stuttgart (http://www.menschen-recht-wasser.de)

Barlow M, Clarke T (2002) Blue Gold, New Press, Toronto

BBU, Brot für die Welt, EED et al. (2006) Gemeinsame Stellungnahme zum Entwurf des Sektorkonzepts Wasser des BMZ, Bonn, Forum Umwelt und Entwicklung (http://www.forumue.de)

BMZ (2004b) Menschenrechtsaktionsplan, Berlin, Juni

BMZ (2006) Sektorkonzept Wasser, Entwurf August, Bonn

Bread for the World, COHRE, Heinrich Böll Stiftung (2005) Monitoring Implementation of the Right to Water: A Framework for Developing Indicators, Global Issue Papers HBS No. 14, Berlin, March

Bread for the World, FIAN-International (2004) Investigating some alleged violations of the Human Right to Water in India. Report of the international fact finding mission to India, January, Heidelberg (www.fian.org or www.menschen-recht-wasser.de)

Brugger F (2004) Some water for all or more water for some? Financing the MDG needs shift in resource allocation, Bread for the World, Stuttgart (http://www.menschen-recht-wasser.de)

CESCR (1999) General Comment No. 12. The right to adequate food. UN-Doc: E/C.12/1999/5, 12 May

CESCR (2002) General Comment No. 15. The right to water. UN-Doc: E/C.12/2002/11

Craven M (2001) The UN Committee on Economic, Social and Cultural Rights. In: Eide A, Krause C, Rosas A (eds) Economic, social and cultural rights, Dordrecht, Boston, London, 455–472

COHRE (2004) Legal resources for the right to water, Geneva

Dubreuil C (2006) The Right to Water: From Concept to Implementation, World Water Council, Mexico

FAO (2004) Final report intergovernmental working group for the elaboration of a set of voluntary guidelines to support the progressive realization of the right to adequate food in the context of national food security, Rome

FAO (2005) Voluntary guidelines on the progressive realization of the right to adequate food in the context of national food security, Rome, adopted in the FAO-Council in November 2004

GKKE (2004) Halbierung der Armut. Der Beitrag des Aktionsprogramms 2015 der Bundesregierung zu den Millenniumszielen, Schriftenreihe der GKKE 35, Berlin

Gorsboth M (2005) Identifying and addressing violations of the human right to water, FIAN-International and Bread for the World, Stuttgart (http://www.fian.org and www.menschen-recht-wasser.de)

International Freshwater Conference (2001) Ministerial declaration and recommendations for action of the International Conference on Freshwater in 2001, Bonn (http://www.water-2001.de)

Liebenberg S (2006) Security of tenure. In: ESC-Review: Economic and Social Rights in South Africa, 7 (3), 1–2

ODI (ed) (2004) Briefing paper: right to water: legal forms, political channels (written by Peter Newborne), July, Sussex

Public Citizen (2004) Will the World Bank back down? Water privatization in a climate of global Protest, Washington, DC

Riedel E (2006) Einführung: Die Allgemeinen Bemerkungen des Ausschusses für Wirtschaftliche, Soziale und Kulturelle Rechte. In: Deutsches Institut für Menschenrechte (ed) Die ‚General Comments' zu den VN-Menschenrechtsverträgen, Baden-Baden, 160–173

Riedel E, Rothen P (eds) (2006) The human right to water, Berliner Wissenschafts-Verlag, Berlin

Ruspekhofer S (2006) The human right to water and food: insights from the work of the special rapporteur on the right to food, Völker- und Europarecht Aktuell. In: Jusletter 27.03.2006

Scanlon J, Cassar A, Nemes N (2004) Water as a human Rights? IUCN Environmental Policy and Law Paper No. 51, IUCN – The World Conservation Union, Gland

United Nations (1987) The Limburg Principles on the Implementation of the International Covenant on Economic. Social and Cultural Rights (E/CN.4/1987/17), Geneva
UNDP (2006) Human Development Report. Beyond scarcity: Power, poverty and the global water crisis. New York, New York, Oxford
UNESCO (2003) Water for People, Water for Life – The UN World Water Development Report. C.S. Knighton (ed) World Water Assessment Programme (United Nations)
Windfuhr M (2006) Freiwillige Leitlinien zum Recht auf Nahrung, Bedeutung, Reichweite und Chancen des neuen Rechtsinstruments. In: Deutsches Institut für Menschenrechte (eds) Jahrbuch Menschenrechte, Frankfurt a. M., 227–235
World Bank (1993): Water Resources Management: A Policy Paper, Washington, DC
World Bank (2003) Efficient, sustainable service for all? OED-Report Nr. 26443, Washington, DC
World Bank (2006): Development Outreach, Special Report October: The Legal Aspects of the World Bank's Work on Human Rights, Washington, DC
World Bank (2007): Issue Briefs, Water, March, Washington, DC
WDM et al. (2007) Down the drain – how aid for water privatisation could be better spent, London
WHO (2003) The right to water, Geneva
WSSD (2002) Plan of implementation of the World Summit on Sustainable Development (http://www.johannesburgsummit.org)
WWAP (2006) Water – a shared responsibility. The United Nations World Water Development Report 2, UNESCO/Berghahn Books, Paris

Rethinking IWRM Under Cultural Considerations

Manfred Matz

Abstract Water management and culture are obviously interlinked. The way access to water is organized in a nation reflects the values and thinking of its people. Common approaches in Integrated Water Resource Management (IWRM) that are based on the Dublin-principles would make us believe that most of the solutions for good water management are rather generic and applicable wherever you are in the world. This article shows how cultural differences have shaped completely different approaches in water management set-ups in France and Germany, both of which are countries where water management has been performed with quite good results over the past decades. Based on this insight, the article questions whether the Dublin Principles and the derived approaches coming from these principles can be applied like blue prints. The analysis shows attempts to show how the Dublin principles can sometimes lead to misunderstandings and even counterproductive effects. The article invites the reader to rethink approaches in IWRM that may be well meaning but are most probably ineffective while giving hints for ways we can create new thinking on water management.

1 Introduction

When we think about water management, we can not ignore its cultural context. Numerous scientists studying cultural differences, e.g. Hofstede/Hofstede (2004) and Trompenaars (1998), have looked beyond the ritual or religious aspects of societies, to define culture as our systems of living and working together. Each country – indeed each region inside larger countries – has different characteristics that influence how its people behave and society is shaped. This refers to everything from the system of schooling to the way people communicate. Certainly, national and local legal, political, administrative and regulatory systems within country, as well as

Manfred Matz
e-mail: manfred.matz@siwi.org

W. Scheumann et al. (eds.), *Water Politics and Development Cooperation*,
doi: 10.1007/978-3-540-76707-7_8,

the way property rights are defined, assigned and enforced are all crucial displays of cultural differences that exist between nations. Many social scientists, especially Hofstede, have shown how cultural differences are translated into the political reality of a country. One very sensitive cultural phenomena related to the internal organization of a country is the way we manage water resources. Others have, for example Huppert (2006), made attempts to relate systems of water management with systems of governance.

To many non-water experts, water management might be thought of as irrigation, water supply or water resource management. Here we talk about water as a resource and the way it is managed; e.g. the way it is allocated to different users and for different purposes. We talk about the 'right' to use water through things like licenses and the duties related to this right. The focus of this article is on water resource management and is different from the management of water services for water resource uses such as water supply, irrigation and others.

In recent years, Integrated Water Resource Management (IWRM) has been promoted by many organizations in the water sector as the best approach to managing water resources. Looking further into the discussion surrounding IWRM, we can see that the tenants translated through the Dublin Principles are said to be somehow universal. After 15 years of following and trying to apply theses principles – with mediocre success only – we have to ask ourselves: Are IWRM principles universal enough that they can be applied anywhere in the world?

According to the number-one promoter of IWRM, the Global Water Partnership (GWP), IWRM is not considered a 'one-size-fits-all' approach but rather a framework in which many cultural specificities can be fitted into to give management practices national or regional shape. However, are these principles sufficiently universal to cope with the different cultural set-ups? Is participation and watershed orientation a precondition for good water management? In which way can a participatory system of water management have a precautionary vision in water resources management? In which way can it compromise it? What about the role of government institutions and non-governmental actors within in different cultural set-ups?

A recently published report on implementation of IWRM in about 60 countries worldwide suggested that good water management relates directly to good governance (GWP 2006). Are the criteria for good governance always the same? This article shows how development aid is applied universally across different cultural settings in the world without an analysis of the cultural and organizational settings that largely determine the applicability and feasibility of many projects.

Therefore, the first part of the article will present the concept of Hofstede and Hofstede (2004). Hofstede argues that the way a society organizes itself is very closely linked to its cultural characteristics. For example how the legal, political, administrative and regulatory systems are implemented is heavily impacted by cultural aspects. Modern history shows that while political systems can be overturned fairly quickly the underlying cultural characteristics of a society remain more constant. These characteristics are hard to change, and if they do change it requires long periods of time.

'You can always buy new trousers, but the crease always starts developing in the same place'[1] – is a message which some might consider discouraging when we think of the importance of major changes necessary to shape future challenges. Globalization is bringing companies, societies, management systems and even the behavior of people closer together. Will this mean, however, that we are somehow heading for one single global cultural shape, which defines the way we behave, create our organizations, manage and use natural resources? The answer is certainly: no!

In order not to indulge too much in cultural aspects of different countries, the author concentrates on two different water management systems, the French and the German, in order to show their impact on development assistance. Mali and Jordan serve as country cases used to analyze the impact of development assistance in water reforms by both German and French official development assistance (ODA).

German ODA has always been very focused on delivering advisory service to the water sector but there is an ambiguous situation. On the one hand Germany has a water management culture, which is viewed as contradicting the Dublin Principle based IWRM – at least as it relates to participation and basin oriented water management. On the other hand, German development organizations place much emphasis on the Dublin Principles. However, the shift towards IWRM in German technical assistance (TA) came only after years and decades of working on sophisticated and seldom-used water management plans, which were often shelved and never be looked at after their completion.

Water management in Germany appears to be a contradiction in and of itself. Although it is highly decentralized, it is not necessarily participatory. It is quite effective but not organized according to basins. This is not in line with current thinking on IWRM. It is possible for the German system to be highly decentralized, because it is based on strict framework of regulations.

Which German water management experiences are, then, relevant for other countries? Ambitious water management plans are part of German advisory service in the water management sector and are largely inspired by the high degree of regulation that defines the German management of water resources. Often those plans have proven to be unworkable. Even less ambitious aspects of these plans – such as licensing procedures for abstraction, protective zones around wells, etc. – are rarely implemented with success in partner countries. Why not? As will be discussed later in the article, I believe their applicability of different water resource management plans is closely linked to culture. Unfortunately, this link is either unknown or ignored within German ODA. As a result, it continues to proceed along the same 'beaten track' without really knowing what the result will be. In order to stay in the business, unfortunately, it seems political correctness often overrules impact-oriented effectiveness.

The French System is considered the shining example and archetype for water management according to IWRM, and has for the most part been the inspiration for the Dublin-Principles based IWRM. The main characteristics of the French system are the basin orientation of the *Agences de l'eau* and its participatory structure. In

[1] Quote from Max Frisch: 'Mein Name sei Gantenbein' (1964) (English title: Wilderness of Mirrors), own translation.

the eyes of many water managers around the world, these two defining aspects of the French water management system make it the 'leader' in good water management. A closer look at the French water management system, however, reveals that there are deficiencies in sustainable water management. As is the case in Germany, the state is the owner of the water resources. Yet, since the state administration for allocation and protection of water resources is weak, many responsibilities have been shifted over to the *Agences de l'eau*. As will be taken up in this article, this has resulted in a number of negative consequences. Despite weaknesses in the French water management system it still remains the model for the Dublin-Principles based IWRM approaches and is followed in several countries in the world.

This article will begin by arguing the serious need for a critical review of the IWRM concept as developed from the four Dublin Principles. The impetus for IWRM development was the fundamental weakness of water management institutions and platforms, which is both in part caused and aggravated by poor legal and regulatory enforcement. IWRM therefore placed emphasis on greater stakeholder participation as a way to replace government structures that simply were not working. However, while greater participation is an important and welcome form of democratization, the idea of introducing participatory allocation of natural resources raises many red flags. For one, precautionary principles run the risk of being buried under the avalanche of personal interest based access claims to water resources. As weak governments do exist, however, an entire 'replacement' of directive by participatory approaches appears to be under certain circumstances a reasonable and justifiable strategy, despite the fact that it is by no means a one-size-fits-all solution for any governance reality. The main message of this article is two-fold. First, culture must be considered in water management through appropriate culturally specific and practically relevant measure and mechanisms. One-size-fits-all measures and mechanisms too often lead to predictable patterns of inefficient and failing water management. Further, we must be sure to establish a system which defends the precautionary vision.

2 Cultural Differences

2.1 What Distinguishes Cultures?

Many experiences show how behavioral aspects divide us (or on occasion even bring us together) and illustrate our differences. So, what are the main characteristics of culture? There is no proven system that can identify what exactly such characteristics are. However, the Dutch cultural management specialists Hofstede and Hofstede (2004) have developed a system which is easy to use, and which gives answers in a simplified way to evaluate the most important criteria needed to distinguish between different cultures. To make the system easier to use, they equate cultures with nations. This means of categorization is problematic for younger nations in the developing world, whose national borders were created out of colonial interest and not through self determination principles based upon shared ethnic groupings and cultural history. This was less a problem in his studies as countries from the developed world were mainly focused on and the indicators found were mostly based on widespread consensus. Unfortunately, for many developing countries only regional studies are available (West Africa and Arab World; see Table 1).

In their study, four indicators were identified that proved to be significant for making cultural specifications: The Power Distance Index (PDI), which reflects the importance of hierarchies, Individuality (IDV), Masculinity (MAS) and Uncertainty Avoidance Index (UAI). Table 1 shows the scores for the countries focused on in this article.[2] Unfortunately, specific scores were not available for Jordan nor for Mali. They are summarized in the categories Arab World and West Africa.

A higher score denotes a stronger result. A high Power Distance Index (PDI) means that such a country is strongly hierarchically in its structure and criticizing superiors is neither common nor appreciated. High PDI countries have strict hierarchical structures in which the boss' word is generally more important than existing rules and regulations. Individuality IDV (the opposite end is called collectivism) shows how strong individual behavior is accepted or rejected. Usually in those countries scoring low in Individuality (IDV) one can easily offend people in public by making him 'lose face'. Protection from in-groups is predominant but ambition and success on the other side is interpreted as a pretension of 'leaving the in-group' and sometimes strongly withheld. Masculinity MAS translates mostly through 'showing off success' and sympathy for the weak. Countries which score low in masculinity have societies where it is well to be seen as unpretentious and to sympathize with the weak, translating for example in higher development aid, as it is the case in Scandinavian countries. The uncertainty avoidance (UAI) index shows the degree on which a culture tends to defend itself from others and avoids ambiguous situations. Usually this is visible in the way the society organizes itself with rules and regulations in private and public context. Countries scoring high in UAI have more rules one is supposed to follow. The strong desire of avoiding ambiguous situations does not automatically mean, however, that regulations are enforced. Ironically, in these countries especially, official regulations are followed less and individuals disobey more often. Hierarchies – some times motivated by the interest of powerful groups – can also overrule easily those regulations. In countries that score

Table 1 Based on Hofstede and Hofstede (2004) scores for selected countries and regions

Country/region	Power distance index	Individuality	Masculinity	Uncertainty avoidance
Arab World	80	38	52	68
France	68	71	43	86
Germany	35	67	66	65
West Africa	77	20	46	54

Based on Hofstede 2004

[2] The whole list of scores of around 50 countries can be found in the aforementioned reference.

high in UAI, thinking against the mainstream is more difficult. The educational system describes a high UAI very good. In countries scoring high in UAI teachers and professors usually are authorities whose expertise is not questioned, who more often teach to ask 'what' instead of 'why'.

For the theme of this article, however, only the PDI and UAI index are focused on as the combination of both indicators describes the way institutional and administrative structures are set up very well. For administration, both the degree of hierarchy and the presences of rules and regulations are highly important.

Figure 1 shows how countries are located with their specific UAI and PDI index. The graph can be divided into four quarters 1–4. The first quarter mainly consists of Scandinavian and Anglophone countries, while in the fourth quarter consists of countries belonging to the Roman language family where catholic religious faith is predominant, as well as the Arab world and most countries in Asia. What are the cultural characteristics of these countries with regard to governance? Certainly, a more authoritarian style with many – often unrealistic – regulations. This might seem contradictory, because it is exactly those countries located in the fourth quarter, which are commonly believed to be more flexible and less strict with regard to regulatory aspects. Meanwhile, countries like Germany known to be strict, inflexible and over-organized (the French calls German to be 'carré'= 'squared') do not score that high in UAI. The mystery lies certainly in the combination with the PDI.

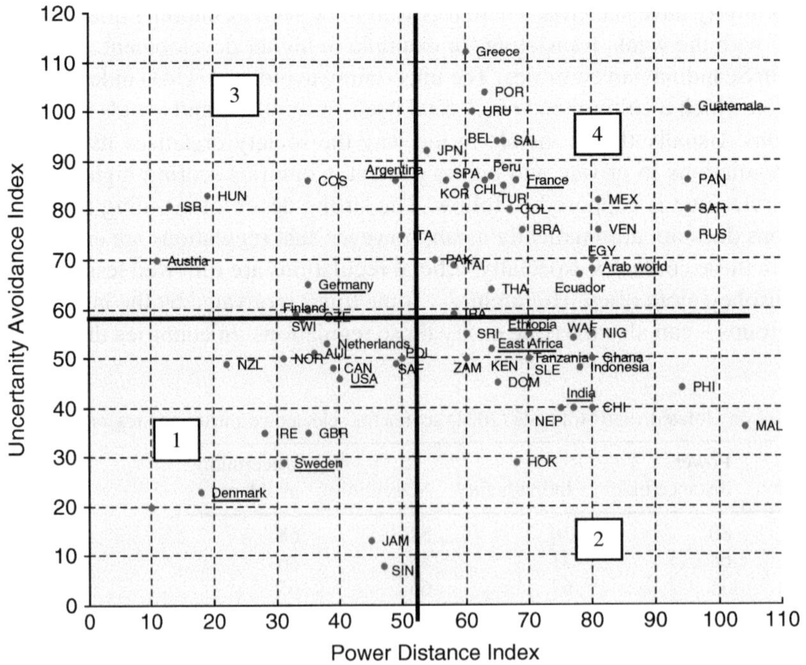

Fig. 1 Ration power distance *Uncertainty avoidance index (selected countries from table 1 have long names); own presentation based on Hofstede (2004)

While countries scoring high in both the PDI and UAI indices have a higher regulatory tendency, this regulation can easily be overruled by hierarchy. The slogan for these countries could be: power is more important than rules. For people originating from countries with lower PDI but relatively high UAI (e.g. Germany) the slogan could be: defend the rule even against your boss. These two visions are stunningly controversial and advisory firms working on German–French intercultural management are very often confronted with exactly this problem of the misinterpretation of the importance of rules stated in agreements. Meanwhile Germans have a strong tendency to stick to the agreements entered into.

French–German joint ventures are thus said to be the most challenging in the world. A very high Individuality (IDV) score in France helps individuals to find solutions through avoiding the inevitable dictation of rules represented in their high UAI. This is another way – besides the overruling by hierarchy – that decreases this dictation of rules. The author of this article experienced a perfect example of this system when he looked for a kindergarten for his 3-year-old child in a French nursery in Bamako, West Africa. While visiting the place the teacher pointed at a sheet of paper on the wall, explaining that this is the timetable for all French nursery schools. The author asked whether this is really the timetable for all French nursery schools all over the world. The reaction from the teacher was one of incredulity: this timetable has of course to be adapted to every single situation. The author asked himself, so why do they put it there. The answer is that they wanted to show their cultural roots. In reality, however, the teachers applied their own system. In a German Kindergarten, nobody would hang a regulation on the wall, which has nothing to do with the place where it was being put. Germany has therefore a very high degree of adapted countrywide regulations into decentralized structures. Decentralizing translates in German culture into having rules down to the lowest level, and then everyone follows these rules. Such decentralized adaptations are – at least by definition – strictly followed up and enforced, and bring the entire nation together like ‘a well-oiled motor’ in which every part has its function and acts according to this.

This example shows that there are general ways of dealing with rules: either you follow them and ‘protect’ them somehow even against superiors (example Germany) or you find ways of getting around them. Here there are two ways: the overruling of rules by an authority and the individual finding their own solution, which put more directly means that they do not obey the rules. According to the scores in PDI and IDV (see Table 1), in France both the two solutions might exist at the same time. In countries of extreme high UAI, there is always a tendency to get around the dictation of rules, because it is more than a member of the society can accept without getting mad. One expression of this is the fact that in countries belonging to the fourth quarter, ‘participatory’ aspects in governance is very popular. Apparently, this can be interpreted as an overcoming of an existing but probably not functional or not wanted strict regulatory system. Participatory approaches and regulatory-oriented (can as well be referred to as directive/repressive) systems are therefore not necessarily opposing; they are to some extent two sides of the same coin. The great difference between them lays not that much in the mere existence of such rules but much more in the implementation of the rules. The balance between the desire for mechanisms avoiding ambiguity and the way they are applied, however, shows much about the cultural aspects of governance.

2.2 The Conflicting Reality of Directive and Incentive-Based/ Participatory Water Management Systems in France and Germany

In order to look deeper into aspects of directive and participatory approaches related to water resources allocation and management the examples of France and Germany are very useful. From a constitutional point of view, both countries organize their water sector quite similarly. Water resources are considered a common good, which consequently means access to water must be controlled by the government through regulatory and licensing procedures. In this respect both countries seem to be very similar. Yet their historic developments have been quite different and the two nations have ended up taking quite opposing ways of water management.

As economic development took shape in Europe during the 1960s its negative impacts on water resource became palpable which gave rise to 'ecological politics'. France and Germany were then in more or less in the same situation. Pollution started to become the visible negative side of economic development, and water resources were specifically in focus.

While Germany developed a regulatory system based on the *Wasserhaushaltsgesetz*, or national water management law, and implemented this system in decentralized local and district authorities, France developed the basin-oriented para-statal system of the *Agences de l'Eau* composed of self-financed quasi-federal organizations. The systems differ most not because of the way they are financed or because one (*Agences de l'eau,* ADE) takes a watershed management approach while the other does not, but they differ rather in how the positive effects on water quality and quantity are intended to be achieved. The German system translated the governments' responsibility – to guarantee interest a sustainable water resource among the public – into a strict regulatory system, where abstraction of water is defined, licensed and controlled. In many cases, so called 'water-and-soil' associations, composed of water suppliers and farmers, apply commonly for licenses and manage the allocated water commonly. Sometimes they manage secondary water courses and assume important management tasks, but depend always of the water administration. The main responsibility for water resource allocation, however, lies within the water department in the administrative district (*Regierungsbezirk*)[3] with its higher authority or in the rural district (*Landkreis*) with its lower authority. Due to the federal system, this can change between the different *Länder*. These authorities grant licenses for the abstraction of water resources according to existing data on groundwater or surface water availability. The work of water authorities, however, is not a

[3] Each *Land* is divided in several of such administrative bodies.

black box. All decisions are subject to publications and open hearings in the concerned water areas.

The development of water management plans as such started only recently as a requirement of the Water Framework Directive (WFD). Subject to such licenses are all abstractions beyond a certain limit, usually 1 cm/day for groundwater. These authorities establish, at the same time, water protection areas around abstraction points like wells and define wastewater treatment plants effluent qualities. All permissions or licenses are subject to strict obeisance and offenders are fined in case of non-compliance.

Although in France a similar regulatory system exists, it is only partially effective. Many discussions by the author with representatives of French government institutions showed that, due to the weak structural set-up of the ministry of environment, licensing of water abstraction is done more erratically than systematically. According to 2002 estimates,[4] only 30 percent of groundwater abstracting wells had a regular license. Until the year 2000, one of the main suppliers of water for Paris, SAGEP, did not have any license for one of its main well fields located approximately 150 km outside Paris.

When discussion began organizing the environmental sector in France, French politicians were aware of the fact that a directive water management system as developed in Germany would probably not work in France.[5] Therefore, the French shaped a system of participatory and incentive based, rather than prescriptive, approaches. The main organizational structures exist within the *Agences de l'eau.* In the incentive-based system, levies for water abstraction – one of the major sources of financing of the ADE – were not compulsory. Not surprisingly, collection was very low in the beginning. As they are now, the levies on raw water abstraction and pollution were intended to subsidize investments that improve water quality and quantity. Industries and communal water suppliers can reduce their fees to the ADE by using water-saving or pollution-preventing technologies. The *Agences de l'eau* had, however, great difficulty in becoming fully effective. After 20 years, collection finally started to increase when industries and water suppliers realized the advantage of using the agencies as a source of funds for investing in more efficient water use technologies or better wastewater treatments, for example.

Since the ADE collected both the user and polluter fees and granted subsidies, the agencies became more of a finance-incentive-motivated environmental investment bank rather than a water manager. This funds-oriented logic was contrary to an impact-oriented logic, which was the intention of the French Ministry of Environment. When it became clear that with this system it was difficult to achieve set targets, the Ministry started to get suspicious about the ADE. Therefore, the Ministry made several attempts to restructure the ADE to become more results-oriented. Since the agencies had become over the last 20 years very powerful and rich institutions, these attempts did not succeed. The most recent development was the passage

[4] Information received during an interview with representatives of the French Ministry for Environment in June 2002.

[5] Thanks for this internal view to M. Pierre-Alain Roche, the former president of the *Agence de l'eau Seine Normandie* and a connoisseur of German culture.

of a new water law in parliament, but it made only minor changes to the agency system. This attempt reflects, nonetheless, the Ministry's desire to be the driving force for water management in France and to overcome the problems related to this parallel structure of the ADE beside the Ministry with its directive organized but ineffective sub-structures.

During the meetings in international watershed basins according to the WFD between French and German representatives, the German commented that the French system was 'rather shady'. From the German viewpoint, surface water management appears to be managed with both the Ministry of Environment and the ADE having a role to play. When it comes to underground water this appears not to be the case. Here the ADE play a less important role and the French national Bureau of Geologic Research and Mines – the BRGM – appears to have much more say. Not surprisingly, a German participant of one such meeting commented that it was completely unclear how 'measures can be applied in such a system'.[6]

One water management story illustrates the French dilemma. In 2001, nitrate concentration in the drinking water of a town in the region of Rennes (Brittany), was high. Consumers complained to the mayor, who referred them to the private company in charge of the municipal water supply. The company itself blamed the French government for not protecting the raw water quality; this, they said, made it impossible to produce good drinking water. The story ended with a condemnation of the French state for not having enforced its own laws. In response, an article in the ministry's internal newspaper blared the headline: 'The Ministry was condemned for not being listened to'. Ironically, this showed the structures impotence at getting into the drivers seat and finally enforcing its own existing laws.

The agencies therefore – in the absence of an efficient regulatory system – took over an important role of the water management in France. The approach developed by the ADE is based on a financing system (as is explained earlier in the paper) and a high degree of participation among users through water committees. On local and regional levels, these user committees make proposals about water and elaborate local and regional water management plans (so called SAGE and SDAGE). These plans are legally binding under French law. The highest public forum is a so-called 'water parliament' that votes on proposals made by the management board and on aspects concerning the agency as a whole.

The French and German cases illustrate the significant differences in how water resources are managed by using either more regulation based or more incentive based models. Both the German and French systems work within their cultural set-ups because their management and governance practices and structures were shaped by and indirectly made to work within their specific cultural contexts. The first analyses under the European Water Framework Directive (WFD), the so-called article five reports, acknowledges that both countries have – through different systems – had a similar level of success in managing water quality and quantity. At the same time, they both face similar problems in areas such as reducing diffuse pollution from agriculture. Interestingly, the WFD will require both systems to become more similar. France will need to become more impact-oriented and Germany more

[6] Member from a water authority in German Middle-Rhine basin.

participatory in water management. For both countries, this challenges the established system much more than it does the achievement of results, because it means overcoming a culturally determined set-up.

The question that we must address is what one can learn from the fact that cultural aspects influence water management? And how does this translate to other countries? In Messner and Scholz (2005) several examples and research results are stated. In the international discourse the participatory system, which – as explained before – has been developed in France, plays a more important role than the directive German one, although the French one is partially developed as a response to a less-than-effective system of governance. This may be, however, simply a German bias. Internationally, the German water system has the reputation of being overly burdensome and old fashioned, since it contains few aspects of participatory approaches and because its water administration is not organized along hydrological boundaries. The long running desire of some German water protagonists to show that participation exists – through the established procedures as explained before and in the watershed orientation practiced in the North Rhine-Westphalian *Wasserverbände* since the beginning of the 20th century – have mostly be in vain. There are two traits that are responsible for the German water management being viewed this way internationally. First, the water administration being organized according to administrative boundaries, and second, relatively little participation exists.

3 IWRM in the International Context

3.1 Revisiting the Dublin Principles

The Dublin Principles discussed and adopted in 1992 mark the beginning of an era designed to overcome the unsatisfactory progress made during the un-mandated International Drinking Water and Sanitation Decade in the 1980s when universal access to water could not be achieved. In the 1992 Dublin and Rio conferences, the main problem identified was the lack of good water management, and not the lack of technology, which was so often stated before.

With the Dublin Principles (see Box 1), a major step was taken to move away from the technological view of water services towards a managerial view. The principles translated rapidly into Integrated Water Resource Management (IWRM), making the terms 'polluter or user pays principle', 'participation', 'watershed management' common terms of water managers worldwide. Many governments started building their policies, strategies and water laws around these key words as well. Several institutions jumped on this train and some, like the Global Water Partnership (GWP), were created especially to spread the IWRM vision around the world. France has been very active in promoting the elements of the Dublin Principles-based IWRM since – as largely described in the last chapter – it has with the *Agences de l'eau* a water management system that displays many of the IWRM elements. But since the French water management system has been described as a system with

Box 1 The Dublin Principles

1. Water is a limited and vulnerable resource, necessary for life, development and the environment.
2. The development and the management of water should be based on a participatory approach, including users, planners and politicians at every level.
3. Women are at the center of the process of water supply, water management and water conservation.
4. For all these – mostly competing – usages, water has an economic dimension. Therefore water has to be considered as well as an economic resource.

some ambiguous elements, it is interesting to analyze which of these culturally defined aspects translate into the general Dublin Principles based IWRM approach, and to what extent French 'problems' find its impact in the approach.

Therefore, in the light of the only mediocre impact in improved water management worldwide it is pertinent to revisit IWRM and its principles. Further, we must evaluate to what extent their influence was positive towards their ability to respond and to overcome the problems of the past. The following chapter will concentrate on these issues, which are related to governance approaches that take into account cultural conditions utilizing the cultural analysis system of Hofstede.

3.2 Water Resource Versus Water Services Management

Until the 1990s, water management was understood mainly as a technology-oriented management of water services with near exclusive focus on areas such as water supply and irrigation. The change from sector orientation to an integrated view, and the consideration of water resource management as the overarching theme, emerged first with the Dublin Principles in 1992 but resulted only quite recently with a separation of water resource from water services management.[7] The 1996 German strategy paper of the BMZ for the water sector was mainly oriented to water supply with few water resource management elements. In the summer of 2007 they made available a revised form with a distinct separation of the water management and water use areas. Water resource management has always been a stepchild of sorts and played a rather insignificant role. Up to now, a kind of competition has existed between water services delivery – mainly water supply – and the IWRM, which is sometimes seen as an ivory tower-like approach. Advocates of the water supply sector approach complain that IWRM is a 'solution in search of its own problem',[8] forgetting that good water management aims at guaranteeing sufficient water for 'their' sector of interest both for today and for future generations. Water reforms in

[7] Refer to World Bank.

[8] Quote form a World Bank expert who prefers to be unknown.

developing countries have therefore concentrated on improved production efficiency in the water delivery sectors rather than in allocation efficiency of water resources. This has to do with the enormous challenge that constitutes changing water management patterns towards a sustainable water management which ensures equal access of all populations.

The most important criteria for good water management are, of course, debatable. One universally acceptable criterion, however, is that effective governance is important. The following country cases describe several aspects of water institutions and water governance. The countries are displaying the same problems but serve as reference to other countries in the world. The description of the country cases is done by considering the following questions:

- Does the governance system acknowledge the importance of water services versus water resource management?
- Is there institutional overlap between water service delivery and water protection functions?
- To what extent is participation a part of the water governance system?
- What mechanisms for water preservation and protection exist?
- How do cultural realities in the country influence water management?

3.3 The Mali Case

This case description shows how overlapping functions and parallel water sector reform processes jeopardize the focus on sustainable water management. The author was an advisor to the Malian water sector between 2000 and 2003.[9]

Mali, in West Africa, is one of the countries with controversial developments in the water sector. The organizational responsibility for water management lies with the *Direction National de l'Hydraulique* (DNH), the water authority, belonging to the Ministry of Energy and Mines. According to the existing laws and decrees the DNH[10] had two main functions: the first was water resource management, which was very much a stepchild since the budgeted funds were not sufficient to undertake relevant activities; the second was the implementation of water supply and sanitation (WSS) projects in small towns and villages. The DNH became the main partner for rural and semi-urban water supply and sanitation projects for many donors. As it happened, tasks related to water supply and sanitation became very attractive for the DNH and its personnel, since cars and office equipment, along with payoffs for key personnel, often sweetened the deal. It was widely said inside the DNH that those without a project were not 'well off'. Meanwhile, a huge number of water supply projects existed but there were relatively few targeting the water resource sector. This was due to the bias of donor agencies towards water supply.

[9] The author was an advisor to the Malian water sector between 2000 and 2003.

[10] Recently a process is underway to split the DNH according to the two functions water service delivery and water resource management.

DNH's economic situation was thus directed by and mainly dependent upon water supply and sanitation projects. However, what could be done with the unrewarding water resource sector? The departments of the DNH related to it were seriously underfinanced. Budgetary allocations from the government were extremely low across the board – not only to the DNH – because of the desperate financial situation of the Malian government. All governmental institutions suffered and were largely dependent on foreign financed projects, with their positive financial impact on the institution as a whole. However, it can be assumed that the reputation of being a 'rich' – one having many projects – government structure translated into even fewer governmental allocation towards the DNH. Although there was a cross-subsidy device in place for water supply projects towards water resource activities, the financial situation did not allow for any activity in water resource management that merits mentioning. Water resource management was therefore minimal.

The DNH enjoyed long-standing support by the French government through several generations of advisors. Those advisors usually had very flawed job descriptions and acted more as personal support to the DNH director. Since there was no real objective-oriented planning in place, the range of activities was rather wide and depended very much on the personal relationship between the director of the DNH and the advisor.

At the end of the 1990s, a GTZ water sector reform project started with the author of this article being the advisor assigned to it. The GTZ project aimed at improving both sectors of activity of the DNH and focused on building up a structure. This allowed conducive and fundable water resource management alongside the improvement of water supply coverage. Over the course of the work, it became clear that structural adaptations were necessary in order to cope with the growing demand in water supply and the need to revive water resource management.

The possibility of separating the water supply from the water management sector was one structural adaptation discussed within the internal working groups installed in the project. But it was easy to note that – if it came to a separation – few personnel would wish to join the water resource sector for the aforementioned reason of the small amount of attention it received by the donor community. Due to this competing situation, the pace of the structural and regulatory reforms was rather low. This changed suddenly and the internal workgroups of the DNH started working vigorously on proposals for structures, bylaws and decrees related to water resource management, making full use of the all of the key terms used in IWRM. Why did this happen?

A new French cooperation project had embarked on a program to support the water sector reform towards establishing IWRM in Mali. The partner, however, was not the Ministry of Energy, but the Ministry of Environment. It was planned and accorded with the Malian Government to build up an *Agence du bassin du fleuve Niger*, in line with the French *Agences* concept. The orientation towards IWRM in West African countries in water politics, however, began already in 1998 with a ministerial conference resolution, held in Ouagadougou (Burkina Faso), to adopt IWRM as a general policy. The Malian Government created the *Agence* and appointed a Director, and the French Cooperation appointed an advisor. It was not surprising that this move provoked a major earthquake inside DNH. The IWRM process, which was slowly but steadily progressing, was challenged suddenly by this parallel structure and activities. The reason for starting to work so fast on the relevant structural and elementary bylaws and decrees was because the newly created water *Agence du basin du fleuve Niger* challenged the existence of the DNH. Several meetings between the Agency and DNH did not result in any agreement about cooperation or division of tasks between the two of them.

The described process overlapped with another donor-driven process, which made reform of the water agencies inevitable. Since 1995, Mali was in a process of decentralization. One of the tasks to be transferred to local communities was water supply and sanitation, which up to now was the mandate of DNH. Therefore, there were actually two battlefields for the DNH, where it had to defend long existing rights. Recently, DNH took one of the first steps towards structural reform of the water sector by hastily started a process of creating a water supply agency. However, donors were not very enthusiastic about this proposal as it was understood to be a move by DNH to save its WSS to the disgrace (and disadvantage) it would face as a result of the empowerment of local communities through the ongoing decentralization process. The position of donors in this regard is, however, ambiguous. For large investment projects, it is always easier to deal with a centralized technical body than with many local partners who have little technical knowledge. This made partners with large sums of money somewhat reluctant towards decentralization processes because it made life for them more complicated.

The Mali example shows some negative aspects in initiating and undertaking water sector reforms, which includes, for example the ignorance of already existing national structures. The process of pushing through the creation of a river basin agency, and thereby ignoring the existence of ongoing reform processes of other institutions, was counterproductive as it leads to double work and institutional competition. Ignoring existing structures and ongoing processes is certainly one of the negative impacts of the approach, which IWRM reform processes tend to have in several countries where they are implemented. This is clearly more a problem of how IWRM is implemented and has little to do with the correct IWRM visions itself. The conflict linked to this will be explained in a later chapter of this article.

With regard to cultural aspects of water governance, we could learn that parallel set-ups were not a problem for the French government supporting two different organizations in quite similar ways. For the Malian DNH it constituted a challenge and a question of survival and was resisted vigorously. For the German technical assistance, meanwhile, this constituted a major logistical problem.

According to the Hofstede and Hofstede system, a high UAI index in the West African Region's and French culture does not mean an inability to live with parallel organizational structures. This is true for neither the French nor the Malians, as long as a controlling force to put the ambiguity in order exists (i.e. power). Since Germans do not rely too much on the importance of power; structures, rules and regulations have to be more precise, consistent, and respected, thus making them much more important than in many other countries. Although this pragmatic way of dealing with regulatory structures is different from the more bureaucratic approach of the French, the UAI index with 65 for Germany is relatively high.

3.4 The Jordan Case

Although several tools for good decision-making – many of them put in place through German TA – do exist in Jordan,[11] this case description also shows how overlapping functions make effective and sustainability-oriented water management systems almost impossible to achieve.

Jordan is one of the countries with the highest water stress. Only 180 m^3/year/percapita of renewable freshwater resources is currently available. The Jordanian government has, with the help of numerous donors, embarked upon improving the efficiency of water uses as well as designing and implementing a water resource regulatory framework. A demand-oriented water policy was issued which referred to some extent to the IWRM principles.

The main user of water resources is the irrigation sector. Although the primary portion of raw water used for irrigation is surface water (in the Jordan Valley), in the highlands only groundwater is used for irrigation. Water supply of the main cities relies on groundwater and the water of the King Abdullah Canal using the water of the Jordan and Yarmuk Rivers. A widespread piping system supplies water for this purpose to the center of the country, where the largest cities are located.

Water supply in Jordan is very irregular. Due to insufficient quantity of water delivered by the water works, most parts of the cities have only one to two service days per week. Loss of water through leaking pipes and water theft aggravates this situation. Underground and roof-tanks have to be built in every house in order to store water to cover the period without water supply. As usual in water scarce situations, supply among the different strata of society is far from being equal. Both in water supply and irrigation, there are privileged users, who usually belong to better-off levels of society. With regard to the equal distribution between current and future generations, there is a strong priority given to the current one. In general, the overuse of renewable groundwater resources for both drinking water and irrigation is about 80 percent in 2004[12], which is leading to dramatic drawdown of groundwater tables.

With this drastic situation in mind, there are two questions to be asked:

1. How does this fit with the positive water resource policy of Jordan?
2. How do the people react to the scarcity?

First, let us consider the people. An observer gets the impression that people are accustomed to that situation. There is no strong tendency to save water, neither in household consumption nor in industrial production. A consciousness toward saving water is virtually non-existent. Second, the government's behavior: the law that controls the use of groundwater for irrigation in the highlands is virtually ineffective. The reason is that it is difficult to enforce this law, because of the strong and well-connected irrigation farmer lobby. The law imposed fees on abstractions from groundwater but only for quantities larger that 150,000 m^3/year. Beside that, the fees

[11] The author was team leader for the design of a new water program in the end of 2005.

[12] According to a GTZ water management project (October 2005).

were insignificantly low. Nevertheless, there was protest in the irrigation lobby and fees had to be brought down to completely negligible levels.

Responsible for collecting those fees is a department inside the Water Authority of Jordan (WAJ), the biggest (and state-run) water supplying company. WAJ belongs to the Ministry of Water and Irrigation and runs the widespread piping system for bulk water to the city water supply schemes. When the author asked for details about funds (amount and destination) he discovered that the total amount was so small that it did not even appear in the department's accounting system. Only those funds paid by a semi-privately managed water supply and irrigation system in the Red Sea town of Aqaba was identifiable. Making matters even more controversial, in this case the WAJ applied the raw water abstraction fee, while it did not do so for its own abstractions. However, this would not have made much sense since WAJ used the funds out of its own budget and for its own expenses, and did not specifically earmark them for water protection or preservation measures.

When the author stressed some of the issues mentioned above to the Jordanian authorities focusing on stronger law enforcement in the framework of a new German Water program, their reaction was somehow rude. The proposal to the water program – for its development the author acted as a team leader – of an additional component of technical advisory service for the water policy sector was rejected. The reasoning was that foreign countries should not interfere with internal political affairs of the country. Discussions about what to do to improve the growing water demand were instead focusing on increasing water production. Water supply driven vision had gained again over water demand management.

This was not at all surprising, given the structure of the Jordanian water sector. As previously indicated, water supply, water for irrigation and water resource management were located in one single ministry. Enforcing the groundwater abstraction law would have required the head of the Ministry for water and irrigation to act against the interests of his own clientele! Since this was unlikely, the groundwater decree was bound to fail. A solution proposed by the author to the Jordanian-German water program to separate water service delivery functions and water resource protection was rejected as completely unrealistic. One possible actor who could assume this role of advocacy of a precautionary vision in water management – the young ministry with a rather inexperienced Minister of environment – was said to be much too weak to play a role with this regard.

With the Jordan case, the lesson learnt is that the directive approach – consisting of water laws, regulations, decrees, bylaws water management information systems and plans, for long a part of the German technical assistance tradition – was completely overruled by influence-dominated, powerful decision-making. Despite the efforts of the German Technical Assistance (TA), Jordan's own policies, regulations and management information tools were erratic and unable to contribute to a more sustainable water resource management. The Jordanian side constantly tries to direct foreign aid into production efficiency, where significant success has been achieved in reducing water losses in the water supply systems in several cities. German TA made various attempts and agreed finally with the Jordanian side in

2005 on a mutual plan for implementing better production efficiency and allocation efficiency. Time will tell if this agreement is successful.

The Arab World score for Power Distance is 80, quite far from Germany's score of 35, while the scores for UAI are very close (65 and 68). This shows the similarly high desire for structures, a similarity then overruled by power (which is likely linked to in-groups). IDV (see Table 1) of the Arab world is very low, at 38, compared to Germany, at 67. This means that collective reasoning and behavior is quite significant and influences widely decision making processes. Well thought through ideas, translated into politics without links to influential strata of society – and future generation and environmental requirements definitely do not belong to this group – do not have much chance to become effective. It is therefore not surprising that the prospects of long-lasting management of the scarce water resources are not bright.

3.5 *Lessons Learnt from Country Cases with Regard to IWRM Processes*

3.5.1 The Wrong Understanding of Integration

In the last chapters we have seen that the Malian DNH 'integrated' two sectors: water supply and water resource policy development and implementation; and that the Jordanian Ministry of Water and Irrigation 'integrated' three: water supply, irrigation and water resource policies and implementation. One of the central aspects of the Dublin Principles-based IWRM is indeed 'integration'. Everybody working in the international water sector knows that this symbolizes the overcoming of the former sector-oriented view. Figure 2 shows the three steps from sector to integrated and sustainable water management.

To the left we have a situation characterized by everybody making use of the available water resources as needed. There is neither coordination nor control, and

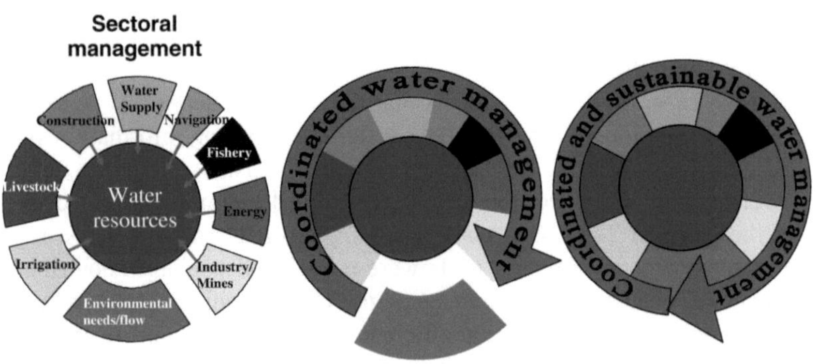

Fig. 2 The three levels of water management.

hardly any management in allocation. In the center, we find the situation that is characterized by a coordinated action among the users (bowed arrow). Users participate and coordinate to reconcile opposing interests around the same resource. The right circle demonstrates a coordinated and sustainable vision, because environmental interests are part of the integration.

This role of advocating for these environmental needs is in most countries given to the government. This holds as well true for both country cases, Mali and Jordan. The question is how the relevant institutions can play this role of being numbering opposition to those in the water service delivery function (water supply and/or irrigation), when at the same time they represent exactly that sector. The interests of those serving water service functions and the requirements of sustainable water management can be, are often are in complete and direct opposition. Therefore, it does not make any sense; indeed it has often proven to be counter-productive, to 'integrate' any of the water service functions structurally with water protection functions. It would make discussions much clearer to talk in terms of 'coordinated' rather than 'integrated' water management. That would not so easily lead to the creation of overlapping and contradictory functions in individual water institutions. This will be, however, quite hard to change.

In developing countries, there is certainly a stronger need for economic development, which more easily justifies the neglect of natural resources protection. Comments from people in developing countries can be interpreted in this way: we are not wealthy enough to care for the environment as much as rich countries can do. Although understandable, this is a fatal and short-sighted position for the economic development of a country, since a sustainable development is based on the vision not to overexploit natural resources in order to guarantee the living conditions, through the precautionary use of natural resources, of future generations as well. This is no new insight but it is still topical.

An analysis of the organizational set-up of Mali and Jordan shows this overlap of functions in the same institutions. This means that one institution has to advocate for water users, as irrigating farmers, water suppliers and so on, while at the same time controlling – which in most cases means limiting – their use of water resources. According to the experience of the author, precautionary approaches to conserve and protect water sources always lose out when they are represented by an institution with overlapping functions that make it compete against water service delivery functions. In Germany, there are in some *Ländern* certain overlapping functions between water management and agriculture, but due to the strictly defined regulations for each sector and the cultural aspect of Germans sticking to those regulations more typically than the French, the problem does not show. In France, however, environmental functions at the *département* level are sometimes taken up by the industrial units with quite negative consequences, as representatives of the Ministry of Environment continuously blare.

Surprisingly, the current IWRM debate does not address this issue at all. This is probably because the current IWRM approach does not focus enough on the governments' role and due to the structural dilemma in water management described above.

3.5.2 Why a Participatory Approach Alone does not Work

Another key word of the Dublin Principles-based IWRM approach focuses on participation. Participation is often linked to decentralization or democratization processes. However, there are several problems that can arise. For one, participation does not always empower the population. Neubert and Herrfahrdt (2005) describe an experience in Uzbekistan where a participatory set-up was 'high jacked' by local leaders who acted in an authoritarian way. Decentralization can therefore create 'smaller kingdoms' which differ little from the ancient and (likely) authoritarian way of governance. Decentralization could also result in more corruption (Easterly 2001), since the responsibility for the common interest gets lost more easily. This can be dangerous for sustainable water management or any other natural resource management. Where participatory systems are put in place to replace somewhat less effective governmental water management, they may also be counterproductive. Who, if not government structures, can ensure the common interest of managing water resources in a sustainable way? The state and government have, therefore, the most important role in allocating water with consideration of sharing the benefits emerging from water among all the users. Sharing benefits means trying to achieve a win-win situation for all users involved, including those of the 'voiceless'. Participatory structures can only play this role to a certain extent because stakeholders look mainly at their own interests. This means that the 'environmental flow requirements' ('environmental needs/flow' segment in Fig. 2) are not represented, or – in other words – that a precautionary vision is unlikely. Fauna, flora and future generations cannot take up 'their' interest in a participatory discussion and defend it against the other users. Protection of water resources thus becomes a side effect of user participation. The 'voiceless water users' need a speaker or an advocate. This could be environmental groups, but in developing countries they have very little power and influence and thus cannot be a strong advocate.

Governments in developing countries are often attracted by participatory approaches. Yet it seems they are often attracted to these approaches without being ready to make the real reforms necessary, in which they themselves have to play an effective role, to make good participation and democratic structures work. This applies to the majority of countries that are located in the fourth quarter of Fig. 1, which according to Hofstede and Hofstede's cultural assessment system displays the relation of power distance and UAI index. It appears that – similar to the situation in France – ironically the parallels of hierarchical structures and tendencies towards participatory approaches seem to coincide quite well. Possibly the latter is an internal cultural respond to the first. It seems as well that the higher the desire to avoid ambiguous situations through regulatory and hierarchical structures is in a system, the more likely the systems tend to fail. This implies a situation that is difficult to bear from a German point of view: a highly overregulated but inefficient or even ineffective regulatory framework that somehow needs participatory grass root-structures to at least have some success.

The focus on participation – although welcomed in the light of increased democratization – under these considerations can become dogmatic and even irrational if somehow accepts the withdrawal of government administration of assuming its role in resource management. Current processes in implementing IWRM approaches in several countries fall way short for this reason. The problem is that too few result-oriented sustainable water management have been identified as one of the critical problems of the French water management system by the *Agences de l'eau* – as described before – and therefore those shortcomings are repeated exaggerated and ignored when translated in IWRM.

4 Critical Review of the German Role in Water Sector Reforms

4.1 The German Dilemma

German ODA can however not plays the role of being an alternative to the deficiencies described above. It struggles are on the other extreme of the scale. Assistance to the water sector through German institutions has concentrated over long time on the whole range of regulatory tools for water management, as well as advising on the elaboration on water master plans, national water policies, job description, technical trainings and strategies. This was very much related to the belief that rules and regulations work everywhere just as they do in Germany itself. Water master plans became for some years, and in some cases still are, the synonym for German water sector policy advisory service. Millions of DM and EURO went into the most sophisticated databases and GIS systems. Few developing countries, however, were able or willing to use these systems and to draw conclusions from the aggregated and analyzed data. Even fewer were able to do what is much more challenging: to enforce the relevant decisions emerging from the available information.

Although the administrative system in Germany is highly decentralized and formally allows a high degree of political involvement of a normal citizen in political decisions, Germany does not have a high reputation in participatory processes internationally. This can very well observed in the IWRM discussion, where Germany is often blamed for being an old-fashioned country with no competence since it lacks river basin orientation and participatory aspects. Though the North-Rhine Westphalia river basin organizations and the wide spread associations culture in the agriculture sector are examples of both river basin orientation and participatory mechanisms that can disprove this myth, they are not, however, well known internationally.

With the upswing in support of the Dublin Principles-based IWRM approaches, however, German TA shifted strongly towards this mainstream instead of jumping on the train with its own experiences. Some of the German experiences with regard to the importance of a well structured government system are, according to the author, very valuable because, as the article illustrates, some elements of the Dublin Principles-based IWRM approach need to be revisited in order to incorporate elements that are more effective.

4.2 The Main Messages for a Culturally Sound Water Management or Revising the Dublin Principles-Based IWRM Approach

The following points resume the message of this article in addressing the currently widespread Dublin Principles-based IWRM approach in water resource management:

Apply policy approaches in the light of the historical and cultural reality of a country. It is critical to spread an approach without analyzing existing structures and their capacities and capabilities. Building up parallel structures in order to overcome less-to-effective governance systems only leads to double work and frustration. Cultural characteristics should be assessed using a system like the one developed by Hofstede system described in the first chapters or any other relevant systems (example: Trompenaars 1998). Although the importance of organizational development – usually equaled as an approach to cultural diversities – has been recognized for a long time, it is often done very schematically and only using cultural assumptions on organizational aspects that do not apply in many partner countries, and sometimes do not even apply in the home country. So, do not make partner countries a playground for principles, which would never work in your own country, even if they are fancy and fashionable. Training, job description, plans and other typical German TA outputs without considering cultural realities are useless. Or with other words, TA that ignores cultural motivated behavior patterns must fail (example Jordan).

Avoid a conflict-of-interest situation in water institutions. Organizations responsible for water preservation should not be at the same time responsible for any water use services like irrigation, water supply, energy production etc. This aspect is ignored in many developing country organizations. Integration of water uses should not be understood as merging water responsibilities in one ministry or organization. The impact on water resources will always be negative since water service delivery aspects – excluding environment as a user – are usually more 'attractive' for developing countries. Problems of future generations seem to be far away, since so many problems already exist today. The optimum solution would be to find a body responsible for IWRM above technical line agencies or departments (inter-ministerial commission, prime ministry, national water commission, etc.), that coordinates the work of the line agencies with regard to the use of water resources and deals with the sometimes conflictive trade-offs of interests. Among these bodies representing the user must be one representing the 'voiceless users' environmental needs.

In negative regulatory and governance framework conditions, it should be investigated whether it makes any sense to continue with the cooperation. If the demand of the partner focuses only on production efficiency and rejects allocation efficiency improvements, the cooperation should be thoroughly analyzed and probably halted. A serious threat to any positive impact of water policy advisory service could be high degrees of corruption and/or nepotism that virtually deprive the central government from any regulatory power.

Do not make participation a dogma. As described in this article, focusing on participatory stakeholder processes alone while ignoring the advocacy of the 'voiceless users' will jeopardize the precautionary vision of water management. Although most welcome this under the pretense of improving democracy, water user interests typically have short-term and personally motivated visions. The leveling out of participatory and directive/repressive elements should be done on the ground and based on the regulatory capacities and possibilities of a country. It is critical to establish participatory approaches without government involvement if one wants to replace an inefficiently working existing government structure. However, under certain conditions of so-called failing states, there might be no other chance other than to work without any government structures since they simply does not exist. In all other cases, interaction between government and participatory user structures should be well defined and part of water management procedures.

Do not make watershed orientation a dogma. The second Dublin Principle is widely interpreted as principle that water management is best done according to natural boundaries, the watershed. Although hydro-'logically' logical, it is not a pre-condition for good water management (see the German example). It helps to increase understanding of the direct link between upstream and downstream users even beyond national boundaries. Overemphasizing this aspect can bind efforts in other more important aspects. It is instead necessary to enhance management concepts that allow for an overarching and implementable concept of sustainability. Watershed orientation is a useful add-on but no means in itself.

The way forward. The main questions to ask in the light of the mediocre impact of water management processes around the world are whether technocrat based water governance approaches are replicable anywhere in the world and whether blue print oriented approaches on participation, watershed management, polluter pays principle and others are replicable in any society?

In order to look forward the following system is proposed, that is based on an analysis that characterize the different approaches in management of public goods and therefore can be used as well in water management. It is based on a three edge characteristics of governance approaches (see Fig. 3). These three ones are schematically displayed in the triangle on this page, with the extremes, 'legal-based', '(dis)incentive based' and 'knowledge-attitude-practice based' approaches, that are described as follows:

- **The knowledge–attitude-practice based approach**: This approach targets a comprehensive and well informed stakeholder society that takes water related decision on water bodies they are using. The synonym for this is the participatory approach that is reflected in the second Dublin Principle. Participatory approaches are widely believed to be a key to good water management and appear therefore as a cornerstone in most IWRM approaches. The 2000 adopted European Water Framework Directive has for example a strong cornerstone in promoting the participation of the general public in water affairs. There are several other countries that – following donor assistance – have embarked on participatory approaches although their governance systems do not provide sufficient democratic elements allowing for that.

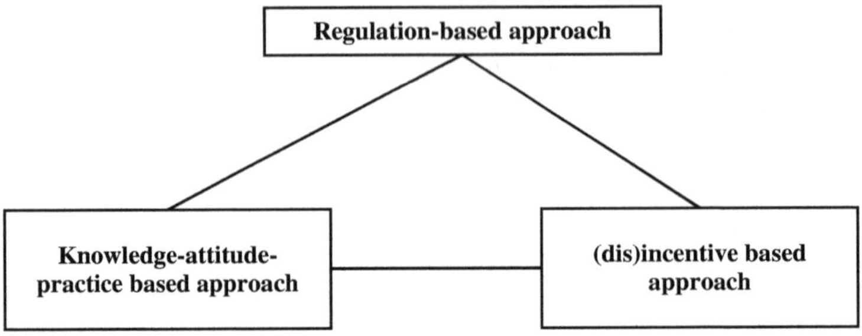

Fig. 3 The approach 'triangle' for water governance

- **The incentive or disincentive driven approach:** Widely reflected in the 'polluter pays-principle' this approach tries to manage water resources by making abstraction and pollution expensive while rewarding water saving or less polluting water uses monetarily. It is reflected in the forth Dublin Principle 'Water is as well an economic resource'. Application of more cost efficient water uses and cleaner water based production therefore gains monetary attractiveness which in turn effectively influences individual, industrial and political water related decision-making. The French *Agences des Eaux* work mainly with this approach.
- **The regulation-based approach**: This means water management is considered a government task and its distribution subject of laws, by-laws and so on. Non-compliance is fined. A typical representative of this technocratic approach is the German water management system although it shows as well aspects of participatory and (dis)inventive approaches.[13] The first Dublin Principle could be indirectly understood as an interpretation of this, but is certainly not widely understood that way.

According to their cultural reality, countries might tend more or less to one corner of the triangle. The system allows locating countries somewhere in between the corners. An important precondition has to be clarified, however, before using this systematic approach for helping developing a water management system in a specific country: a clear political commitment for the process and a security that continuing engagement in those procedures, which are deduced from that exercise and are not object of disturbing factors such as nepotism and corruption, are critically needed.

5 Final Comment

IWRM is a means to an end and not a means in itself. However, the religious ferocity with which it is nowadays sometimes promoted, nourishes the critical voices of its conceptual basement. It is time to review some of the ivory-tower like criteria that

[13] Water supply is one of the most expensive in the world due to extreme water saving measures with results in water losses of less than five percent.

are used in an inflationary way. It is necessary to revise criteria – the mediocre impacts of Dublin-based IWRM approaches worldwide are the best arguments for doing so. Continuous concentration on doing the things right prevents us from asking ourselves whether these are the right things to do and not only 'are we doing the things right?'

The right balance of directive and participatory approaches is crucial, as are organizational structures that do not hinder themselves by unsolvable internal conflict of interests. Cultural aspects play a long-ignored, crucial role in making governance systems work. No system is a silver bullet.

Cultural differences, and the ways to get around them, are meant to bring us together, not separate us. The more we understand the specific behavior of other cultures, the more we are able to appreciate, enjoy them and work successfully together. No one should try to forget his roots and be somebody else. Remember Max Frisch: 'Creases always appear in the same place, although in new trousers.'

References

Easterly W (2001) The elusive quest for growth. Economists' adventures and misadventures in the Tropics, Cambridge, Mass

Frisch M (1964) Mein Name sei Gantenbein, Suhrkamp, Frankfurt a. M

GWP (2006) Setting the stage for change. Second informal survey by the GWP network giving the status of the 2005 WSSD target on national integrated water resources management and water efficiency plans. Stockholm

Hofstede GH, Hofstede GJ (2004) Cultures and organisations: Software of the mind, New York

Huppert W (2006) Situation conformity in water governance, Presentation held at the Center for Development Research (ZEF), Bonn, May

Messner D, Scholz I (eds) (2005) Zukunftsfragen der Entwicklungspolitik, Nomos, Baden–Baden

Neubert S, Herrfahrdt E (2005) Integriertes Wasserressourcen-Management. Ein realistisches Konzept für Enwicklungs- und Transformationsländer? In: Messner D, Scholz I (2005) Zukunftsfragen der Entwicklungspolitik, Nomos, Baden–Baden

Trompenaars F (1998) Riding the waves of culture. Understanding diversity in global business, 2nd edn, New York

Part III
Politics of Water Supply and Sanitation

Private Sector Participation in Water Supply and Sanitation

A Contribution to Attaining the Millennium Development Goals or Merely the Export of Old Solutions?

Thomas Kluge and Ulrich Scheele

Abstract Billions of people in developing countries do not have access to clean water and sanitation services. Against this background, the MDGs represent very ambitious targets for the coming years. In the light of the economic and financial problems of the developing countries, strengthening the role of the private sector in the provision of infrastructure seems to be an important part of any strategy to fight poverty. In recent times however, the privatization strategy has increasingly come under pressure, resulting e.g. in a declining interest of large international corporations. Apart from regulation problems the question is, whether the existing centralized network based infrastructure models found in industrialized countries can form the basis for a successful international water policy strategy. Alternatives can be found in decentralized systems which are adaptable to local conditions and flexible enough to adjust to new situations, and at the same time being sustainable in an ecological sense. To deal with future challenges not only in developing but also in industrialized countries, German water utilities should concentrate on such infrastructure models with a focus on sustainability instead of exporting traditional solutions to infrastructure problems.

1 Introduction

"Pipe dreams" is the title of a current report dealing with the privatization of water supply and sewage disposal in developing countries which addresses, among other things, the unfulfilled expectations of water companies operating internationally.[1]

The obvious problems that have occurred in many cases of privatization in recent years are also reflected in a marked decline in private investment in the water sector. The growing skepticism towards privatization takes on special significance,

Thomas Kluge and Ulrich Scheele
e-mail: kluge@isoe.de

[1] Public Services International (PSI) Pipe Dreams: The Failure of the Private Sector to Invest in Water Services in Developing Countries, London.

W. Scheumann et al. (eds.), *Water Politics and Development Cooperation*,
doi: 10.1007/978-3-540-76707-7_9,

especially against the background of the very ambitious Millennium Development Goals (MDGs). Concrete targets for the development of water supply and sanitation were first set at the Johannesburg Summit on Sustainable Development.

The private sector, despite all criticism, can and should retain a prominent role in attaining these goals. However, the results of previous privatization practices in both developing and industrialized countries, which have generally been very ambivalent, necessitate a critical review of past approaches at the very least.

This also appears to be of importance from a (German) national perspective, not only because the Federal Republic of Germany is the third largest donor in the water sector, but also because the demand for stronger international engagement by German supply companies has become more vocal in the course of the current debate about the future development of the German water industry's organizational framework. This call for a strengthening of German companies in the global water market is, among other things, a central component of the so-called modernization program; the German water sector sees this program as a way to position itself with respect to far reaching privatization and liberalization demands (Bundesregierung 2006).

One can ask with justification whether an organizational model that was rejected in the home market is at all capable of being an export success. However, it appears to be more important in the longer term to clarify the question of what exported solutions might look like if they are to respond to the previous privatization experience while simultaneously making use of the special advantages of the German municipal supply and disposal systems. Decentralized solutions are most likely to play a role here, as are special institutional structures that safeguard the public interest. This article attempts to address this problem by relating the international debate on privatization to the national debate about the future of the German water industry.

The first section will briefly consider the coming challenges presented by the international water crisis and the Millennium Development Goals, which should be the primary determinants of investment requirements. With this as a starting point we will explore the background, motives and models for privatization in order to then address the deficits and problems that have been experienced to date in greater detail. The final section will then consider the potential role of German companies in the global water market. This will serve primarily to throw light on new, alternative systems.

2 Water Crisis and the Millennium Development Goals

The lack of or inadequate management of water resources is considered one of the critical factors that will negatively affect the chances of sustainable economic and social development in many parts of the world in coming decades. Many regions are characterized by a shortage of water, with the problems being aggravated by pollution and excessive use of existing water resources. The growing demand for drinking water and the demands on water from industry and agriculture are causing conflicts over water resources to increase. Furthermore, general political stability in some parts of the world is deteriorating as a consequence of conflict between countries that must use common water resources.

The special problems related to water are primarily the result of unequal regional distribution of this useful resource and, particularly in the developing countries, the lack of or inadequate financial, institutional and technical foundations for the construction of water supply and sewage systems.

According to the United Nation's World Water Development Report, approximately 1.1 billion people lack access to an adequate water supply, and around 2.4 billion have no access to sanitation (WWAP 2006; UNDP 2006). Water-related illnesses are among the most common causes of death in developing countries. Each year 1.8 million children die from diseases caused by unclean water and insufficient sanitary facilities. "At the start of the 21st century, unclean water is the world's second largest killer of children." (UNDP 2006, Foreword).

Given the status quo, the problems will become more acute and will require an enormous political and financial effort to turn things around. The World Summit for Sustainable Development in Johannesburg in 2002 declared the sufficient supply of water to be one of the central challenges of the 21st century and set very ambitious goals for the first time. These include, among other things, cutting the number of people without access to water supply and sewage disposal services in half by the year 2015.

Table 1 summarizes the importance of an adequate water supply for the attainment of the Millennium Development Goals.

Meanwhile, a multitude of studies, reports and guidelines have been published containing recommendations for implementation and action (UN Millennium Project 2005). Greater participation on the part of the private sector is seen as necessary not only to provide financing but also as an important step in the direction of more efficient and transparent organizational structures. In particular, the public utilities serving cities and agglomeration areas in developing countries are typically viewed as performing poorly (inefficient structures, high leakage rates, poor quality and poorly assured supply, no cost-covering pricing policy).

The achievement of the Millennium Goals is, moreover, not merely a technical problem; it encompasses legal as well as institutional reforms, and, in particular, the question of financing. The financial implications of the Millennium Goals were first comprehensively discussed at the International Conference on Fresh-water in Bonn in 2001. A report, written under the auspices of Michel Camdessus, the former director of the IMF, was presented at the 3rd World Water Forum in 2003. This report dealt systematically with all aspects of financing (UN Millennium Project 2005). Although the report confronted the question of the necessary political and institutional reforms and left the question of privatization to the individual countries, the Camdessus Report met with considerable criticism from the very beginning (Hall 2004).[2] The report pointed out that reaching the Millennium Goals would require

[2] Cf. also: Stakeholders' responses to the recommendations of the Panel. http://www.worldwatercouncil.org/index.php?id=552. Cited 20 Jul 2007.

Table 1 Importance of water supply for the attainment of the MDGs

MDG	Contribution of improved drinking water supply and sanitation
Goal 1: Eradicate extreme poverty and hunger	- Higher productivity results when sickness in adults and children is reduced due to safe drinking water and adequate sanitation - Reduction in health costs which are a big burden to poor families - Healthy people are better able to absorb nutrients in food - Higher productivity due to reduced water collection time
Goal 2: Achieve universal primary education	- Better school attendance due to better health and reduced water collection time - Separate school sanitation facilities for girls and boys increase girls' school attendance
Goal 3: Promote gender equality and empower women	- Improved health, less care-giving for sick children and reduced water collection time give women more time for productive activities, education and leisure - Water and sanitation closer to homes reduce risks of assault for women and girls while collecting water or searching for privacy
Goal 4: Reduce child mortality	- Improved sanitation and drinking water sources reduce child mortality
Goal 5: Improve maternal health	- Improved water sources reduce labor burden and health problems, reducing maternal mortality risks - Safe drinking water and basic sanitation at home and in health-care facilities ensure basic hygiene following delivery
Goal 6: Combat HIV/AIDS, malaria and other diseases	- Improved sanitation and drinking water sources, and water management in human settlements, reduce water-bound diseases and transmission risks of malaria and dengue fever
Goal 7: Ensure environmental sustainability	Adequate treatment and disposal of waste water contributes to ecosystem conservation and to the reduction of the pressure on freshwater resources Careful use of water resources prevents contamination of groundwater
Goal 8: Develop a global partnership for development	Development agencies and partnerships recognize the fundamental role of safe drinking water and basic sanitation in economic and social development

Sources: Kraehenbuehl and Johner 2004, 3; Willoughby 2004, 3

a doubling of yearly investments over a 20–25 year period;[3] it also dealt with individual financing instruments, but did not go into any details. Central problems considered in the Camdessus Report are the numerous risks faced by investors in developing countries; a description of strategies and mechanisms for risk reduction follows. As a result, critics see the report as a prescription for privatization. However, the report has been criticized since it is primarily concerned with large-scale infrastructure projects and the construction of water and sewage systems in urban areas and in doing so ignores the central problems: how, first and foremost, the poorest countries in the world can gain access to financing sources, and how project financing for the poorest population strata in rural regions can be guaranteed. Since the report's submission numerous initiatives have been started which could be used to fill these financing gaps.[4]

At the beginning of the 1990s, the call for the privatization of companies within state grid-bounded infrastructure industries met with wide interest and found direct support in the policies of international development organizations and donor institutions. However, the call for private solutions cannot be explained by economic reasons alone, but only within the context of the underlying political-economic debate of the 1980s and the 1990s. This totally new orientation may be described with a few catchphrases: deregulation, in the sense of a reduction of state influence; a new balance between market and state; a new allocation of risks; and a long, slow farewell to the Keynesian model of the welfare state.

These developments with regard to policy for developing and newly industrializing countries can be seen, for example, in the so-called Washington Consensus. This doctrine, preached for the most part by the World Bank and the International Monetary Fund, calls for comprehensive reforms aimed at macroeconomic stability, primarily through the dismantling of protectionist measures and the opening of commodity and capital markets; the aim of these reforms is to promote not only economic growth in these countries but, at the same time, to contribute significantly to fighting poverty. Key building blocks of this consensus include deregulation and reduction of bureaucracy, dismantling of subsidies, strict budget discipline, as well as a very explicit privatization policy and the opening of the countries involved to foreign investment. As a result, the field has been prepared for large international multi-utilities, who see in water management a huge market that promises long term profits.

However, common goods must first be turned into marketable products before infrastructure markets can be opened up. This commodification has provoked an intensive and controversial discussion, particularly in the field of water management, concerning whether water should be seen as an economic, marketable commodity or as a human right. Many organizations, including some in the UN, but above all

[3] Investment needs are estimated on approximately 180 billion US$; this sum also covers investments in new programs for the improvement of industrial and agricultural water use (Winpenny 2003, 13). Several cost studies were analysed in a recent study by the World Water Council (cf. WWC 2006).

[4] The focus is on such models exploiting all financial sources at the local and regional level (cf. van Hofwegen 2006).

church groups, environmental organizations and opponents of globalization have called for a human right to water and are united in their opposition to privatization in this sensitive area of provision of basic human goods. They are demanding a human right to water that is legally enforceable and the allocation of public funds to guarantee that even the poorest groups have access to a minimum water supply level and sewage disposal facilities.

However, this fundamental conflict between the two positions, which has continually marked many international water conferences over the last few years, can be resolved, if one is prepared to see the human right to water as a complementary building block of any effort to attain the Millennium Goals and not just in opposition to the economic interpretation.

Privatization programs will have to place more emphasis on the associated need for regulation; at the same time, questions of allocation will become central, and indeed starting with the design phase of the privatization model. This will necessarily have an effect on price and fee policies, the design of contracts, the choice of technology and the design of regulations. A strong emphasis on a human right to water will also mean that public financing will remain essential even where private solutions are favored. We will now look more closely at some of these points.

There was a marked increase in the number of cases of privatization and associated investment by the mid-1990s. This development is also reflected in the water sector, mainly in countries in which state or municipal companies provided the water supply and sewage disposal services exclusively. Since the mid-1990s, there has been a strong decline in privatization projects, whereby a not insignificant part of the remaining investment was accounted for by the prosperous developing and emerging markets.[5] The search for the reasons for this development, especially in the water sector, requires intensive examination of privatization concepts and experiences resulting from them.

3 Privatization and Liberalization in the Water Industry

The privatization of infrastructure services formerly provided by state, i.e. public companies or administrations, had become a worldwide phenomenon by the beginning of the 1980s. Privatization became the centerpiece of the global economic-political reform.

In spite of the large amount of literature on this subject and the extensive list of documented examples of privatization, the definition of the concept of privatization often remains unclear and ambiguous. Privatization of the infrastructure services, in particular, can take very different forms; the models differ mainly depending on

[5] For the development of privatisation in the network industries of the developing countries, cf. among others: Private Participation in Infrastructure (PPI), Project Database of the World Bank (http://ppi.worldbank.org/); with a somewhat limited geographical area: http://www.rivatisationprivatisationprivatisationbarometer.net; for summaries, cf. also Kikeri and Kolo (2005); with special reference to water cf. Balance and Trémolet (2005).

which functions and elements of the value creation chain are provided by the private supplier, who owns the facilities and how the facilities are transferred following expiration of the contracted term (Table 2). Privatization is sometimes spoken of when the only change involved is a change in the form of organization, i.e. a private legal form is selected but the capital shares remain entirely in the hands of the State.

The concept of privatization is also discussed in another context in water supply. The provision of water services presupposes access to water resources. It is frequently suggested by critics of privatization that it is the water resource itself that is privatized, and that this life-giving resource, essential for the life of a society, is placed under the control and subject to the interests of a profit-maximizing private company. This critical evaluation is justified, although management of the resource is usually under state control, independent of whether the water supply is provided by a private or a public company. In fact, there are cases in which (private) utilities seek direct influence over the management of the resource and the allocation of user rights when privatization of water supply is discussed. What is normally meant, however, is that services are supplied by private companies.[6]

The privatization of water supply services raises several special problems that result from the special characteristics of the sector and do not occur in other network industries (cf. Winkler 2005):

- From an economic perspective, the water supply is considered a classic natural monopoly with characteristics such as subadditivity in the cost function, economies of scale and sunk costs. This means that a single company can usually supply the market more cost-effectively than any larger number of companies.
- Water supply is characterized by a high degree of external effects; water is a basic good that consumers have little or no substitute for.

Public monopolies, with few exceptions worldwide, have traditionally provided water supply services due to the special characteristics of this sector. Ownership allows the State, and in many cases municipalities, to directly influence the provision of services and to safeguard public interests. Under these conditions, the transfer of a public company into the private sector simply means transformation from a public to a private monopoly.

The problem of monopoly can be solved by a simultaneous liberalization of the market, in the sense of opening up a previously protected market. Competition in this branch would discipline the companies and assure that they cannot exploit their position of market dominance at the cost of the consumers. The introduction of competition in the telecommunications or energy industries is difficult but basically possible. Competition models quickly confront their limits in the water industry. Competition that has been applied in other sectors through the construction of additional supply networks or through third party access models is not economically feasible in the water industry and is often technically difficult to realize. Potential competition has no influence in the water industry either: if it were possible for new

[6] One could also refer to "privatisation of the water supply" to describe alternative supply solutions in the urban areas of developing countries (water kiosks, private water vendors).

Table 2 Institutional options for water supply

	Operator Tasks	Operator Risks	Operator's Income	Ownership of Facilities	Ownership of Infrastructure Network
Management contract	Provision of services for payment of a fee	Low, dependant on bonus system	Fixed fees and bonuses	Contracting office	Contracting office
Affermage	Operation of supply facilities for a fee; no investment in the infrastructure	Significant, operational risks, economic risk	Affermage fee x volume of water sold	Operator	Contracting office
Lease	Operation of the supply facilities: income from sale of services to customers; payment of a fee to the office that issued the contract; no investment obligation	Significant, operational risks, economic risk lower than by affermage since the leasing fee is fixed	Customer payments minus payment of a leasing fee	Operator	Contracting office
Concession	Operation of the supply facilities, financial investment, does not secure ownership of infrastructure	High, operational risks, economic risk, investment risks	Customer payments minus payment of a concession fee	Operator	Contracting office
Divesture	Operation of the supply facilities, financial investment, secures ownership of the infrastructure	High, operational risks, economic risk, investment risks	Customer payments, license fees	Operator	Operator

Source: Kessides 2004, 228

companies to enter the market at any time and in this way contest the position of the previous monopoly supplier, this could, at least theoretically, contribute to disciplining the monopolist. However, this type of solution is doomed by the high sunk costs.

The privatization of the water supply alone does not represent an alternative to the traditional state provision of services unless the monopoly problem is solved at the same time (public monopolies constitute similar challenges and can also be used as "cash cows"). In principle, there are numerous solutions worthy of consideration: private supply companies can be subjected to state regulation; this regulation can apply to both quality and supply standards, investment commitments and, above all, can be oriented towards the regulation of prices. Such a model has been used in England and Wales. The results have been ambivalent after 15 years of experience; fundamental problems remain, especially with respect to price regulation (cf. Vass 2004).[7]

The most common organizational model worldwide in the water industry is based on the principle of market competition. Each individual service is tendered for a specific period of time in this model. In the auction process that follows, the companies bid for contracts in which they can offer various "price – quality packages". The company offering the most favorable terms with regard to given selection criteria is awarded the contract. This company is not required to compete during the term of the concession and the service is offered for tender again upon expiration of the contract. In theory, this model will have the same results as a competitive market.

This model of inviting bids initially appears very simple; however, practical implementation requires the clarification of a number of fundamental decisions:

- What should be offered? Contracts for specific service components or for complete services could be awarded through the bid process and the contracts could require appropriate investment in the facilities and the network.
- The term of the contract is closely related to the above determination: the shorter the contract, the more intensive the competition. On the other hand, longer terms are necessary if private companies are to be offered the opportunity to recoup the investments they have already made.
- Which of the actual supply and disposal tasks can be regulated in the contracts? How specific can these requirements be without unduly limiting the private company's scope of decision-making?
- How can concession contracts assure the specified quality standards and who is responsible for monitoring them?
- Price regulation is an important part of every concession contract. A company with interest in acquiring a concession would include water prices in its offer that it believes would allow it to cover the costs of providing the service in addition to a profit.
- At the same time, very long-term contracts require agreement on specific pricing formulas that allow the water tariff to be adjusted in response to changing general conditions. The prices are often connected to the index of specific costs or can be

[7] For regulation details, cf. also: Office of Water Services (http://www.ofwat.gov.uk).

adjusted with regard to the investment obligation. It is also necessary to regulate the question of whether and under what conditions the contract can be renegotiated. However, if the contracts are renegotiated too often, the basic principle of this competition model must be questioned.

The engagement of private water supply companies in developing and emerging markets is primarily based on the concession model. This privatization form was decisively propagated in the past and promoted by international organizations such as the World Bank. The fact that many global players in the international water market are of French origin and can draw on over 100 years of experience with such models in their home market contributed to its spread.

4 Experience with Privatization Models in the Water Industry

When compared to the overall extent of privatization in the network industries, the share of water supply and sewage disposal is small, and private water supply systems worldwide are the exception rather than the rule. In spite of this, the debate over privatizing water supplies is at the centre of the very critical debate over reforms in the network industries. The special weight accorded the water sector here is clearly related to the specific characteristics of the sector. Controversy over such models has become more heated in the course of the globalization debate and the yet unfinished discussion of the GATS agreement (Grosso 2005; Gerber and Stütz 2005; Kirkpatrick and Parker 2004).

How is this often severe criticism of privatization in the water industry to be judged? In the light of previous experience, does increased private engagement in this sector represent a solution at all? A systematic examination of the practice of privatization is necessary before these questions can be answered. One is confronted by a series of fundamental methodological challenges since the effects of other influencing factors must be excluded when assessing the effect of privatization. Case studies of individual instances of privatization currently dominate the literature. These allow a detailed analysis that takes into account all of the relevant local and regional idiosyncrasies; naturally, the transferability of the results remains limited for this reason. In contrast, statistical, econometric procedures permit a systematic analysis of the interrelationship of effects. The results depend not only on the analytical method selected, but above all on the availability of adequate data. There is also an abundance of studies on the overall economic effects of privatization and liberalization in network industries; however, the number dealing with the water industry is limited. The work of Galiana et al. (2005) is an example of the econometric approach. They use Argentinean data to demonstrate a positive relationship between the increase in privatization and the decline in infant mortality.

The literature has documented numerous cases of privatization in recent years. Given the political volatility of the subject, it is clear that the problematic cases predominate. In one of its reports, the World Bank examined the number of failed infrastructure projects and the reasons for their failure (Harris et al. 2003). According to this report, only 48 of the approx. 2,500 projects failed between 1990 and 2001. Only 7 projects were identified in the water sector and these accounted for 11 percent of the total investment volume. However, this review may be too superficial and may consequently present a too optimistic picture in that it does not account for projects that have not officially failed but have been inadequately implemented.

There are a number of projects worldwide involving "private sector participation" that may certainly be described as successful. Some cases have involved projects that have drawn on national funds or capital markets. Most of the literature, however, is concerned with projects that have undergone privatization in a more "classical" form, where global players or their subsidiaries have been involved.

Two of these many projects will be looked at below. These projects are not untypical and at the same time they display the fundamental problems involved. One case involves a project in South America that attained a certain amount of popularity. While this project failed in the end, the second example, from Africa, is still in its difficult implementation phase.[8]

5 The Case of Cochabamba

Privatization of water supply in the Bolivian city of Cochabamba is considered a prime example of a failed privatization model (Clarke et al. 2004; Hall and Lobina 2002; Lobina 2000). This case received a great deal of media attention because the opponents among the affected population were violently attacked by security personnel. The water supply, in a region of around 500,000 inhabitants, was considered to be completely inadequate. Privatization of the supply also appeared to be an appropriate solution given the experience in other parts of the country.

In April of 1999, a contract with a term of 40 years was concluded with Aguas del Tunari, a consortium made up of International Water Ltd. (55 percent),[9] Riverstar International (25 percent) and four Bolivian companies. The concession was not awarded through a competitive process since only one company applied for the contract. The concession contract contained a list of provisions, which in the end caused the project to fail and led to dissolution of the contract at the end of September, 2000.

The contract included specific conditions related to the targeted development of connection levels for the supply and disposal networks. A 9-category increasing-block tariff was selected as tariff structure; the price was increased over 35 percent shortly after the conclusion of the contract and notification was given of a further planned increase of 20 percent.

The contract was particularly controversial due to the fact that it guaranteed the company a specific level of profit. The consumers were therefore clearly burdened with the greatest share of risk: a fall in demand would not affect the company since

---

[8] Cf. for further case studies: Public Services International Research Unit (www.psiru.org).

[9] Until November, 1999, the US-American Bechtel Enterprise Holdings was in the possession of 100 percent of International Water Ltd.

the contract guaranteed a rate of return (ROR) of 15–17 percent. The contract could be renegotiated to adapt the tariffs where appropriate. Risks associated with the exchange rate were also eliminated since the concession contract was remunerated on a US dollar basis.

On the other hand, the concession contract left open a multitude of questions and great uncertainty, especially in respect to the control of access to water resources and the question of financing the new development of water resources. The company's information policy was also totally inadequate; it refused, for example, to disclose the basis of its pricing policy.

6 The Case of Mali[10]

Mali counts as one of the world's poorest countries, with corresponding deficits in its supply of infrastructure services. Following a bid competition in 2000, the water supply was transferred to Energie du Mali (EDM) per concession contract. The main shareholder of this company is the French SAUR Group with 60 percent of the shares; the remaining 40 percent are held by the Malian government. Regulation of this private company, which is also responsible for the supply of energy, was entrusted to an independent public authority, the CREE. The concession contract bound the company to a series of obligations and concrete investment requirements. However, the EDM was initially very slow in meeting its obligations, and the regulatory body did not possess the authority or means to force the company to fulfill the investment requirements.

However, the decisive conflict was triggered by the question of setting tariffs. The concession contract contained a tariff adjustment clause. It was clear that this formula would lead to a significant increase in prices even before the conclusion of the contract; however, the government agreed to the contract, presumably in the hope of renegotiation. The Malian government considered the tariff increases that were already pending in the very first years of the contract to be politically unenforceable. The government "solved" the problem by using tax income to subsidize the company in the amount of the planned tariff adjustment.

However, the regulatory body saw this solution as a practice that was against the interests of the consumers, i.e. the taxpayers, and froze the tariffs. The supply company then declared that it was no longer in a position to fulfill the obligations agreed upon in the concession contract. In the renegotiation that followed between the company, the regulatory body and the government, the tariff formula was not adjusted; rather the concession contract was transformed into a so-called affermage, i.e. a services contract that requires no investment by the company. This made the Malian government responsible for all future investments; in view of the country's precarious economic and financial position, the effect of this change of contract on the infrastructure is obvious.

[10] Cf. for details Pollem (2007).

7 Conceptual Problems Related to the Implementation of Privatization Programs

A number of general conclusions can be drawn against the background of the large number of privatization cases that are now on hand. The critical reservations apply more or less to all network industries, but especially to the water industry. The problems are related both to fundamental deficits in the conceptual design and to the lack of implementation experience.

Trebing and Voll (2006) identified the inconsistencies between the goals defined by international financing organizations and the political goals that individual states hope to achieve through reform of the sector as an important flaw in many privatization programs. While the programs of international organizations (World Bank, IMF, OECD) normally reflect the macroeconomic goals of a free global market for goods and capital, the goals of reform programs at the national level are often determined by specific political, legal-institutional and social conditions.

On the other hand, national governments are not always in a position to deal with new structures or prepared to take on new challenges. Limited administrative capacity, unclear lines of responsibility, an insufficient data base, generally weak state structure, high levels of corruption, as well as a lack of decentralization of governmental power and administrative competence are just some of the conditions under which reforms within the water sector must be implemented. New forms of regulation, namely those which depend on a transfer of authority to independent institutions, face additional problems in countries which have highly centralized structures. The possibility of political influence here is clear. Reference can again be made to Trebing and Voll (2006, 311) in this context:

> Neither party truly accepted that a model that relied on private investment required compensatory tariffs, including a return adequate to the risk of the investment. Also, for reform to be successful, government needed to recognize its new role and set policy, not tariffs, pay its own bills in a timely fashion, not interfere with service and personnel decisions, and to demonstrate a sustained commitment to the sector.[11]

The following are some of the factors that negatively affect implementation: unrealistic timelines; the creation of irreversible structures, from which future governments can only separate themselves with difficulty; an often unmistakable fascination for complex market models and regulatory institutions; insufficient consideration of market size; a failure to consider the transfer costs bound up with each step of a reform; underestimation of what was needed to change from one bid system to another; an especially decisive lack of experience with the concepts and instruments for directing and monitoring private operators on the part of the new regulatory institutions.

Concession models represent the most common form of private sector participation. The specific problems of this model arise primarily as a result of the divergent interests of the different actors: while operating companies are interested in minimizing project risks, assuring profits and therefore favor long-term contracts, they

[11] For critical consideration, cf. also: Kessides (2005); Lopez-de-Silanes (2005); Gleick (2004).

must take into consideration the desire of the communities and regions awarding the contracts to establish a competitive environment in order to motivate the companies to provide their services more efficiently. This speaks more in favor of short-term contracts and more frequent calls for bids. France has had decades of experience with this model of privatization and has recently reacted to the obvious deficits of the model (limited competition, corruption, high prices, etc.) with changes in the regulatory framework. These problems are turning up more and more in post-colonial countries that do not have the necessary formal and informal governmental institutions to deal with such problems.

Concession models are used in all infrastructure sectors; however, their deficiencies can be seen most clearly in water supply and sewage disposal.

- Concessions are often not awarded in a competitive process; this can be traced back to a conscious avoidance of this requirement with corruption remaining an unresolved problem; this may also be due to the fact that there were insufficient applicants;
- The concession contracts are not sufficiently specific; this allows the companies a great deal of flexibility in the interpretation of the contract;
- Regulatory institutions often lack the authority and capacity necessary to effectively enforce the companies' obligation fulfillment;
- The relationship between a company and the authorities is not infrequently characterized by a certain asymmetry of information, usually disadvantageous to state institutions, which do not possess the necessary information to guarantee compliance with the contract. This is compounded by the fact that the companies have little interest in fulfilling their obligation to provide information;
- Contracts are renegotiated too often; pressure comes not only from the companies but also from the regulatory institutions, i.e., strategic behavior can be identified in both partners to the contract.

However the greatest problems are revealed with respect to the contractually regulated pricing policy. The fundamental demands on designing tariffs are many:

- Cost coverage: income derived from tariffs must be sufficient to cover the cost of the system.
- Economic efficiency: prices must be set in such a way that they provide signals for the efficient use of the resource; prices should be an expression of the total costs of customer decisions.
- Equity: consumers with similar characteristics must be dealt with equally.
- Social aspects: water must be made available at minimal cost to the poorer groups within the population.

The implementation of such a water price policy in the developing countries is inevitably confronted with problems; for example, technical-institutional requirements for charging such tariffs generally do not exist.[12] In many cases, the very first

[12] Discussions related to the implementation of the European Water Framework Directive show that the demand for cost-covering prices has not been implemented in many European countries either; cf. also Hrovatin and Bailey (2001).

charges for the provision of water services were levied in the course of the privatization. It is ultimately the alleged negative social implications of such policies that have incited massive opposition in the countries concerned and have caused many projects to fail. The demand that water prices be subsidized as a means of easing the impact of these social consequences is understandable; however, such a strategy may also be subject to criticism (Mitlin 2004).

The question of where the funding could come from remains. More important, however, is the danger that subsidization could actually reinforce existing social inequity. The population groups already connected to a central network, generally those with higher income, would benefit from such subsidies, while the inhabitants of unserviced areas remain ignored. Consideration must also be given to consequences for the supply company itself, which would then lack the necessary funding for the expansion of its supply system.

The discrepancy between positive results observed micro-economically – many studies refer to more efficient structures – and the social consequences of such policies has especially contributed to the current crisis in privatization. The World Bank recently began to re-evaluate its former infrastructure policy, a policy that to date has not reached the poorest of the population in the affected countries (Infrastructure Network 2006a, b; Estache 2004). The OECD has also readdressed the question of the future form of its infrastructure policy (OECD-Development Co-operation Directorate 2006; Kraehenbuehl and Johner 2004). There is the demand for a strategy of pro-poor growth that should ensure that the investment in infrastructure actually contributes to a reduction in poverty and social inequality.

While the Camdessus Report called for a doubling of the financing stream in order to achieve the ambitious Millennium Goals, the funds coming from private sector participation and flowing into the construction of water supply and sewage disposal systems have in fact tended to decline (Kürschner-Pelkmann 2006). This results in part from the growing resistance to the privatization of the water supply from non-governmental groups, as well as from some politicians. At the same time, however, companies have become more and more aware of the financial risks they are facing, above all with large scale projects in developing countries. As a result, the large operating companies that are still active have been concentrating their business activities on more low risk projects.

However, it appears that the gap which "traditional" international corporations have left behind is currently being filled, at least in part, by a growing number of new players in the market. According to an overview by the OECD, more than 130 new operating companies have entered the water supply market in the last several years. This involves companies, both state-owned and privately owned, from industrialized countries as well as from developing countries. For the first time, the Global Forum on Sustainable Development, held by the World Bank and the OECD at the end of 2006, concerned itself more broadly with this "new landscape of water and sanitation operators in developing countries" (OECD 2006) and with the "end of the oligopoly" (Marin 2006).

The OECD and World Bank see these as welcome developments since the appearance of new companies on the scene is expected to pressure current state-owned companies (the incumbents) to raise their efficiency. However, closer analysis shows that classical private participation is not a complete withdrawal from the market and that a more nuanced view of financing is required (Marin and Izaguirre 2006). The large international corporations will also play a role in the future; however, it appears that they are going to concentrate on fewer countries, smaller projects and, more and more, on management-only contracts; that is, projects with small investment commitments. On the other hand, the "new entrants" will have the prospect of less political resistance, access to local and regional financial markets and of becoming active in those cities and rural areas that were previously not covered or covered only to a limited extent.

An important advantage, moreover, is that the operating companies will be in a better position to strengthen local and regional economies. In particular, many of the operating companies from developing countries have only been active in other sectors (mainly construction) so far and do not always have the necessary expertise in the water supply field. However, developing such know-how seems to be decisive for the long term, if financing from international donor organizations is to flow into these projects. The World Bank sees a particularly good chance for cooperation between these new entrants and international companies in this field. However, these partnerships should be organized in such a way that the local operator is not always the "junior partner"; rather, the explicit goal of such cooperation must be to develop such operators into independent actors capable of surviving (Saghir 2006). An important precondition for such a strategy is the attractiveness of this kind of cooperation for foreign operators.

However, in view of the fact that basic problems of supply infrastructure in developing countries have not yet been solved, the urgent question remains as to what an alternative strategy might look like.

A report for the OECD Infrastructure and Poverty Reduction Task Force summarized the most significant approaches discussed as follows (Kraehenbuehl and Johner 2004, 11f.; OECD-Development Co-operation Directorate 2006):

- "Involve the private sector primarily in all areas where competition exists, i.e. in engineering, construction, manufacture, installation and maintenance.
- Avoid private companies as owners of natural monopolies.
- Promote the local private sector.
- Involve private operators, international and local, in competition and short or medium term management or service contracts only.
- Promote ownership and management responsibility of municipal and communal water organisations."

All of the suggestions conclude that in spite of shortcomings, private financing and management solutions cannot be forgone in the future and that priority must be given to the consideration of local conditions and requirements and to the implementation of locally adapted technical and institutional solutions.

German water supply companies have not played a significant role in the international market to date. However, the attempt to position itself as a successful competitor among the large water companies in this market seems to be neither useful nor realistic, even in the future. The reorientation of the infrastructure policies described should rather be taken as an occasion to shift the focus towards alternative system solutions and to transfer the experience gained here.

The final section attempts to sketch some of the basic system alternatives possible.

8 Alternative Approaches to Solving Water Crises

The debate concerning the construction of water supply and sanitation systems assumes the participation of foreign contractors building centralized systems in urban areas. To be sure, great importance is placed on alternatives which are tailored to take into account the needs and conditions of rural and suburban areas (ecosan) in reaching the Millennium Development Goals; however, the global players active in the international water market have so far given priority to centralized solutions based on standard technology. The operating companies come from countries where centralized systems usually provide country-wide infrastructure service of a generally high level of quality and safety. The companies view such systems as technically feasible and economically risk-free. For the most part, international donor organizations and the developing countries themselves also favor systems that rely on known procedures and approaches.

A new coolness towards the old centralized solutions is growing, both in the industry and in developing countries. Alternatives are being sought in the construction of increasingly decentralized and flexible systems for water supply and sewage disposal, alternatives that are superior to centralized systems with regard to sustainability and that are at least on par with them in terms of economic considerations.

While the infrastructures of developing countries continue to be characterized by growth, the water sector in Germany and Western Europe, in comparison, is characterized by stagnant, if not declining, demand. The question is the extent of the implications that such different starting points have for the question of organization in providing water services.

Skepticism towards water industry privatization has also become widespread in industrial countries. The parliament in the Netherlands decided to prohibit privatization in the water industry. In England, to date the only Western European country with a completely privatized water industry, the first thoughts of re-nationalizing the industry are emerging. Among other things, these ideas are fostered by the private company experience that investment, especially in network structures with a long lifespan, is associated with increasing financial risks and declining profit. Even France is experiencing increased criticism of the "French" concession model. The criticisms of this model include the lack of competition, long terms and a lack of transparency.

All of this more than begs the question of whether developing countries are being sold a model with privatization that is increasingly the subject of debate in the "exporting" industrial countries themselves. And in addition to these points, the question must be asked as to whether the – very ideological – debate about legal form and regulation does not overlook and obscure some of the basic requirements for the provision of water services.

Conventional supply and disposal strategies developed in industrialized nations do not appear to offer promise as single, successful solutions to the achievement of the Millennium Development Goals; these strategies are also very expensive solutions. Growing economic pressure is increasingly apparent in the metropolitan regions: Europe and North America are characterized by aging societies; this demographic profile is expressed in the reduced consumption of infrastructure services and climbing costs in infrastructure systems (Kluge et al. 2006). The situation in the developing countries is a complete contrast: They are very young societies demographically (with high migration rates into cities) and demands on the infrastructure display a correspondingly high growth rate. The consequence is an even higher investment requirement, especially where investment is to be in classic, country-wide systems. In contrast, the challenges confronting the industrial nations of the North are a consequence of increasing system costs. One speaks of the so-called fixed costs trap because the fixed costs in water supply and sewage disposal systems can lie at over 80 percent and these costs must be borne by increasingly fewer consumers when the population is decreasing (Rothenberger 2003).

Therefore, the infrastructure systems of the industrial countries are marked by an ever-increasing need for adjustment to demographic change. The loss of functionality in these systems is expressed, in particular, where functional threshold levels are not met due to a decline in consumption. For example, undesirable decay processes and stagnation can occur within the sewage system if an adequate flow rate cannot be assured. Expensive sewer flushing must be carried out in these cases, which also contributes to continually increasing costs (Koziol 2006).

There is another fundamental problem with former conventional sewage treatment technologies (mechanical, biological, denitrating or dephosphating), the unavoidable penetration by so-called micro-pollutants through the various filter systems in spite of the high level of this technology. Moreover, the conventional European/North American sewage treatment systems are characterized by the consumption of important resources such as energy and nutrients (phosphate, nitrate). These treatment systems require large amounts of energy; at the same time, an elaborated process is needed to remove nutrients such as nitrate and phosphate for oxygenation of the surface water (Schwarz 2005). The sustainability of these supply and disposal systems is increasingly in doubt in spite of the level of supply achieved (Balkema et al. 2002; Hellström et al. 2000; Kärrmann, 2001).

This situation is made more acute by the fact that the sewage and water supply systems of most industrial countries are in great need of modernization. A study by the German Institute of Urban Affairs (DiFu) has forecast that the German water industry alone (supply and disposal) has an investment requirement of more than € 110 billion over a period of 10 years (Kluge et al. 2006).

However, real alternatives already exist to this complex, expensive and inflexible traditional supply and disposal system. In some semi-central system designs, for example, sewage is regarded as a resource and is processed together with biological household waste to produce methane gas, and, in other processes, to extract nitrate and phosphate (Otterpohl 2002). These modern alternative systems are made possible by the application of separation technologies based on new filter technologies (nanotechnology), and by innovative control and modeling techniques (artificial neuronal networks, fuzzy logic, etc.). These intelligent system solutions also focus on the integration of rainwater as a raw water source; the prerequisite for this is the possibility of processing it into drinking water quality through special separation techniques (Schwarz 2005).

These few examples demonstrate the special potential of these alternative systems: they allow a clear increase in resource and energy efficiency. However, other aspects of these alternative systems are also notable: Decentralization through the use of modular design adapted to local conditions allows the development of complex, adaptive system solutions. The high adaptability of these structures lies in their ability to adjust to fluctuations in the external environment, e.g., the changing population density, resulting in part from of the portability of the modular components. This ability to rapidly adjust to changing general conditions has clear advantages over the classical system design.

To summarize the above, it may be argued that technology has developed significantly beyond that which is currently represented in existing infrastructures and through management. A "window of opportunity" has opened, especially where the classical systems have reached the end of their technical and economic lives and the decision-makers face the question of whether to invest in renovation or to install a new system. The conditions for a complete change of system could not be more opportune in terms of the system's lifetime (Koziol 2006).

However, it is an open question whether this opportunity will be taken: Opposition to reform, of greater or lesser intensity, can be observed in many industrial countries in spite of the still predominantly public legal structures. So, for example, Germany continues to adhere to its accustomed "level of technology", i.e., "generally recognized technological precepts," with respect to the selection and installation of sewage handling systems. These technological norms are promoted and enforced by branch-specific specialist associations (DVGW/DWA). These established European technological models are particularly in demand in developing countries; conversely, this means that it is only the currently established system structures in industrial countries that enjoy the preconditions for being transferred to other countries (Schramm and Kluge 2002).

However, the new system alternatives outlined above would ease the situation greatly by virtue of their high resource efficiency, their modularity and their adaptability to other environments not only in Germany, but also in so-called third world countries.

Realization of the great potential of these alternative systems requires pioneers, i.e., cities and municipalities that consciously decide in favor of installing semi-decentralized facilities in response to new challenges. There have been no such pioneers in Germany to date, which is all the more regrettable since German industry is in possession of numerous patents in the area of components for semi-decentralized facilities that are just waiting to be implemented. At the same time, these systems could provide Germany with interesting private-public partnership organization models.

If the establishment of local, regionally adapted integrated systems is seen as an important approach to solving the world's water problems, it is also necessary to overcome opposition to development and innovation in the Northern industrial countries. This could also mean removing Germany from constant competition with England and France for the provision of conventional solutions. A paradigm change would not only be required in respect to technological systems; it would also be necessary to discard the focus on "blue" water. The focus would increasingly be concerned with developing cross-sector and cross-media supply and disposal networks (water, energy, nutrients) in semi-decentralized form. It only appears to be possible to develop long-term, fundable, organizationally sound and socially acceptable solutions to the worldwide water problem by turning away from policy that relies exclusively on conventional technology models.

References

Balance T, Trémolet S (2005) Private sector participation in urban water supply in sub-Sahara Africa, GTZ, KfW Bankengruppe, Frankfurt a. M., November

Balkema AJ, Preisig HA, Otterpohl R, Lambert DJD (2002) Indicators for sustainability assessment of wastewater treatment systems. In: Urban Water, Vol. 4, No. 2, 153–161

Bundesregierung (2006) Bericht der Bundesregierung zur Modernisierungsstrategie für die deutsche Wasserwirtschaft und für ein stärkeres internationales Engagement der deutschen Wasserwirtschaft, Berlin, März

Clarke GRG, Kosec K, Wallsten S (2004) Has private participation in water and sewerage improved coverage? Empirical evidence from Latin America, World Bank Policy Research Working Paper 3445, Washington, DC, November

Estache A (2004) A selected survey of recent economic literature on emerging infrastructure policy issues in developing countries. Background paper for the October 2004 Berlin meeting of the POVNET Infrastructure Working Group, World Bank, Washington, DC, August

Galiana S, Gertler P, Schargrodsky E (2005) Water for life: The impact of the privatisation of water services on child mortality. In: Journal of Political Economy, Vol. 113, No.1, 83–120

Gerber L, Stütz J (2005) Privatisierung und Liberalisierung der Wasserversorgung auf Druck von internationalen Finanzinstitutionen und WT – sind die Vorwürfe der NGOs berechtigt? In: Aussenwirtschaft, 60 Jg., H. III, 305–345

Gleick PH (2004) The Millennium Development Goals for water: crucial objectives, inadequate commitments. In: Gleick PH (ed) The World's Water 2004–2005. The biennial report on freshwater resources, Ecosystem Studies, Island Press, Washington, DC, 1–15

Grosso MG (2005) Liberalising network infrastructure services and the GATs, OECD, Trade Directorate Trade Committee, Working Party of the Trade Committee OECD Trade, Paris

Hall D (2004) Water finance – A discussion note, Public Services International Research Unit (PSIRU), Paper commissioned by Public Services International (PSI) World Social Forum Delhi/Mumbai, January

Hall D, Lobina E (2002) Water privatisation in Latin America, 2002, Public Services International Research Unit (PSIRU), London

Harris C, Hodges J, Schur M, Shukla P (2003) Infrastructure Projects. A review of canceled private projects, public policy for the private sector, Note No. 252, World Bank, Washington, DC, January

Hellström D, Jeppssion U, Klärrman E (2000) A framework for systems analysis of urban water management. In: Environmental Impact Assessment Review, Vol. 20, No. 3, 311–321

Hrovatin N, Bailey StJ (2001) Implementing the European Commission's water pricing communication: cross – country perspectives. In: Utilities Policy Vol. 10, No. 1, 13–24

Infrastructure Network (2006a) Infrastructure: lessons from the last two decades of World Bank Engagement, Discussion Paper 35199, Vol. 1, Washington DC, January

Infrastructure Network (2006b) Scaling up infrastructure: building on strengths, learning from mistakes, Discussion Paper 35199, Vol. 2,Washington DC, January
Kärrmann E (2001) Strategies towards sustainable wastewater management. In: Urban Water Vol. 3, 63–72
Kessides I (2005) Infrastructure privatization: gone too far? Or too early to tell? Privatization Barometer, The PB Newsletter, No. 3, July, 21–31
Kessides IN (2004) Reforming infrastructure: privatization, regulation and competition, World Bank Policy Research Report, Washington, DC
Kikeri S, Kolo AF (2005) Privatization: trends and recent developments, World Bank Policy Research Working Paper 3765, November
Kirkpatrick C, Parker D (2004) Regulation and the privatization of water services in developing countries: assessing the impact of the General Agreement on Trade in Services (GATS), Centre on Regulation and Competition, Working Paper Series No. 67, University Manchester, June
Kluge Th, Libbe J, Scheele U et al. (2006) Der netWORKS-Ansatz zur integrierten Strategiebildung. In: Kluge Th, Libbe J (eds): Transformation netzgebundener Infrastruktur: Strategien für Kommunen am Beispiel Wasser, Deutsches Institut für Urbanistik, Berlin, 34–63
Koziol M (2006) Transformationsmanagement unter den besonderen Bedingungen der Schrumpfung. In: Kluge Th, Libbe J (eds) Transformation netzgebundener Infrastruktur: Strategien für Kommunen am Beispiel Wasser, Deutsches Institut für Urbanistik, Berlin, 402–451
Kraehenbuehl J, Johner O (2004) OECD Infrastructure and Poverty Reduction Task Team: water infrastructure for poverty reduction, final report, commissioned by Swiss State Secretariat for Economic Affairs (seco), Zurich, 22 November
Kürschner-Pelkmann F (2006) Der Traum vom schnellen Wasser – Geld. In: Aus Politik und Zeitgeschichte Nr. 25, 3–7
Lobina E (2000) Cochabamba – water war, Public Services International Research Unit (PSIRU), PSIRU reports, London
Lopez-de-Silanes F (2005) Dos and Don’ts in Privatization: Evidence from Less Developed Countries. In: Fondazione Eni Enrico Mattei – FEEM, Privatization Barometer, The PB Newsletter, No. 2, January, 36–42
Marin P (2006) Opportunities and challenges from the emergence of new private operators in developing countries, OECD Global Forum on Sustainable Development, Paris, November 29–30
Marin Ph, Izaguirre AK (2006), Private participation in water. Toward a new generation of projects? GRIDLines Note No 14, September
Mitlin D (2004) Beyond second best: the whys, hows and wherefores of water subsidies, Centre on Regulation and Competition, Working Paper Series Paper No. 93, University of Manchester, February
OECD-Development Co-operation Directorate (2006) Guiding Principles on using infrastructure to reduce poverty. Task Team on Infrastructure for Poverty Reduction (InfraPoor), Paris, March
OECD (2006) A new landscape of water and sanitation operators in developing countries. OECD Global Forum on Sustainable Development / World Bank “Public-Private Partnership in water supply and sanitation – recent trends and new opportunities”, Paris, 29–30 November, Press Release
Otterpohl R (2002) Perspektiven für die dezentrale Abwasserreinigung: Das abwasserfreie Haus und dezentrale Abwasseranlagen in der Innenstadt? In: Wasser & Boden, 54 Jg., H. 5, 12–15
Pollem O (2007, forthcoming) Die Regulierung des Wassersektors in Entwicklungsländern, PhD thesis, Carl von Ossietzky Universität Oldenburg, Oldenburg
Public Services International (PSI)/World Development Movement (2006) Pipe dreams: The failure of the private sector to invest in water services in developing countries, London, March
Rothenberger D (2003) Dynamik der Veränderungskräfte in der Abwasserentsorgung. EAWAG-news 57, 11–13
Saghir J (2006) Policy implications taking the agenda forward, OECD Global Forum on Sustainable Development, Paris, 29–30 November

Schramm E, Kluge Th (2002) Ein nachhaltigerer Umgang mit Wasser im Norden – Chancen für den Süden. In: DGU Nachrichten, Nr. 25, Juni, 9–13

Schwarz St (2005) Kostbares Naß. In: Fraunhofer Magazin 2/2005, 62–63. http://www. fraunhofer.de/fhg/Images/magazin 2.2005-62_tcm5-18722.pdf. Cited 20 December 2005

Trebing HM, Voll SP (2006) Infrastructure deregulation and privatization in industrialized and emerging economies. In: Journal of Economic Issues, Vol. XL, No. 2, 1–9

UN Millennium Project (2005) Preparing national strategies to achieve the Millennium Development Goals. A handbook, New York

UNDP (2006) Human Development Report 2006: Beyond scarcity: power, poverty and the global water crisis, New York

van Hofwegen P (2006) Task force on financing water for all. Enhancing access to finance for local governments. Financing water for agriculture. Chaired by Angel Gurria, World Water Council

Vass P (2004) The development of water regulation – A collection of reviews, Collections series 6, Centre for the study of Regulated Industries (CRI), University of Bath

Willoughby CH (2004) Infrastructure and the Millennium Development Goals. Session on complementarity of infrastructure for achieving the MDGs, Berlin, 27 October. . http://www.oecd.org/dataoecd/22/31/36567911.pdf. Cited 20 July 2007

Winkler J (2005) Wettbewerb für den deutschen Trinkwassermarkt: Vom freiwilligen Benchmarking zur disaggregierten Regulierung, Baden-Baden

Winpenny J (2003) Financing water for all. Report of the world panel on financing water infrastructure, World Water Council, Chaired by Michel Camdessus, Marseille, March

WWAP (2006) Water – a shared responsibility. The United Nations World Water Development Report 2, UNESCO/Berghahn Books, Paris

WWC (2006) Costing MDG target 10 on water supply and sanitation: comparative analysis, obstacles and recommendations, March, Marseille

Sector Reforms for Sustainable Financing of Water and Wastewater Services

Franz-Josef Batz

Abstract Achieving the MDGs in the water sector requires a substantial increase of investments. Funds for investments have to be mobilized from all sources. Mobilizing local resources is hereby key for sustainable financing. This chapter argues that for mobilizing local resources for the water sector, sector reforms are imperative. These include reforms of the water sector, in the financial sector and in public administration. Measures in the water sector alone will not be sufficient. Sector experts and policy makers have to broaden their view and work across their sectors to achieve sustainable financing of the water sector. Moreover, the governments of the developing countries are responsible for creating the enabling environment. Development cooperation can assist in this task. Provision of Overseas Development Assistance (ODA) alone may not be sufficient. Capacity Building is equally important and should accompany all development measures in the water sector.

1 Introduction

With the discussion on the Millennium Development Goals (MDGs), the issue of development financing enjoys greater priority on the political agenda. The European Union (EU) has committed to increase its ODA up to 0.56 percent of its Gross National Income (GNI) by 2010 – with half of the additional € 20 billion going to Africa – and to 0.7 percent of GNI by 2015 (Council of the European Union 2005). The water sector has also benefited from this development with the introduction of new instruments, such as the EU Water Facility, the African Water Facility, or the European Infrastructure Partnership. With the help of these commitments and instruments, transfer payments to the African water sector under ODA are supposed to increase. One important question arising out of this is whether the additional ODA funds transferred will be used efficiently and effectively. And can the water sector

Franz-Josef Batz
e-mail: Franz-Josef.Batz@gtz.de

W. Scheumann et al. (eds.), *Water Politics and Development Cooperation*,
doi: 10.1007/978-3-540-76707-7_10,

in African countries absorb these additional financial flows? Will the funds make a sustainable contribution to the MDGs or will they only have a brief impact? What role do sector reforms play in providing sustainable finance to the water sector in Africa?

2 Water Sector Reforms Are Key for Sustainable Financing

To the best of our knowledge, worldwide, the annual funding shortfall for achieving the MDGs in water supply and sanitation is estimated at around US$ 9–30 billion (EUWI 2003). WaterAid estimates that public sector investment reaches 69 percent of total investment; external aid flows 17 percent; international private flows 9 percent; and domestic private sector investment 5 percent (Annamradju et al. 2001). These figures show that ODA covers only a minor part of the financial burden in the water sector, and even doubling ODA inflow will not change this. This makes it all the more important for African countries to make greater efforts to raise funds from their own resources in order to ensure a sustainable financing of their water sectors. However, sectoral analyses and benchmarking studies show that the water sectors of most African countries remain very inefficient as long as sector reforms are not implemented.

Water tariffs do not cover costs: In most African water utilities (approximately 60 percent), revenues do not even cover operating and maintenance costs (Table 1). There are many reasons for this poor figure: many public water supply utilities are required to keep tariffs low for political reasons. Further, water tariffs do not reflect actual consumption. Generally there is also no enforcement mechanism for payment either. Over a long period, water users have become accustomed to pay very little or nothing for water, which has led to a strong subsidization mentality in many countries. In addition, water users are often unwilling to pay, as the services are of poor quality and tariff increases have not necessarily been accompanied by improved quality (PPIAF and WSP 2002). However, substantial improvements are feasible, as the benchmarking data reveal: The best 25 percent of the African water utilities studied by the International Benchmarking Network achieved average cost recovery rates (O&M) of up to 160 percent, approaching the average cost recovery rate in industrialized countries.

Water utilities are inefficient: Data of the International Benchmarking Network show average values for non revenue water of 32 percent (reaching as high as 70 percent!). Average values for staff per 1000 connections are also extremely high as compared to other regions and the values for average billing period and collecting efficiency indicate poor performance. Again, there is potential for improvement: the top 25 percent of African water utilities reach values in all of these indicators that are comparable to industrialized countries.

Regulation and monitoring are marginally developed: Of the 55 countries in Africa studied by the World Bank (2005) 31 countries have neither governmental nor autonomous regulatory agencies. Only five countries have independent

Table 1 Water sector indicators in Africa compared with other developing countries and industrialized nations

Indicators	Africa (all countries)	Africa (top 25 %)	Developing countries (top 25 %)	Industrial countries (average)
Cost recovery (O&M%)				
- average	94	160	180	180
- min	2	120	139	98
- max	400	400	400	353
<100 % (% of utilities)	60	0	0	1
Non revenue water (%) average	32	13	20	16
Staff/1000 connections average	20	4	6	2
Billing period (months) average	8.8	2.2	1.2	1
Collecting efficiency average	71	100	100	98

Source: International Benchmarking Network 2005

institutions responsible for regulating the water and wastewater sector. Regulation is essential for setting standards with respect to the tariff structure and the quality of services. It provides utilities with incentives to operate efficiently and to be socially responsible. Regulators have an important role to play as neutral brokers between policy players and operators. At the same time, they have to allow an adequate return of investment. Thus regulation helps to attract capital to the sector, including from private sources. Not existing or ineffective regulation may have a negative influence on the revenue stream of a utility and on its ability to cover costs and to finance new investments.

3 State Financing Systems Are Equally Important

Ultimately, user fees and tax revenue are the basic source of finance for water infrastructure. As most water utilities in Africa are publicly owned these two sources are closely related. However, state financing systems are poorly developed in Africa.

Low public investment in the water sector: Worldwide, it is estimated that the water sectors absorb 1–3 percent of the state budgets. African countries on average provide less than 1 percent of their annual budget to the water sector. Investment in sanitation is even less. Reasons include high debt service, which takes up to 60 percent of national budgets, low state revenue due to ineffective tax systems (EUWI 2003) but also a low priority for investments in the water sector (Care et al. 2004). Moreover sub-Sahara Africa has a tax ratio of 17 percent (Asche 2005), which is the lowest in the world. Unproductive and antiquated tax systems (emphasis on import duties, little taxation of domestic economic activities and assets), weak tax administrations, importance of the informal sector and the resulting narrowing of the tax assessment base all sharply limit tax revenue (ATPC 2004).

Fee revenues are too low: In Africa approximately 90 percent of the water utilities are publicly owned (Calaguas and Cann 2006), i.e. operation, maintenance and expansion must be financed by public institutions through fees. Fees are a primary source of revenue for local authorities. However, collecting fees (including water tariffs) is not very popular among local functionaries, and so tariff policy is being abused for political purposes. This is a major reason for the failure to cover costs.

Weak fiscal decentralization: In many African countries, decentralization has shifted responsibilities from the central level to subnational levels. In many cases though fiscal decentralization has not yet been implemented. African central governments have difficulties to relinquish important sources of tax revenue to subnational levels of government (Fjeldstad and Rakner 2003). Traditional subnational taxes such as land tax generate too little revenue to meet the financial needs. Financial transfers (interstate and from the central government) are generally inadequate.

Lack of technical and human capacity: At many subnational government levels, the tax potential is nowhere fully realized. Estimates for African countries suggest that improvements in tax assessment and collection (tax administration) could increase tax revenue at subnational level by often more than 50 percent (Steffensen and Trollegaard 2000).

Lack of access to credit markets: In many African countries, regions, and provinces local authorities have no access to credit markets. Reasons are partly legal restrictions, motivated by fiscal considerations: access to loans could weaken fiscal discipline (soft budget constraint), moral hazard problems from a possible accumulation of liabilities could increase, as the central government would have to take responsibility in cases of insolvency, complicating macroeconomic management by the central government. Other reasons relate to the creditworthiness of subnational government levels: the main reasons for problems herefore are low and widely fluctuating revenues from fees and taxes, lack of tangible security, lack of repayment culture, poor financial management and inadequate regulation of local authorities as well as lack of credit rating (Winipenny 2003; EUWI 2003).

4 Weak Local Financial Systems and Capital Markets Do Not Channel Capital into the Water Sector

Given the low levels of international capital flows to Africa and to the water sector, local financial and capital markets have to play a particularly important role in financing water infrastructure. These markets should be in a position to meet the financial needs of the actors in the water sector on a demand-driven basis.

For example:

- by giving individual households access to the formal banking system so that they can borrow at market rates in line with their bank ability: for purposes such as financing household connections, operating and maintaining their own wells and sanitation facilities etc.

- by giving smaller utilities (private or public) and small scale operators access to loans to finance new investments, operating and maintenance of smaller piped systems, financing investments for water kiosks, tankers, public toilets, waste water trucks etc.
- by ensuring that water utilities have access to a capital market capable of mobilizing local savings (e.g. pension funds etc.). Local sources of finance have a particularly significant role to play here, as most water utilities do not have access to financing in international capital markets. International financing also involves risk from currency fluctuations.

Sub-Saharan Africa has a deficit in terms of widespread provision of financial services to the population. DfID estimates that Africa-wide the unbanked population is between 80 and 90 percent (Applegarth 2004). This primarily affects poorer population groups, particularly in rural areas, and Small and Medium Enterprises (SMEs) in all sectors of the economy, including the water sector. As a result, the financial and investment potential of these households and enterprises cannot be adequately realized. The reasons for the shortfall in the provision of financial services include the poor environment for sustainable financial intermediation (e.g. weak financial market regulation and supervisory agency). The same applies to institutions and structures at the meso level of the financial system, e.g. associations, credit information agencies, Apex institutions, which are either absent or have serious weaknesses.

Capital markets in most African countries are either not existing or are inadequately developed. This is why even large enterprises are only able to access domestic refinancing through the banking sector. This leads to a situation where commercial banks focus on larger enterprises and neglect other customer groups perceived as too risky. These commercial banks generally only mobilize short term financing (Applegarth 2004).

In consequence, there is a general shortage of long-term capital needed for investment in the water sector. This is due to the low level of local savings which the formal banking sector can mobilize. Reasons for this include the high capital flight from Africa: 39 percent of private wealth produced in Africa is invested outside Africa. Africa has the lowest savings ratio (ratio of bank deposits to GNI) of all regions of the world, only 22 percent of GNI compared with 40 percent for Southern Asia and the Latin American Countries. However, particularly people in rural areas save primarily real assets such as livestock, housing, informal businesses etc. which by far outweigh ODA grants (Applegarth 2004).

These figures show that developing efficient financial systems and capital markets could mobilize considerable funds from local resources. To meet the need for investment in the productive and social sectors, Africa must at least double the (monetary) savings ratio. For a long time Africa has neglected the mobilization of local savings. In line with this, financial sector and capital market development play a major role in mobilizing and channeling capital to the water sector. Reforms relate not only to the banking system but also to developing pension and insurance systems, social security systems, which offer secure investments for savings and also function as instruments to mobilize local capital.

5 Increased Financial Transfers Outstrip the Absorption Capacity of the Countries

There is agreement in principle that ODA promotes economic growth. However, ODA also has decreasing marginal returns extending to negative impacts on growth. Studies show that there might be a saturation point for ODA between 15 and 45 percent of Gross Domestic Income (Applegarth 2004). Many African countries have already reached this point (Table 2). It is thus probable that the impact of ODA decreases and turns negative. Rapidly increasing funds could even accelerate this process. Reasons can be found in:

Macroeconomic constraints. Rapid increases in ODA transfers can result in macroeconomic instabilities. They have an inflationary impact, and if they fluctuate they have an adverse impact on public and private investment. They can distort the labor market, e.g. through a sudden increase in demand for qualified skilled labor etc.

Crowding out the private sector. This is particularly relevant in LDCs and Poverty Reduction Strategy (PRS) countries with a high ODA share in their Gross Domestic Income and in connection with budget financing. The external flow of funds can lead to excess liquidity, which is soaked up in order to avoid inflation by relatively simple monetary policy instruments, such as state bonds. This can lead to crowding out of the private sector, as it is more profitable for banks to invest their funds in low-risk, high-interest government bonds than to lend to small and medium-sized water utilities or other population groups perceived as high-risk.

Weak institutions: Many countries lack the institutional and human resource requirements to use growing aid effectively, with little or no regulation and monitoring instruments, inefficient public administrations, inadequate financial management at local level, lack of public control (incentives to corruption). The concentration of program-oriented aid (SWAP, budgetary support) and an increase in ODA may also lead to overload the respective institutions in tendering, processing and implementing infrastructure measures.

Table 2 Share of external aid payments in Gross Domestic Income in Africa

Countries	1997	1998	2002	2003
Congo DR	5.5	2.2	14.7	n.a.
Ethiopia	8.4	10.2	21.7	22.8
Malawi	13.8	25.6	20.2	29.9
Mali	17.7	13.6	15.1	12.7
Rwanda	12.5	17.7	20.8	20.0
Sierra Leone	14.3	16.3	47.0	39.0
Tanzania	12.5	12.1	13.2	16.3
Sub-Saharan Africa	4.5	4.7	6.3	6.0

Source: World Development Indicators 2004 and 2005

Conclusions for Policy Advice

We do not solve problems by throwing money at them: Policy must not surrender to the hypothetical "silver bullet" of increasing financial transfers in order to reach the MDGs. ODA alone will never reach sufficient levels for solving the financing constraints in the water sector, let alone the problems associated with ODA in general. Financing the water sector will continue to depend on the ability of the respective country to mobilize local resources and create suitable conditions for investment (local or international). This takes time (decades in some cases) and requires political commitment by the partner country to initiate and implement reform processes. It is therefore imperative that increased financial transfers are accompanied by sector reforms, which improve the absorption and implementation capabilities of the national structures.

Bilateral and multilateral development cooperation must press harder for self-responsibility: The governments of the developing countries are responsible for creating the environment for sustainable financing of the water infrastructure. Politicians in the South must modernize their economies to achieve higher economic growth. They must invest in water and make their country attractive to private investors – not least to domestic private investors who currently prefer to invest their money "safely" abroad.

For the *water sector* this means first of all, formulating and implementing financing policies. These include tariff systems, which reflect the needs and capabilities of different social groups, which are transparent, reliable and simple, and which are based on cost recovery. Tariff policies need to ensure that water supply and sanitation (WSS) costs are covered with the aid of user fees, connection fees and subsidies. This also includes effective regulation and monitoring which creates incentives and offers security to investors (cost recovery, financial sustainability, profits, protection against political influence, incentives to invest in poorer areas etc.) and is independent of the service providers and government. This policy creates incentives for transforming current inefficient water utilities into enterprises, which are financially sustainable and customer oriented.

Reforming state financial systems requires giving greater priority to state investment in the water sector within national development policies. This means not only increasing its priority within the state budget (taking into account macroeconomic stability criteria), but also to give greater attention to water in the national PRS. In the course of decentralization, the transfer of additional tasks and burdens to subnational regional and local authorities must be accompanied by the transfer and development of capacities and mandates for cities and local authorities to finance investments in the water sector. In many developing countries this means starting with the reform of internal financial relationships (e.g. reforming the tax system, direct taxes, equalization payments between subnational governments, transfers by central government), with the goal of ensuring that cities and local authorities have reliable sources of income. It also requires enabling regional and local authorities to meet their capital needs from internal and external markets. Besides the necessary adjustments to the legal and institutional framework, it also requires the creation of financial capacities, for example, local authorities' financial control and management, and systems for local authority borrowing or debt management. This also includes strengthening the willingness to collect taxes, fees etc.

Creating a *sustainable financial system* is the basic requirement for durable and demand-oriented provision of financial services by local financial markets. This requires a common understanding of competitive financial sector promotion on the part of the actors involved, and specifically between the policy areas involved, e.g. financial sector policy and water sector policy. Of central importance here is the regulatory and political environment in order to establish a transparent and competitive framework. This could create greater confidence for demand and supply of financial services. The challenges to financial institutions at the institutional level are primarily to strengthen the institutional capacity, mobilizing local resources, liquidity management, and managing risks and costs in developing innovative – and in particular – long-term financial products.

Greater weight must be given to capacity development: The current popularity of infrastructure financing needs to be matched by support in other respects, or else there is the risk that funds will be used inefficiently. This could fall back on development cooperation and infrastructure promotion. Development cooperation institutions must extend their actions beyond the water sector: sectoral reforms must take into account measures and reforms in other sectors than the water sector. Therefore the profile of the development cooperation adviser must offer a broader range of competences. This also means greater transsectoral and interdisciplinary cooperation with the partner countries and among development cooperation organizations. Coherence needs to be established between policy areas, e.g. new financing instruments must be compatible with the financial system and must not distort local financial markets. ODA should not replace national and subnational budgets (e.g. in PRS).

Acknowledgements Paper presented at the thematic group meeting on water at German Development Institute (DIE), Bonn, October 2005. The conclusions expressed are those of the author and should not necessarily be attributed to GTZ or German Development Cooperation as a whole.

References

Applegarth P (2004) Capital market and financial sector development in sub-Saharan Africa. A report of the Africa Policy Advisory Panel, Center for Strategic and International Studies (CSIS), Washington, DC

Annamradju S, Calaguas B, Gutierrez E (2001) Financing water and sanitation, key issues for increasing resources to the sector, WaterAid, London, United Kingdom

Asche H (2005) Öffentlicher Haushalt und Haushaltsprozess in Subsahara-Afrika. In: Herausforderungen für die öffentliche Entwicklungsfinanzierung in Subsahara-Afrika: Aktuelle Debatten über Instrumente und Optionen, Bericht zum DIE-Afrika-Fachgespräch am 1. Juni, Bonn

ATPC (African Trade Policy Center) (2004) Fiscal implication of trade liberalisation in African countries, Economic Commission for Africa, Addis Ababa

Calaguas B, Cann V (2006) Reforming public utilities to meet the water and sanitation goal, World Development Movement and WaterAid

Care, Oxfam, Freshwater Action Network, Green Cross, Tear Fund, Water Aid, WWF (2004) A scorecard assessment of developing country and donor progress

Council of the European Union (2005) The EU and Africa: Towards a strategic partnership, 15961/05 (367), Brussels

EUWI (2003) The European Union Water Initiative: Final report of the financial component, Brussels

Fjeldstad OH, Rakner L (2003) Taxation and tax reform in developing countries: Illustrations from sub-Sahara Africa, Chr. Michelsen Institute, Bergen

International Benchmarking Network (2005) IBNET Indicator Search Facility. http://www.ib-net.org. Cited 17 Oct 2005

PPIAF, WSP (2002) New designs for water and sanitation transactions, Washington, DC

Steffensen J, Trollegaard S (2000) Fiscal decentralisation and sub-national government finance in relation to infrastructure and service provision. Synthesis report on 6 sub-Saharan African country studies, World Bank, Washington, DC

Winipenny J (2003) Financing water for all. Report of the world panel on financing water infrastructure, Kyoto

World Bank (2005) International Directory of Utility Regulatory Institutions http://www. worldbank.org/html/fpd/psd/ifur/directory. Cited 17 Oct 2005

The Political Economy of Water and Sanitation Services in Colombia

Matthias Krause

Abstract Adequate domestic water supply and sanitation (WSS) services are a crucial element for escaping poverty. Lacking access to these services is one important factor causing infectious diseases and decreasing the productivity of households. During the last decade the academic and political debate on WSS reform has focused on the polemic issue of private versus public ownership of service providers. The paper argues that this focus is too narrow to explain success or failure in the delivery of WSS services. Instead, the paper shifts the attention towards the role of governance and explores the influence of governance issues on WSS services for the case of Colombia. Achieving broad access to good-quality and low-cost services presupposes a complex mix of poverty- and efficiency-oriented WSS policies. Therefore, the political and administrative governance which shapes the actions of the main actors involved (politicians, regulators, private and public service providers, and users) has a relevant effect on the poverty-orientation and the efficiency of service delivery.

1 Introduction

Adequate domestic water supply and sanitation (WSS) services are a crucial element for escaping poverty. Lacking access to these services is one important factor causing infectious diseases and decreasing productivity and economic possibilities of the household. The provision of equitable and efficient WSS services presupposes effective policies in order to (i) implement affordable solutions for the poor who lack access as well as to tackle (ii) the significant health and environmental externalities and (iii) the natural monopoly characteristics associated with these services. This is a complex issue because expanding services and improving their quality typically cannot be reached without improving efficiency and financial sustainability of service providers. Moreover, neglecting environmental sustainability will put in danger the availability of safe water which in its turn will hurt especially the poor.

Matthias Krause
e-mail: Matthias.Krause@die-gdi.de

W. Scheumann et al. (eds.), *Water Politics and Development Cooperation*,
doi: 10.1007/978-3-540-76707-7_11,

In Chap. 2 of this paper the main elements of a "good" WSS policy are outlined based on a discussion of the poverty- and efficiency-related problems of WSS services. Subsequently, and using a political economy perspective, the relation between weaknesses in WSS policy and failures in political and administrative governance is described. During the last decade the academic and political debate has focused too narrowly on the privatization issue. It is argued that a broader governance perspective is better suited to analyze policy weaknesses and to further develop research on WSS sector reform.

Chapter 3 presents empirical evidence from Colombia. The hypothesis explored is that weaknesses in WSS policies can be explained by weaknesses in political and administrative governance. The Colombian WSS sector is very suited to explore this hypothesis. Starting in the late 1980s it has been deeply transformed with the goal of improving achievements in equity and efficiency. The reform included the promotion of private sector participation (PSP) which allows assessing how governance weaknesses influence both public service providers and providers with PSP. Preliminary evidence indicates that reluctance of politicians to grant full independence to the regulator and to service providers is detrimental to efficiency, that there is a discrimination of the rural poor in subsidy policy that may be related to their political voicelessness and that inefficiencies in the use of publicly provided grants are related to a weak administrative governance. In Chap. 4 some general conclusions are drawn.

2 Conceptual Approach

2.1 Social Goals and Implications for WSS Policy

The academic and public debate on the organization of domestic WSS services centers around three social goals: poverty alleviation, efficiency improvements and environmental sustainability. In what follows the former two are going to be discussed (environmental aspects are treated under the heading "efficiency").[1]

2.1.1 Poverty Alleviation

Poverty can be broadly defined as a multiple and pronounced deprivation in well-being.[2] Poverty manifests itself in different dimensions of well-being: (i) material

[1] This is a rather limited perspective on environmental sustainability. In the literature that centers on sustainable water resource management, environmental aspects are treated in a much more elaborated manner. Cf. Neubert (2005) for an in-depth discussion of theoretical and practical aspects of the Integrated Water Resource Management concept that integrates environmental, economic and social goals from a resource management perspective.

[2] According to a widespread consensus in development literature, the term poverty should not be limited to lack of income or monetary wealth. The following concept is based on World Bank (2001, Chap. 1).

deprivation, (ii) low achievements in education, (iii) ill health, (iv) vulnerability, (v) political voicelessness and powerlessness. Having this broad poverty concept in mind, it can be argued in the first place that lacking access to improved[3] WSS services itself is a symptom of poverty as it is a kind of material deprivation. At the same time, lacking access to these services is one important factor causing deprivations in other dimensions. It increases the risk of falling ill with the consequence of decreasing productivity and economic possibilities of the household.[4] Productivity and economic possibilities do not only decrease due to the consequences of ill health but also due to time and money spent in getting safe water, e.g. from a dug well two miles away or from an expensive source like water vendors. Finally, lacking adequate WSS services is often also a consequence of other poverty dimensions: From a conceptual point of view, there are two exemplary institutional settings (out of a continuum of combinations) for allocating WSS services: (i) free markets that allocate services according to the price mechanism and (ii) state organizations that allocate services according to a political logic. In general the poor are handicapped with respect to both allocation mechanisms because of (i) their low purchasing power and (ii) their political voicelessness and powerlessness.[5]

The close interrelation of lacking WSS services and being poor is a strong argument for a poverty-oriented WSS policy.[6] Policies should promote access of the poor to adequate services and make sure that these services are affordable for low-income households. This can be achieved e.g. by subsidizing access to (and possibly consumption of) WSS services of the poor. A free market allocation based on the price mechanism alone that is not complemented by any specific poverty-oriented policies which might include subsidies, will not use the potential of WSS policy to contribute to poverty alleviation.

---

[3] Improved water services: Household connections, public standpipes, boreholes, protected dug wells or springs, rainwater collection. Improved sanitation services: Connections to public sewers or septic tanks, pour-flush latrines, ventilated improved pit latrines, simple pit latrines. Cf. WHO and UNICEF (2000, annex A). For a more in-depth discussion of definitions and characteristics of improved (or adequate or safe) WSS services cf. UN-HABITAT (2003, 2–5).

[4] The health and economic benefits of WSS services to households are well documented. Cf. UN-HABITAT (2003, Chap. 2), WHO and UNICEF (2000, 1–3), Hutton and Haller (2004), Fuest and Laube (2004, Chap. 3.2). For an in-depth empirical analysis and discussion of health benefits from WSS services cf. e.g. Jalan and Ravallion (2001), Esrey (1996).

[5] For an in-depth analysis of the disadvantage of poor people to influence government decisions on general pro-poor policies and to obtain benefits from such policies cf. Johnson and Start (2001).

[6] An argument closely related to this is to highlight the human rights character of WSS services. The General Comment No. 15 of the International Covenant on Economic, Cultural and Social Rights has reinforced the human rights character of water services. It states that "[t]he human right to water entitles everyone to sufficient, affordable, physically accessible, safe and acceptable water for personal and domestic use" (United Nations Economic and Social Council 2002; as cited by Budds and McGranahan 2003, 94).

2.1.2 Efficiency I: Market Failure

An important strand of (economic) literature is concerned primarily with efficiency.[7] The network supply of WSS services shows two important characteristics that prevent markets from working efficiently, i.e. from serving the greatest number of consumers according to their preferences at the lowest possible cost (cf. e.g. Noll 2002):

Natural monopoly[8] – in a free market setting this phenomenon means that consumers are faced with a single supplier that can charge high prices for low-quality services because he has not to fear any competition due to the cost characteristics of the supply technology. This leads to losses in consumer welfare.

Environmental and health externalities[9] – in a free market setting these phenomena lead to situations where too much water is extracted from a water source as compared to its capacity to replenish, too few sanitation facilities are installed or the water body is polluted because the environmental and health costs associated with these activities are not reflected in the market price and are thus not accounted for by the individual households or firms.

Both kind of market failures are good reasons for the state to take an active role in designing specific policies and regulations for promoting low-cost and good-quality services as well as for internalizing external effects. It does, however, not necessarily imply state ownership of service providers.

2.1.3 Efficiency II: State Failure

The traditional model for organizing WSS services in most developing countries until the 1990s has been public provision by state-owned monopolies with politically set tariffs that in most cases contained some sort of cross-subsidies between consumer classes. The choice of this model has often been justified on equity and efficiency grounds. Unfortunately, in many cities in developing countries this traditional model has been associated with disappointing achievements both with respect to the

[7] It can be distinguished between the narrow concept of (i) operational efficiency which means that a given output (here quantity and quality of WSS services) is produced at the lowest possible cost and the encompassing concept of (ii) allocative efficiency which means that the maximum number of consumers are provided with goods and services according to their preferences at the lowest possible cost. Cf. e.g. Rosen (1992, Chap. 4).

[8] A market is a natural monopoly when – due to economies of scale and/or scope – the lowest cost alternative to satisfy a given demand is supply by one single firm. For a technical discussion of the natural monopoly cf. Braeutigam (1989).

[9] According to Rosen (1992, 66), “externality” can be defined as an “activity of one person affecting the welfare of another in a way that is outside the market.” For an in-depth analysis of externalities and public goods cf. Cornes and Sandler (1996).

equity and the efficiency goal.[10] According to an important strand in literature,[11] the main reasons for the bad performance lie in the negative consequences of (i) state ownership and (ii) absence of competition on the operational efficiency[12] of service providers. As there are no private shareholders that press the management to use inputs in the most efficient way to maximize profits, and as customers cannot sanction bad service by switching the supplier, managerial slack occurs. Moreover, management decisions are not taken according to business logic but according to a political logic with the consequence of a low-level equilibrium.[13]

Inspired by this state failure argumentation, International Finance Institutions and important donor countries during the 1990s promoted private sector participation[14] (PSP) in developing countries, usually combined with the building up of regulatory agencies for monopoly regulation. Assessing the success of this strategy has been a matter of controversy. As far as the academic debate is concerned, the picture is inconclusive.[15] Most analysts acknowledge improvements in operational efficiency but underline that overall success has been diminished by ineffective regulation and insufficient poverty orientation. An undisputed result is that private investments in the WSS sector in developing countries have materialized only to a small fraction and have lagged far behind the expectations (cf. Izaguirre 2004). In the political arena, PSP in WSS has been highly controversial from the beginning and it has in several cases encountered strong political opposition.[16] Typically it is claimed that PSP is inequitable and worsens the situation of the poor.

---

[10] Low coverage especially among low-income households, intermittent and bad service, high water losses, failure of the subsidies to reach the poor, average tariffs below average costs of service (which means that the resulting financial losses have to be covered by state transfers), unnecessarily high costs of service provision (e.g. due to overstaffing). Cf. Kessides (2004, 2–3 and 220), World Bank (1994, 25–36).

[11] Cf. Shirley and Walsh (2000) and the overview of literature given there. Applied to the case of the WSS sector, cf. Foster (1996, 1–5).

[12] See footnote 8.

[13] As far as the author knows, the term "low-level equilibrium" in the context of WSS has been coined by Pablo Spiller. Cf. Spiller and Savedoff (1999).

[14] The term private sector participation (PSP) is used to describe a range of different ways of involving the private sector in the provision of WSS services: service contract, management contract, lease, concession, build-operate-transfer schemes (BOT), joint venture, divestiture. These different models imply different types of articulation between private and public actors with differing divisions of responsibilities, risks and rights. For a description of different forms of PSP cf. Shirley (2000, 150), Budds and McGranahan (2003, 89–90), Foster (1996, 12).

[15] Whereas several case studies show improvements in coverage, operational efficiency and water losses following PSP – cf. e.g. Shirley and Ménard (2002) – a recent analysis using a quantitative approach finds no significant difference in coverage improvements between PSP and public provision (cf. Clarke et al. 2004).

[16] Cf. Kessides (2004, 6). Generally, we can observe that the global political debate on water issues is marked by ideological polarization. Cf. Urquhart and Moore (2004) and Budds and McGranahan (2003) for a survey of the different positions and arguments in this debate.

2.1.4 Elements of a "Good" WSS Policy

There is no such thing as a "unique state-of-art blueprint" for poverty-oriented and efficiency-enhancing WSS policy. With respect to the polemic issue of PSP recent proposals are less dogmatic (cf. World Bank 2004; Kessides 2004, Chap. 5). In what follows the main lines of an encompassing WSS reform are sketched that probably many analysts would endorse.

1. Organizational separation of the functions policy-making, regulation and service provision aiming at making harmful political influence on regulation and management of WSS services more difficult and at encouraging actors to concentrate on their core functions: public officials on policy-making, regulators on tariff and quality regulation, and service providers on managing WSS services.
2. Poverty orientation: To assure that the poor gain access to safe WSS services and that the use of these services is affordable to them is a fundamental task of policy-making. Measures include subsidizing[17] the connection costs (and possibly the current tariffs) of low-income households, defining and monitoring connection targets in low-income neighborhoods, empowering the poor for participating in regulation and planning of WSS in order to enable them to exercise their citizen and consumer rights.
3. Economic regulation: This element aims at protecting WSS service users from the abuse of monopolistic power and consists basically in defining (and controlling) efficient tariff and quality levels that have to be met by service providers and in promoting competition for the market (e.g. competitive bidding of concessions) to substitute for the absent competition in the market.[18]
4. Tariff reform: This element basically is about aligning the tariff level with the level of efficient[19] costs and aims at improving allocative efficiency as well as financial sustainability of service providers. This (together with PSP) typically is the most conflictive measure because in the very most cases it means that tariffs have to be increased.
5. Corporatisation[20] of service providers, i.e. the transformation of service providers into autonomous organizations. This bundle of measures includes both options (a) commercialization under public ownership as well as (b) PSP in one of its various forms. It aims at improving operational efficiency and at insulating management from harmful political influence. Minimum requirements for improvements to occur under public ownership are that service providers are transformed into incorporated enterprises with an own budget and autonomy of the management

[17] The question of how to design WSS subsidies schemes in order to effectively reach the poor without compromising (too much) overall efficiency of WSS service provision is a complex issue on which quite a lot of theoretical and empirical research has been done. Cf. Yepes (1999, 2003), Estache et al. (2002, Chaps. 3, 4).

[18] For the terms "competition in the market" and "for the market" cf. World Bank (1994, Chap. 3).

[19] It is the main task of regulation to assure that tariffs do not rise above the level of efficient costs. See above.

[20] On corporatization and commercialization cf. World Bank (1994, Chap. 2).

to decide on the level and quality of service provision, on manpower and budget allocation.

6. Environmental regulation: Protecting water bodies from pollution and assuring long-term availability of fresh water resources internalizes negative external effects on downstream users and on the environment and is a condition for the sustainability of WSS service provision. Typical measures include quantitative limits and/or fees for the abstraction from and the pollution of the water bodies, subsidies for sewerage treatment plants as well as the building up of monitoring systems for water quality and quantity.

2.2 Governance Matters

The hypothesis explored in this paper is that the unsatisfactory achievements in both the traditional and the market-oriented model may be related to weaknesses in governance.[21] As has been discussed above, the achievements in the WSS sector with respect to poverty alleviation and efficiency improvements are threatened by both state failure and market failure. Even when relying exclusively on private utilities for service provision, the state is necessary for economic and environmental regulation and for poverty orientation. If under the traditional model the state failed to achieve equity and efficiency, being responsible for policy-making, regulation and service provision, why should the overall result improve with the state being necessarily still responsible for the former two functions, unless the underlying reasons for the state failure are explicitly and effectively addressed in WSS sector reform? What market failure and state failure theories have in common is that they describe conditions under which self-centered behavior of individual actors does not lead to a maximization of the social goals equity and efficiency, i.e. conditions under which self-centered behavior can be detrimental to social welfare.[22]

From a normative point of view the main function of a good governance structure is helping to reduce detrimental forms of self-centered behavior and, in general, to shape the behavior of the main actors in the WSS sector (basically politicians, policy makers, regulators, private and public service providers, and users) in such a way that it is conducive to the equity and efficiency goals. Weaknesses in governance can have a negative effect on WSS policies and thus on service performance by two different ways: (i) Policies do not point at the right direction because of failures during the process of policy formulation and political decision-making. This is

[21] In a general way governance can be perceived as "the sum of the many ways individuals and [organizations.], public and private, manage their common affairs"; definition of the Commission on Global Governance as cited in Weiss (2000, 796). In this paper the focus is set on the common affairs with respect to the provision of WSS services.

[22] This is not to say that self-centered behavior is generally detrimental to social welfare (self-centeredness of human behavior is a basic assumption of the theoretical approach chosen here). On the contrary: under the conditions of perfect markets the interaction of self-centered individuals leads to an efficient allocation of resources (which however might be considered as inequitable).

related to weaknesses in political governance. Example: subsidies are only available for those who are already connected to the WSS system leaving all the low-income households that lack a WSS connection without any assistance because the non-connected poor have a weak political voice and thus are not taken into account by politicians and policy-makers when drafting subsidy rules. (ii) Policies point at the right direction but they are not properly implemented because actors use the rooms for maneuver policies leave open for pursuing particularistic goals to the detriment of social goals, i.e. because of failures during the process of policy implementation. This is related to weaknesses in administrative governance. Example: subsidies do not reach the poor because the public administration exercises its discretional power to use the money for the pursue of particularistic interests which in its turn is facilitated by a low transparency of budget processes, a weak accountability of the public administration and a lack of effective sanctioning mechanisms.

In what follows the Colombian WSS policy is analyzed. The focus is on finding evidence for the hypothesis that weaknesses in WSS policies are related to weaknesses in political and administrative governance.

3 Evidence from the Colombian Case Study

Colombia has been chosen as a subject of a case study because it offers both public providers and providers with PSP. Therefore, it allows to explore the influence of governance on both provider types and to analyze whether there are systematic differences between public providers and providers with PSP.

Starting in the late 1980s the Colombian WSS sector has been transformed with the goal of improving equity and efficiency. Two major reforms have shaped the governance structure of the Colombian WSS sector: Decentralization and public utilities reform (cf. Maldonado and Vargas 2001; Fernández 2004; SSPD 2002). Through the decentralization process (started in 1986) the responsibility for service provision and infrastructure assets were transferred from the national level to municipalities. Before, there had been a mixed responsibility with the major cities having been served typically by municipal corporations, and the remaining urban and rural areas by national government-owned providers managed by the Instituto Nacional de Fomento Municipal (INSFOPAL). The public utilities reform (started in 1994) introduced major institutional changes in the WSS sector. The most important changes were (i) the establishment of regulatory agencies at the national level with the mandate to regulate tariffs and enhance competition, (ii) the introduction of the principle of cost covering tariffs, (iii) corporatization of service providers and promotion of PSP, as well as (iv) the creation of institutions for user participation. The public utilities law (no. 142, 1994) confirmed (v) the principle of solidarity prescribing cross-subsidies between socio-economic groups and declaring the covering of unmet WSS needs a priority.

The remainder of this paper is organized as follows. First, a short overview of the WSS service situation in Colombia is given (3.1). The subsequent analysis concentrates on policy weaknesses (not so much on achievements) and potential connections to governance failures. It concentrates on two essential WSS policy elements: organization of policy-making, regulation and service provision (3.2) and poverty orientation (3.3). Finally some general conclusions are presented (4).

3.1 Overview of the WSS Service Situation

Between 1993 and 2003 overall coverage rates have increased from 79.7 percent to 86.1 percent (improved water services) and from 73.2 percent to 82 percent (improved sanitation services) of total population (see Table 1). Coverage in rural areas is much lower than in urban areas (more or less half). Access to WSS services in Colombia varies according to the wealth of households (cf. OPS and OMS 2001, 11–12 and 79). By 1997 only 57.5 percent of the households belonging to the decile with the lowest per capita consumption spending had an in-house water connection whereas among the decile with the highest per capita consumption spending the same share amounted to 98.3 percent. Again rural areas differ markedly from urban. Differences between poor and rich are much bigger on the countryside (poorest decile: 29.2 percent, richest decile: 54.9 percent)[23] than in the city (poorest decile: 91.5 percent, richest decile: 99.1 percent). Differences between poor and rich as well as between rural and urban areas are less pronounced when in addition to in-house connections other improved WSS services (wells, public standpipes) are taken into consideration (cf. Meléndez 2004, 49ff.).

Table 2 presents data for a sample of 26 providers from intermediate[24] Colombian cities, distinguishing between providers with PSP and without PSP. It reports

Table 1 Coverage with improved water supply and sanitation services (percent of population)

Service	1993 (1)	2003 (2)	Change 2/1
Urban			
Water	94.6	97.4	1.03
Sewerage	81.8	90.2	1.10
Rural			
Water	41.1	53.1	1.29
Sanitation	51	57.9	1.14
- Sewerage	14.6	15.2	1.04
- Other	36.4	42.7	1.17
Total			
Water	79.7	86.1	1.08
Sewerage	63	71	1.13
Sanitation	73.2	82	1.12

Source: DNP (2004, 5), modified by the author

[23] These figures relate to the area "rural 2" as classified by DANE (cf. OPS and OMS 2001, 79).

[24] 100,000–700,000 thousand habitants. The four main capitals Bogotá, Medellín, Cali and Barranquilla (more than 1.3 million habitants) have not been considered in order to reduce the heterogeneity of the sample with respect to city size and thus ease comparability of the cases.

Table 2 Performance of urban WSS services with and without PSP (1998–2003)

Indicator	Providers with PSP	Providers without PSP	Total
Coverage water connections			
Average 1998–2003 (%)	90.8	92.4	91.9
Change 1998–2003 (03/98)	1.089	1.029	1.049
Coverage sewer connections			
Average 1998–2003 (%)	75.7	88	83.7
Change 1998–2003 (03/98)	1.025	1.028	1.027
Unaccounted-for water			
Average 1998–2003 (%)	45.8	45.2	45.4
Change 1998–2003 (03/98)	0.919	1.183	1.088
Collection rate			
Average 1998–2003 (%)	84.4	81.6	82.5
Change 1998–2003 (03/98)	1.145	0.995	1.043

Source: Author's calculations based on data provided by SSPD and CRA; sample of 26 intermediate cities: 9 cases with PSP, 17 cases without PSP

average indicators for coverage with (a) water and (b) sewer connections as well as for operational efficiency: (c) unaccounted-for water (percentage of potable water that is produced but not billed: the lower the indicator the more efficient) and (d) collection rate (percentage of the amount billed to customers that is collected: the higher the indicator the more efficient). Figures for the total sample of 26 service providers have developed into the desired direction (change 1998–2003) except for unaccounted-for water that has increased by 8.8 percent on an unsatisfactorily high average level (45.4 percent). Except for coverage with sewer connections, providers with PSP show more significant improvements than those without PSP (the latter actually have worsened with respect to unaccounted-for water and collection rate). When looking at average levels of the indicators however, providers with PSP perform worse except for the collection rate.[25]

In the course of the WSS sector reform tariffs have been increased significantly. E.g. between 1998 and 2001 the sector specific price index rose by 42.7 percent (i.e. 12.6 percent per year) in real terms (cf. SSPD 2002, 30–32). Real tariff increases have coincided with some improvements in coverage and operational efficiency (see above) but also with higher proportional spending of low-income households. The share of household income spent on WSS services of the poorest quintile amounts to 6.8 percent (2003) which lies above the internationally recommended 5 percent (cf. Meléndez 2004, 23 and 53).

[25] This picture for providers with PSP (clear improvements but unsatisfactory average levels) may be related to the fact that PSP in Colombia – in most cases – was initiated when service performance and financial situation of the public provider had become alarming and consequently the pressure on politicians to remedy the situation very high. This means that PSP typically was started in situations where indicators were very bad (information gained in interviews with experts and representatives of organizations of the Colombian WSS sector).

3.2 Organizational Structure of Policy-Making, Regulation and Service Provision[26]

Table 3 shows the organizational structure of the functions policy-making, regulation and service provision which is strongly influenced by the fact that Colombia is a decentralized state (functionally, fiscally and politically).

Although municipalities are fully responsible for service provision in their territories and consequently mayors are responsible for local WSS policy-making (definition of coverage targets and urban planning, decisions on corporatization of service providers, allocation of public funds), the national government retains crucial responsibilities in the realm of WSS policy-making like sectoral planning (including national coverage targets and allocation of public funds) and capacity building. Colombia is a special case as compared to international practice because responsibility for economic regulation is split. The task of defining tariff and quality of service levels is done by CRA (Regulatory Commission for Water and Sanitation), control is accomplished by SSPD (Superintendency for Public Utility Services) and monitoring is a shared responsibility of both agencies. The 1,091 municipalities existing on Colombian territory accomplish their task of service provision by a variety of organizational forms: incorporated enterprises (with or without PSP); directly by the municipal administration or, especially in rural contexts, by different forms of community organizations. The main resources to finance service provision come from tariffs paid by users and publicly funded subsidies (see below).

Table 3 Distribution of responsibilities according to territorial levels of government

Function	National	Provincial	Municipal
Policy-making	National Ministry of Environment, Housing and Development – MAVDT	–	Mayor
Regulation	Regulatory Commission for Water and Sanitation – CRA (*economic regulation*)	Autonomous Regional Bodies (*environmental regulation*)	–
Control	Superintendency for Public Utility Services – SSPD (*economic control*)	Autonomous Regional Bodies (*environmental control*)	–
Service provision	–	–	Municipal administration; Public enterprise; Private enterprise

Source: Based on Fernández (2004, 28), modified by the author

[26] This section is based on Fernández (2004).

In what follows two potential weaknesses of this governance structure are briefly discussed that are both related to the risk of detrimental political influence: (i) independence of the regulatory agency CRA and (ii) independence of service providers.

There is a strong consensus that in order to accomplish their tasks of protecting consumers from monopolistic market power as well as of promoting efficiency and investments in the WSS service sector, regulatory agencies shall be independent (i.e. be insulated from the influence of politically or economically powerful groups and possess an adequate technical expertise).[27] In the case of CRA independence can be questioned because of several features contradicting typical formal safeguards that would support independence:[28]

1. Ministries are directly involved in the decisions of the regulatory commission: 3 out of 7 members are ministers – the remaining 4 are technical experts.
2. Technical experts are appointed exclusively by the executive (President) – the legislative branch is not involved.
3. Appointment of technical experts is not based on a transparent and competitive selection that would assure technical expertise.

These weaknesses in the independence of CRA may have contributed to the delay in the revision of the outdated methodology for tariff regulation that should have been completed by 2001 but actually was still not fully implemented by July 2005, increasing this way the uncertainty in the WSS service industry (cf. also Fernández 2004, 81–83). The revision of the old methodology was very necessary because it contained hardly any incentives for efficiency what means that tariff increases partially reflected rents and inefficient costs (especially in big and intermediate cities where tariff increases have been substantial).

The management of a service provider should be independent from detrimental political influence on management decisions (e.g. concerning staff selection, technology choice, investment) in order to concentrate on operational efficiency within the given policy and regulatory framework that should be conducive to equity and efficiency. The institutional safeguards that prevent detrimental influence of the municipal government are strong in case of a private incorporated enterprise, average in case of a public incorporated enterprise and more or less absent in case services are provided by the municipal administration. There is some empirical evidence indicating that detrimental political influence is a bigger problem in the case of public provision:

One indicator for political interference is the frequency of change of the general manager. Unfortunately, this indicator is only available for a small sample of 4 service providers (2 public incorporated enterprises; 2 private incorporated enterprises) for which it was collected during this investigation. Whereas both public enterprises have had 6 general managers between 1998 and 2005, one private enterprise had 4 and the other 1 during the same period.

---

[27] Cf. Smith (1997, especially 11), formal safeguards.

[28] What follows is based on information gained in interviews with experts and representatives of organizations of the Colombian WSS sector.

As has been already discussed above (see Table 2), providers with PSP on average showed improvements in both indicators for operational efficiency (unaccounted-for water and collection rate) between 1998 and 2003 whereas providers without PSP on average worsened with respect to these indicators.[29]

Many public providers (especially in intermediate and small cities) did not implement tariff increases that would have been necessary to approximate cost covering levels because tariff increases where blocked by municipal governments (cf. Fernández 2004, 81).

3.3 Poverty Orientation

The analysis of the poverty orientation of Colombian WSS policy is limited to the issue of subsidies. With respect to their principal objective, it can be distinguished between subsidies (i) aiming at lowering the cost of connecting to the service for low-income households who lack access and those (ii) aiming at lowering the cost of using the service for low-income households who already have access (in order to guarantee affordability and to prevent withdrawal from the service).[30] In the academic debate the first objective is considered as more relevant, especially in countries where access is low and lack of access correlates negatively with income (cf. Foster 1996, 14–16). Access in Colombia (see Table 1) cannot be considered as very low: as compared to 20 other Latin American countries, in 2000 Colombia ranked on position 5 (water) and 6 (sanitation) (WHO and UNICEF 2000). Lack of access to both services is much more common in rural areas where approximately 86.5 percent of all households without improved WSS services live. In rural areas there is a clear negative correlation between lack of access and income whereas in urban areas this negative correlation holds only for sanitation services (cf. Meléndez 2004, 53).

With respect to the way of delivering subsidies to the target population, one can distinguish between cross-subsidies that are delivered through the tariff scheme (the principle being that some consumer classes pay more than the cost of service whereas other consumer classes pay less) and direct subsidies that can either be delivered through the supply side (e.g. by financing a network extension to a poor part of the town with public grants and thus without extra costs for low-income dwellers) or through the demand side (e.g. by granting cash or vouchers to low-income households so that they can pay for their connection costs).[31]

There are at least three important criteria to assess subsidy schemes: (i) Their targeting quality (i.e. the extent to which subsidies reach those households that are or should be eligible), (ii) the cost of administrating the targeting system and (iii) the extent to which subsidies harm allocative efficiency (cf. e.g. Foster et al. 2000).

---

[29] A higher operational efficiency of providers with PSP as compared to public providers has been confirmed in a regression analysis performed by the author. Cf. Krause (2007)

[30] Cf. Estache et al. (2002, 14–21), who make this distinction and, in addition to subsidies, discuss further instruments aiming at these two objectives.

[31] On these issues cf. Estache et al. (2002, Chaps. 3–4), Yepes (2003), Krause (2003, 24–25).

Direct subsidies to the demand side are typically the most preferred instrument among economists because (contrary to cross-subsidies) they do not distort the price signal and are thus less harmful for allocative efficiency (cf. Yepes 1999, 4–7). In Colombia, subsidies in the WSS sector take two forms: Public grants for supplying services and cross-subsidies.

3.3.1 Public Grants for Supplying Services

Important amounts of public grants are allocated to the WSS sector by the national government (a smaller fraction) and by municipalities (the vast majority) that, in principle, could be used to pursue the objective of expanding access among low-income households. The bulk of these resources come from the national budget. Table 4 shows subsidies stemming from the national budget that have been allocated to the whole environmental services sector (WSS services, waste water treatment, solid waste disposal). It can be seen that earmarked transfers[32] to municipalities within the framework of the decentralization law no. 715 (2001) are by far the most important source with a share of 86 percent. Transfers are earmarked to be used for the environmental services sector but the administration of these funds and the spending decision is made by municipal governments. These transfers are not only the most important source of subsidies, they are important even when compared to the value added of the whole sector: 33.5 percent on average as measured in current prices for the years 1994–2003.[33]

With respect to the important portion of transfers that are used for WSS services, there is no general procedure on the micro level that is designed for targeting expenditures so as to expanding access among low-income households. (However, the formula for distributing transfers among the 1,091 municipalities considers

Table 4 National fiscal spending for WSS, waste water treatment and solid waste disposal, 1998–2002

Source	Thousands US$ 2003		Proportion (%)
	Total amount 5 years	Average per year	
Transfers to municipalities (decentralization law 715, 2001)	1,195,716	239,143	86
National royalties fund	56,234	11,247	4
Ministry for economic development	84,410	16,882	6
Other	54,220	10,843	4
Total	*1,390,579*	*278,116*	*100*

Source: Fernández (2004, 62)

[32] DNP (2004) provide a description of the fiscal transfer system from national to municipal governments with a focus on the WSS sector.

[33] Own calculation with the data reported by DNP (2004, 8).

poverty as one important criterion which means that in practice poorer and smaller municipalities receive higher transfers per habitant than richer and bigger municipalities; cf. DNP 2004, 6; Fernández 2004, 71–72). The targeting of local spending depends exclusively on the priorities and decisions of the municipal government which in turn are influenced by local political governance. There are some indications that, on average, municipal governments do not target public grants to low-income households lacking access to improved WSS services: As has been said above, 86.5 percent of these households live in rural areas. However, just 46 percent of earmarked transfers (1996–2001) have gone to rural areas (cf. DNP 2004, 12). This spending pattern could be related to the political voicelessness of the rural poor.

One could argue that a sort of indirect targeting would be achieved if public grants were efficiently used for expanding general access to WSS services (because of the negative correlation between lack of access and income – see above). However, there are concerns with respect to the efficient use of transfers. According to a recent calculus of the National Planning Department (cf. DNP 2004, 17), national coverage rates in 2003 could have been 100 percent (piped water) and 95 percent (sewerage), if transfers to municipalities (1996–2003) would have been executed efficiently in infrastructure expansion projects (in contrast to 88 percent and 74 percent, respectively, that were actually measured in 2003). Some observed problems related to administrative governance features are:[34] (A) An important portion of transfers earmarked for WSS are not executed for that purpose. (B) Fiscal control and auditing procedures related to transfers are weak. (C) Funds are used for items that have no direct impact on coverage expansion (e.g. for covering deficits generated by the cross-subsidy-scheme, see below). (D) Due to lacking technical expertise and lacking articulation between the planning activities of the municipal administration and the service provider, funds are used for infrastructure projects that are unsuitable or unworkable. (E) Corruption related to the use of public grants for construction activities in the WSS sector is a common phenomenon in Colombia and compounds the before mentioned problems.

3.3.2 Cross-Subsidies

The Colombian cross-subsidy-scheme is organized as follows. All consumers are divided into four classes: Industrial, commercial, official and residential (the latter is by far the most important class by number and consumption). Residential consumers are again subdivided into 6 socio-economic strata (1 being the poorest and 6 the richest). Up to of 20 m^3/connection/month (defined as basic consumption level by CRA), the residential consumer classes 1, 2, 3 pay a subsidized tariff that lies below the cost of service (50 percent, 40 percent, 15 percent, respectively) as regulated by CRA whereas the residential consumer classes 5 and 6, industrial and

[34] For the following points cf. DNP (2004), Fernández (2004, 63 and 70–72), information gained in interviews with experts and representatives of organizations of the Colombian WSS sector.

commercial classes pay a surcharge (20 percent respectively).[35] Classification of residential consumers is a responsibility of mayors and is done based on a standardized methodology provided by the National Department of Statistics which considers basically the characteristics of the housing unit (cf. Meléndez 2004, 25–26).

Apart from the well known negative effects on allocative efficiency, which in some cases can be of considerable magnitude (cf. e.g. Yepes 2003, 7), the Colombian cross-subsidy scheme shows the following weaknesses:

1. Targeting quality is low because the error of inclusion is high: 51 percent of consumers that receive a subsidy through the tariff have an income that lies above the national poverty line (the positive thing about this is that the error of exclusion is extremely low: only 0.7 percent of the connected households with an income below poverty line do not receive a subsidy).
2. Another characteristic of the cross-subsidy scheme is that it does not redistribute from rich to poor but is slightly regressive.
3. The cross-subsidy scheme is not operational in smaller and poorer cities and localities because there exist no consumer classes that pay a surcharge. This means that deficits between the cost of service as regulated by CRA and tariff revenues have to either be covered by the municipal government (that can use earmarked transfers for that purpose) or by the service provider (which means that necessary expenditures may have to be postponed and service quality may decrease).

These three characteristics indicate that – apart from urban low-income households that already have access to WSS services – urban middle classes benefit most from the cross-subsidy-scheme. The losers of the Colombian subsidy policy seem to be the rural poor. This would support the hypothesis that politicians and policymakers allocate subsidies to benefit (urban) constituencies that have relatively more political power and are better organized.

4 Conclusions

Based on the analysis so far, the following tentative conclusions can be drawn:

Achieving the goal of poverty alleviation is interrelated with achieving the efficiency goal. Providers that are managed efficiently and are financially sustainable have a greater capacity to expand services to the non-connected poor and to provide a good quality service than poorly managed and financially strained providers. However, whether providers actually engage in expanding services to poor neighborhoods depends on the application of specific poverty-oriented policies.

In the paper it was argued that achieving poverty- and efficiency-orientation in WSS services presupposes a complex policy mix, the quality of which depends on the quality of the political and administrative governance of the country or region.

[35] Cf. Fernández (2004, 54–57); residential class 4 and official consumers neither receive a subsidy nor pay a surcharge.

The Colombian case study has yielded some empirical indications that several of the weaknesses observed for Colombian WSS policy are related to governance weaknesses:

- Politicians seem to be reluctant to grant full independence to the regulator and to public service providers. This behavior can plausibly be explained by the fact that politicians would lose control of tariff setting and of the allocation of funds by granting full independence, and thus would lose important means to win elections and maintain clientelistic networks.
- Nevertheless, this practice is harmful for the efficiency of WSS service supply. According to the available data, providers with PSP have shown greater improvements in operational efficiency than public providers.[36] This result supports the hypothesis that the institutional characteristics of PSP provide a greater protection against political interference in the management than corporatization under public ownership.
- Those people that have the greatest need for WSS services (the rural poor who lack access to improved services) are disadvantaged with respect to subsidy policy. A plausible explanation for this observation is that the rural poor lack the political voice that would be necessary to introduce reforms to subsidy policy in order to focus scarce subsidies on needy households.

Further research seems desirable to confirm the tentative results obtained in the present paper. An implication of this analysis for policy makers is to keep in mind that adverse governance factors may go against well-intentioned policies aiming at enhancing poverty-orientation and efficiency of WS services and to render them useless in the end.

Acknowledgements An earlier version of this paper has been presented at the International Workshop "Water Poverty and Social Crisis: Perspectives for Research and Action", Agadir (Morocco), 12th–15th December 2005, organised by Institut de recherche pour le développement (IRD).

References

Braeutigam RR (1989) Optimal policies for natural monopolies. In: Schmalensee R, Willig RD (eds) Handbook of industrial organization, Vol. 2, Elsevier, Amsterdam, 1289–1346

Budds J, McGranahan G (2003) Are the debates on water privatization missing the point? Experiences from Africa, Asia and Latin America. Environment & Urbanization, Vol. 15, No. 2, 87–113

Clarke GRG, Kosec K, Wallsten SJ (2004) Has private participation in water and sewerage improved coverage? Empirical evidence from Latin America, World Bank Policy Research Working Paper 3445, Washington, DC

Cornes R, Sandler T (1996) The theory of externalities, public goods and club goods, 2nd edn, Cambridge University Press, Cambridge, MA

[36] This result has recently been confirmed in a regression analysis. Cf. Krause (2007).

DNP (2004) El sistema general de participaciones en el sector de agua potable y saneamiento básico, Bogotá
Esrey SA (1996) Water, waste, and well-being: A multicountry study. American Journal of Epidemiology, Vol. 143, No. 6, 608–623
Estache A, Foster V, Wodon Q (2002) Accounting for poverty in infrastructure reform: Learning from Latin America's experience, WBI development studies, World Bank, Washington, DC
Fernández D (2004) Sector agua potable, Informes de base. Colombia: Desarrollo económico reciente en infraestructura (REDI), World Bank, Washington, DC
Foster V (1996) Policy issues for the water and sanitation sectors, Inter-American Development Bank, Washington, DC
Foster V, Gómez-Lobo A, Halpern J (2000) Designing direct subsidies for the poor – a water and sanitation case study. In: Public Policy for the Private Sector, Note No. 211
Fuest V, Laube W (2004) Konzept einer armutsorientierten Entwicklungszusammenarbeit im Wassersektor, Teilgutachten 1 im Rahmen des Forschungsprojekts des Bundesministeriums für wirtschaftliche Zusammenarbeit und Entwicklung: "Wasser und Armut. Schlussfolgerungen für die Umsetzung des Aktionsprogramms 2015 und das Erreichen der Millennium Development Goals", German Development Institute, Bonn
Hutton G, Haller L (2004) Evaluation of the costs and benefits of water and sanitation improvements at the global level, WHO, Geneva
Izaguirre AK (2004) Private infrastructure: Activity down by 30 percent in 2002. In: Public Policy for the Private Sector, Note No. 267
Jalan J, Ravallion M (2001) Does piped water reduce diarrhea for children in rural India, World Bank Policy Research Working Paper 2664, Washington, DC
Johnson C, Start D (2001) Rights, claims, and capture: Understanding the politics of pro-poor policy, Overseas Development Institute, Working paper 145, London
Kessides IN (2004) Reforming infrastructure: Privatization, regulation, and competition, World Bank Policy Research Report, Washington, DC
Krause M (2007) The political economy of water and sanitation in developing countries: Cross-country evidence and a case study on Colombia, Draft 29 January 2007, Bonn: unpublished manuscript
Krause M et al. (2003) Sustainable provision of renewable energy technologies for rural electrification in Brazil: An assessment of the photovoltaic option, German Development Institute Reports and Working Papers 10/2003, Bonn
Maldonado MM, Vargas Forero G (2001) Evaluación de la descentralización municipal en Colombia. La descentralización en el sector de agua potable y saneamiento básico, Archivos de Economía, Documento 166, DNP, Dirección de Estudios Económicos, Bogotá
Meléndez M. (2004) Subsidios al consumo de los servicios públicos en Colombia. Hacia dónde movernos? Informes de base. Colombia: Desarrollo económico reciente en infraestructura (REDI), World Bank, Washington, DC
Neubert S et al. (eds) (2005) Integriertes Wasserressourcenmanagement (IWRM) – Ein Konzept in die Praxis überführen, Nomos, Baden-Baden
Noll R (2002) The economics of urban water systems. In: Shirley MM (ed), Thirsting for efficiency: The economics and politics of urban water system reform, Pergamon, Amsterdam, New York, 43–63
OPS, OMS (2001) Desigualdades en el acceso, uso y gasto con el agua potable en América Latina y el Caribe, Colombia, Serie Informes Técnicos No. 4, Washington, DC
Rosen HS (1992) Public finance, 3rd edn, Irwin, Homewood, Boston
Shirley MM (2000) Reforming urban water systems: A tale of four cities. In: Manzetti L (ed), Regulatory policy in Latin America: Post-privatization realities, North-South Center Press at the University of Miami, Miami, FL, 147–170
Shirley MM, Ménard C (2002) Cities awash: A synthesis of the country cases. In: Shirley MM (ed), Thirsting for efficiency: the economics and politics of urban water system reform, Pergamon, Amsterdam, New York, 1–41

Shirley MM, Walsh P (2000) Public versus private ownership: The current state of the debate, Working paper 2420, World Bank, Washington, DC
Smith W (1997) Utility regulators – the independence debate, in: Public Policy for the Private Sector, Quarterly No. 12, December, 9–12
Spiller PT, Savedoff W (1999) Government opportunism and the provision of water. In: Savedoff W, Spiller PT (eds), Spilled water. Institutional commitment in the provision of water services, Latin American Research Network, Inter-American Development Bank, Washington, DC, 1–34
SSPD (2002) Acueducto, alcantarillado y aseo 1998–2001. In: Supercifras en m3, No. 6, 1–143
UN-HABITAT (2003) Water and sanitation in the World's cities. Local action for global goods, Earthscan, London, Sterling
United Nations Economic and Social Council (2002) Substantive issues arising in the implementation of the International Covenant on Economic, Social and Cultural Rights, Draft, General comment No. 15: The right to water, Office for the UN High Commissioner for Human Rights, Geneva
Urquhart P, Moore D (2004) Global water scoping process. Is there a case for a multistakeholder review of private sector participation in water and sanitation? Scoping report, Executive summary, ASSEMAE, Brasilia
Weiss TG (2000) Governance, good governance and global governance: Conceptual and actual challenges. Third World Quarterly, Vol. 21, No. 5, 795–814
World Bank (1994) World development report 1994. Infrastructure for development, Washington, DC
World Bank (2001) World Development Report 2000/2001. Attacking poverty, Washington, DC
World Bank (2004) Public and private sector roles in water supply and sanitation services. Operational guidance for World Bank Group staff, Washington, DC
WHO, UNICEF (2000) Global water supply and sanitation assessment 2000 report, Geneva
Yepes G (1999) Do cross-subsidies help the poor to benefit from water and wastewater services? Lesson from Guayaquil, Working paper, Water and Sanitation Program, World Bank, UNDP, Washington, DC
Yepes G (2003) Los subsidios cruzados en los servicios de agua potable y saneamiento, Informe de trabajo, Inter-American Development Bank, Washington, DC

Part IV
Power Plays in Irrigation Reforms

Political Power Play in Bulgaria's Irrigation Sector Reform

Insa Theesfeld

Abstract This article aims at explaining the extent to which the state and international donors intervene in Bulgaria's irrigation sector by enforcing legislation in the transition country. The process of designing national irrigation sector policy reform and, in particular, its implementation in post-socialist Bulgaria is understood as an intended process influenced by the respective political power holders. Four subsequent legislations which determine Bulgaria's irrigation sector policy are therefore analyzed with aspects of the public choice theory of institutional change. Empirical examples emphasize how state authorities extend their decision-making power by the way regulations are implemented and how individual actors extract private short-term rents from the resource system. Commitment of political leaders is shown to be a decisive determinant for effective devolution-oriented policy implementation in natural resource management.

1 Introduction

Recently, common-pool resource scholars call to take distributional aspects into account when analyzing institutional change in common-pool resource management (Meinzen-Dick et al. 2002, 652; Agrawal 2001, 1650–1656). The way benefits are distributed among various actors is decisive and the respective political weight of the latter can influence the likelihood of institutional change (Baland and Platteau 1998, 649). When new rules are implemented, some people benefit more than others. Indeed, some may even benefit at the expense of others. This applies also to Bulgaria's irrigation sector in transition, where power processes have determined outcomes of the national irrigation water policy.

The aim of this article is to explain the process of designing irrigation sector policy reform in post-socialist Bulgaria. This part of the agricultural sector reform was

Insa Theesfeld
e-mail: insa.theesfeld@agrar.hu-berlin.de

W. Scheumann et al. (eds.), *Water Politics and Development Cooperation*,
doi: 10.1007/978-3-540-76707-7_12,

politically neglected for one decade. A political economy driven empirical analysis reveals political intentions and strategies behind designing irrigation sector policies. It provides insights why the formally devolution-oriented reform – which gives the impression of handing over decision-making power in resource management to local communities – is in fact a further concentration of power in the irrigation sector. Empirical evidence is given for pseudo-devolution due to the actual implementation process of Bulgaria's recent legislation in the irrigation sector, which results in a concentration of property rights with the state authorities. In fact, this leads to even stronger influence and control of state authorities over local communities.

Irrigation, until 1990 a major water user in Bulgaria, has been drastically affected by the transition process. Uneven distribution of Bulgaria's natural water resources over time and space makes irrigation necessary to reduce production risk and insures the common-pool resource a continuous high economic importance. Yet, irrigation systems were built to serve large production units during socialism and do not meet the needs of the huge number of small-scale landowners that emerged following the land restitution process. This misfit is described for many post-socialist countries, such as the Kyrgyz Republic (Herrfahrdt et al. 2006, 47). In Bulgaria the facilities have largely deteriorated, the property rights on the infrastructure are ambiguous and the water loss in the system amount to 70 percent due to non-maintained facilities and stealing of water.

Currently, two ministries form the head of Bulgaria's hierarchical organizational structure of the irrigation sector. The Ministry of Environment and Water is responsible for Bulgaria's water sector, including the environmental supervision of the country's natural water resources and the coordination of the overall water balance. The Ministry of Environment and Water coordinates water supply and demand for the various sectoral components: power, industry, municipality, and agriculture. Water consumption in agriculture comprises crop irrigation, livestock breeding, and fish farming. For the irrigation sector in particular, the Irrigation Office, which is affiliated with the Plant Growing Directorate of the Ministry of Agriculture and Forestry, is in charge of. The Irrigation Office can be considered the coordination unit implementing the state irrigation policy.

The Ministry of Agriculture and Forestry delegates the management of the irrigation sector to the Irrigation System Company (ISC) state firm. The ISC has been registered under commercial law as a stock-holding company with the state as the sole owner. It has a monopoly on irrigation water supply. Allocation of irrigation water based on market coordination such as trading water rights or quotas does not exist. Irrigation sector management is centralized. Decisions are implemented top-down, and currently there are no opportunities for the agricultural water users to participate. No other organization besides the ISC plays a major role in the irrigation sector. The ISC is responsible for the management, operation and maintenance of all state-owned irrigation and drainage systems in Bulgaria. Twenty-three regional branches operate semi-autonomously but answer to the head office in Sofia, especially for financial control (World Bank Office Sofia 1999, Annex 1). The hierarchy forms an umbrella structure of the ISC from the head office to the regional branches and onto the employees working in the villages – the water technicians and the water guards.

This article has two main strands and is structured as follows. First, Sect. 2 contrasts the devolution-oriented reform and the de facto power concentration with the state authorities in the irrigation sector. Four subsequent regulative acts aiming at devolution[1] determine the irrigation sector policy in Bulgaria: the World Bank project to set up water user organizations, the Bulgarian Water Law, the Water User Association Act to facilitate water user associations and the latest amending bills to the Water User Association Act. Their main formal objectives and effective impacts are described in detail. Empirical examples serve to underpin the de facto pseudo-devolution. The examples emphasize the way private short-term rents are extracted from the resource system and the way state authorities extend their decision-making power by the way regulations are implemented. Second, Sect. 3 adds a detailed scheduling of the water sector policies from a political economy perspective. In particular, it is referred to aspects of the public choice theory of institutional change focusing on voters' decisions and party competition. It is shown how political actors design the policies in a way that their party can gain votes in the next election campaign and how political actors can misuse the resource system for economic benefit. The latter is the consequence of their short-time horizon in politics. Political and economic hold-to-power-strategies including power misuse to keep control under new legislative conditions and private budget maximizing are analyzed in this section. Section 4 discusses, in the light of the political and economic hold-to-power strategies, the possibilities for legal transplants from non-transition countries. Commitment of political leaders is shown to be a decisive determinant for effective devolution-oriented policy implementation.

2 Post-socialist Irrigation Sector Policy

During the first decade of transition, property rights on water, including irrigation infrastructure, were implicitly regulated or passed by. Political reforms in the irrigation sector started as late as the land restitution process had approached its end and was not as diversified as the land reform.

Besides legislation that builds the legislative impact, there are other formal but non-legislative impacts on the national irrigation policy, such as the activities of the World Bank project facilitating water user organizations (WUOs). Both, legislative and non-legislative impacts in the post-socialist irrigation sector are driven by political strategies and must be jointly considered.

Since 1991, the World Bank has attempted to set up WUOs (World Bank Office Sofia 1999) within the frame of the Cooperative Law. Thereafter, in the light of EU accession, the Bulgarian government enacted two new laws with major impact on the irrigation sector: the Bulgarian Water Law, implemented in January 2000, as

[1] According to Knox and Meinzen-Dick (2001, 3) devolution indicates the transfer of responsibility and authority (here: over natural resources) from the state to non-governmental bodies, particularly user groups. In contrast, decentralization refers to authority and management transfer to lower levels of government.

well as the Water User Association Act, which came into force in March 2001 with its by-laws. Both legal acts claim to reform and decentralize the former centrally planned water sector and increase the involvement of local actors.

2.1 World Bank Project and the Cooperative Law

Roughly, the World Bank project can be subdivided into three phases. The pilot phase of the World Bank project lasted from 1991 to 1995. The objective was to analyze the opportunities for the foundation of WUOs according to the Turkish model. A small group of experts, mainly employees from the ISC, were trained by the World Bank and conducted field visits at WUOs in Turkey. Finally, four pilot WUOs were established in Bulgaria.

The second phase of the project officially pursued devolution objectives in irrigation management and was running from 1995 until 1996. Together with the governmental power holders, the World Bank hired consultants to organize the setup of WUOs. In July 1996, one consultant was employed at each ISC regional branch for at least 3 months. In the World Bank report these consultants were referred to as "local facilitators... who assisted the start of WUOs" (World Bank Office Sofia 1999, 4). The consultants did not have to prove their hydromeliorative expertise. Crucial for the development of the WUOs was that the consultants were paid on the basis of the number of registered WUOs, obviously a strong incentive to found as many WUOs on paper as possible. Thus, by the end of 1997, 206 WUOs were reported to be established, 128 of which were registered at courts (World Bank Office Sofia 1999, 4).

The third phase of the project began with the termination of the regional facilitator's work by the end of 1996. During this third phase, the World Bank attempted to transfer management responsibilities from the ISC state firm to local WUOs. They should take over the responsibility of operation and maintenance of the irrigation and drainage infrastructure at local level. The basic idea behind this project was "to manifest the WUOs readiness and willingness to rehabilitate the irrigation facilities managed by them through their own funds" (World Bank Office Sofia 1999, 6).

In 1999, the World Bank project phased out. In 1999, 172 WUOs were reported to have registered with the court. Most of them, 164 were founded under the Cooperative Law. The Ministry of Agriculture and Forestry reported that out of the total reported number of WUOs about 30 were actively operating. Finally, only eight WUOs were listed and received a World Bank credit for rehabilitating their irrigation facilities. In 1999, the World Bank admitted that only one out of the 172 WUOs meets all requirements for an operational WUO, the rest of them existed only on paper (World Bank Office Sofia 1999, 6).

Most of the irrigation systems under WUOs' command did not represent independent hydrological systems, which means that the ISC sold water to these WUOs. Despite this fact, all WUOs defined the water price for their clients arbitrarily, i.e. independently from the ISC and the state. This resulted in most cases in artificially high prices, instead of lower prices. The latter is a sign for the strong profit orientation of the initiators. Frequently outsiders of the villages managed to receive the use rights for the canal system. By reducing maintenance work to a minimum and not ensuring a reliable water supply, they concentrated on collecting the water fees, partly in advance. Thus, they managed to extract high personal rents, with severe consequences for the already deteriorated infrastructure. There was no financial help from the Ministry of Agriculture and Forestry to repair or maintain the infrastructure used by the WUOs, although some founders had hoped to additionally receive state grants or a World Bank loan when establishing a WUO. Most of the WUOs either terminated their work voluntarily as the expected financial support failed to appear, or their existence was terminated by the Ministry of Agriculture and Forestry after strong protest by the water users who refused to pay the overpriced water fee. In both cases the use rights on the infrastructure fell back to the ISC state firm.

The following observation gives evidence for extracting private short-term rents from the resource system. The empirical material was collected during 3 months fieldwork between 1999 and 2002. Besides expert interviews and 17 village case studies which provided an overview of the irrigation situation, four in-depth village case studies were carried out. In 1999, in one of the case study villages, nonvillagers founded a WUO according to the Cooperative Law. Founders only had to be landowners of plots located alongside the main distribution canal, which serves a number of villages. This foundation was inscrutable for the population of the respective village. For instance, the manager of this organization refused to name the other six founders and members. Most of the villagers were unaware of the possibility to establish a WUO. Likewise, they knew hardly anything about the formal existence of a WUO in their village. The villagers spoke of this organization either as a private water firm or as a tenant renting the canal system. The villagers were only aware that the water guard was from their village, without knowing the other parties involved. The water guard was the father of the manager of the organization. The situation resembled one of open access, with efforts by a formal institution to exert some authority. The effective water ordering and appropriation rules in the village show that the WUO was not an effective company. During spring 2000, the water guard employed five pensioners for five days to clean the canals, which was the only maintenance work in the season completed by the WUO.

The manager of the WUO took advantage of the information asymmetry that existed between him and the villagers. He held a leadership position in the Youth Organization of the Peasant Party, which held governmental power in coalition with the Union of Democratic Forces (UDF) from 1997 until 2001. Due to his political engagement, the manager of the WUO had access to various kinds of information and could participate in a course offered by the World Bank, in which he was trained in establishing WUOs under the Cooperative Law. He used his powerful position, good contacts, and supplement knowledge to establish this WUO. The prestige he had earned by the establishment furthered him in his career in politics. He gained extra income from the collection of water fees and made an additional profit by not spending adequate funds for maintenance work.

2.2 The Bulgarian Water Law

The Bulgarian Water Law entered into force in January 2000 (State Gazette No. 67/ 27.07.1999). It defines the formal ownership structures on the water resources of the country. Further, it specifies formal property rights, especially control rights, on water usage and partly on the water facility management. The property rights specifications – the rights and duties evolving from the ownership and the constraints to ownership – have a larger impact on the irrigation practices than the formal ownership structure. For instance, the law restricts private ownership rights on water through resource quotas limiting the water uptake from wells by landowners to ten cubic meters water withdrawal in 24 hours and not more than 0.2 liter per second (Art. 43 (2)). Beyond this limit water users must apply for permission and pay tax (Art. 44 (1)); yet no control mechanism for the maximum discharges is outlined.

2.3 Water User Association Act and By-laws

In Bulgaria, a special law for water user associations (WUAs) was enforced in March 2001 (State Gazette No. 34/06.04.2001). The WUA Act aims at strong decentralization and devolution of responsibilities. In particular, the main elements of the law are (a) transferring ownership of internal canal systems from the state to agricultural producers and (b) changing the direction of the decision-making process from top-down to bottom-up and thus delegate irrigation system management and property rights to the water users. The introduction to the WUA Act describes the motivation of the law: to adapt the irrigation sector to the farm structures and property rights on land that evolved after 1990, as a result of Bulgaria's agricultural reform. Article 2 outlines the purpose of the WUAs: "Water User Associations shall be voluntary organizations of natural and legal persons, which, in accordance with the interest of their members and society and through mutual assistance and cooperation, shall perform activities related to the irrigation and drainage of agricultural lands and the maintenance of irrigation and drainage infrastructure on a specified territory."

All WUOs, which had been formally established, had to re-register under the condition of the WUA Act within a period of 6 months from 1st of April 2001. WUOs which are not re-registered until the end of this period shall be terminated and taken off from the register. Their irrigation infrastructure shall fall back into the management of the ISC state firm.

The operation of WUAs shall be under the supervision of the State (Art. 5). There are various functions of a supervisory body, which shall be accomplished by the Ministry of Agriculture and Forestry (Art. 5). For instance, amendments in the WUA's statutes have to be approved by the supervisory body (Art. 16). This supervisory body is specified as a Hydromeliorative Agency in subsequent by-laws. These aspects show already the strong influence of the state within the WUAs-formally community organizations.

2.4 *Evidence for Extending Decision-Making Power of State Authorities*

There is divergence between the formal devolution attempts and the actual further concentration of power with the ISC state firm. This is manifested by analyzing actual registration processes of WUAs, which show the competition between the WUAs to be founded and the ISC state firm. First the formal procedure for the registration is described, which rather hinders than favors the foundation of WUAs. Second, the granting of only an opportunity for a use rights transfer specified in the WUA Act is a critical issue.

2.4.1 Formal Registration Process of Water User Associations

In contrast to the registration of a WUO under the former Cooperative Law, the registration of a WUA under the WUA Act is strictly regulated. To start the procedure for the establishment of a WUA a local "Constituent Committee" has to be founded, which comprises at least five persons who possess the title deeds of their land and who are served by the same irrigation infrastructure. This group of people has to apply for the "opening of an establishment procedure"[2] (Art. 8(2)). The Hydromeliorative Agency shall be responsible for the supervision of the restructuring of the ISC regional branches, for the development of the WUAs as well as for the allocation of funds for the rehabilitation of the irrigation system. In addition it shall decide as a supervisory body on the application and eventually ratify the order for the opening of an establishment procedure (Art. 9). In contrast to the legal requirement (State Gazette No. 53/12.06.2001), up to July 2002 no Hydromeliorative Agency was established with the deciding power over the applications to open the procedure to establish a WUA. Especially the latter duty has been taken over by a Temporary Committee. Insiders call this Temporary Committee a "stop block" indicating its purposeful slowing down of the management devolution process.

A first indication for the actual concentration of power with the state authorities is the membership of this Temporary Committee which cannot be regarded as neutral as it only comprises specialists from the ISC and the Ministry of Agriculture and Forestry. There is only one participant who does not represent one of the two closely linked state authorities, who is a member of the union of water users. Yet, this one has no vote right. The Committee's composition represents a crucial aspect in regard to the competition between the ISC state firm and the planned WUAs.

There is rivalry for the midsized infrastructure between the ISC state firm and the WUAs. The ISC state firm wants to keep certain irrigation systems under its control, e.g. the profitable ones or those ones with good established relationships to big individual water users, which ensure reliable side-payments. In particular, the ISC tries to keep those irrigation command areas under its control which incorporate a

[2] The term indicates a two stages process. The procedure to apply for the opening of the establishment procedure is a preceding requirement.

water dam, a barrage or a main distribution canal. These so-called bottlenecks are easy to control and manageable at low costs. For the manager at the ISC head office the advantage of a WUA arises from the fact that his company has to negotiate only with one contractor for a certain command area, assuming that the water dam or the main distribution canal as supplying infrastructure remains under the control of the ISC. The ISC wishes to outsource other irrigation command areas, especially those with a very small-scale ramified and destroyed canal network, to the water users. In general, with a growing number of WUAs the ISC looses its legitimation to exist. The fear of loosing responsibility and hence jobs, triggers off competition between the ISC and the WUAs. In the light of high annual subsidies paid by the state in support of the irrigation sector, the status quo puts the ISC in the position to distribute and decide over the full amount of subsidies. Whereas a decentralization of management would imply a decentralized subsidy distribution, an effect not welcomed by the ISC employees. Regarding the competing situation, the empowerment of the ISC department heads to decide over the WUA's establishments through their representation in the Temporary Committee clearly favors the advocates of the centralized irrigation system management.

Until July 2002, 150 applications have been received for 110 command areas. The Temporary Committee held eight meetings up to July 2002 to decide on applications. It gave advice to the Minister to approve or to dismiss applications, resulting in 70 orders for the opening of the establishment procedure signed by the Minister. In addition to the frequent complaints of WUA initiators referring to the ISC regional branches' hindering legal establishment procedures, another indication for the strong competition is the fact that up to 2002, only one WUA was registered. Moreover, this WUA did not receive water from the ISC but independently from a river.

In several cases two applications referred to an identical command area or one identical Constituent Committee applied for different command areas. In cases where more than one application for opening an establishment procedure is handed in, for an identical command area, the Ministry can decide how to proceed. In the latter case, it takes over the role of the Constituent Committee with the aim to either postpone the decision about the management of the WUA by the water users or to directly decide and adjust the boundaries of the command area.

After a Constituent Committee received the order to open the procedure, the second stage in the application procedure is reached and, in compliance with the WUA Act, further formal steps have to be passed. The initiators have to organize at least two preliminary "Constituent Meetings", which have to be announced in local and national newspapers. The Constituent Meeting is legitimate, if the persons described in Article 6(2) participate (Art. 11). The objective of the meeting is to compile a list of all potential founders of a WUA. The latter refers to the rule that 51 percent of land owners and users who own and use more than 50 percent of the agricultural land on the irrigation command area have to be founding members. This requirement is restrained by Article 12 which states that each founding member may authorize another person to represent him/her in the Constituent Meeting in writing by his/her signature and by witness of a Notary Public. In particular, this article involves opportunities to bypass the law. For instance, large tenants frequently act and sign on behalf of their lessors without notifying them. The participant list of the Constituent Meeting, as well as the statutes of the planned WUA have again to be approved by the Minister for Agriculture in subsequent steps.

2.4.2 Opportunity for a Use Rights Transfer

Another interesting issue of the WUA Act to be highlighted here concerns the use rights of the infrastructure granted to WUAs and the ownership transfer. Article 47(1) states: "The associations shall be entitled to acquire use rights, free of charge, over the irrigation facilities as well as the service equipment on the command area of the association, included in the property of trade associations in which the state is a sole trader [i.e. the ISC]. The terms and conditions for transferring and withdrawing use rights shall be in conformity with an ordinance issued by the Council of Ministers on a proposal from the Ministry of Agriculture and Forestry." A WUA which uses facilities in compliance with the previous statements "shall be entitled, within a period of up to 5 years from use right acquisition, to acquire property rights on them free of charge by a decision of the Council of Ministers on a proposal from the Ministry of Agriculture and Forestry" (Art. 47(4)).

At this point a digression to the Water Law (State Gazette no. 67/27.07.1999) should help to illustrate the deliberate fuzziness of the laws which enable the state at any time to exert an influence on the irrigation sector. For instance, the text of Article 91(1) of the Water Law declares: "Owners of water economy systems shall be able to concede rights to use over the systems or technologically detached parts of them to water user association in connection with the subject of activity of the association for a term no longer than 10 years." The latter regulations, together with the provisions specified in the WUA Act, specify only an opportunity for establishment of use rights. The granting of use rights to WUAs is not a mandatory obligation, but a legal option for the state firm (World Bank Office Sofia 1999, Annex 4 – Legal Aspects). The World Bank conceives the granting only of an option to establish a use right an inadequate legal solution referring to the present crisis in the irrigation sector which additionally hampers investments of the associations. The World Bank concludes, that with the Water Law the rights of the ISC "partially retained in full", because the ISC remains owner of infrastructure used by the WUAs. For the infrastructure that is declared public state property, a use right for the state firm will be established (World Bank Office Sofia 1999, Annex 4 – Legal Aspects).

2.5 Latest Amending Bills to the Executive Hydromelioration Agency's Structural Rules and the Water User Association Act

There are three legal amendments to the Executive Hydromelioration Agency's Structural Rules dating from October 2001, January 2003 and March 2003 (State Gazette No. 88/12.10.2001; State Gazette No. 4/14.01.2003; State Gazette No. 25/ 18.03.2003). The amendments served to strengthen the Minister of Agriculture's role within the Agency. The latest amendment from 2003 further centralized the responsibilities. Accordingly, the Minister of Agriculture was granted the decision-making power while the Agency was subordinated within the hierarchy. The power to decide on investment grants is becoming increasingly centralized as well. Likewise, the Agency lost its decision-making power and was degraded to fulfilling an advisory role, supporting the Minister in his duties rather than executing the duties, as had been indicated in previous versions of the Executive Hydromelioration Agency's Structural Rules. Moreover, in previous versions of the rules, the Director of the Agency had to be installed by the Minister for duration of 5 years. Due to the latest amendment his term of office can be terminated at any time by the Minister, indicating increased instability of the Director's position and one-sided dependence.

Currently, in Bulgaria two proposals to amend the WUA Act are circulated, one submitted by the Council of Ministers in March 2004 and one submitted by a member of the Turkish Party in July 2004. There were a number of first and second readings of these bills in plenary sessions in the National Assembly. However, these amendments have not been adopted by the parliament. With the starting of the pre-election period and the parliamentary elections in Bulgaria in summer 2005 the discussion about the WUA issue seems to be suspended.

As regards the devolution processes and the redistribution of property rights in the irrigation sector, two aspects in the first amending bill are of particular interest. First, the rights and decision-making power of the Minister of Agriculture is foreseen to be expanded (Sect. 15 Art. 66(2)). For instance, the Minister should approve the statutes of the WUA's and changes to it, he can also start an own initiative to found a WUA, and he can issue an order on the rules concerning the infrastructure use. This opposes the decentralization objective and the claimed aim of devolving responsibilities from the state to the communities. In fact, a concentration of power would be the result.

Second, the number of persons necessary to found a WUA should be reduced. This change in procedural rules leads at first sight to simplified practices and a reduction in obligations, but bears the risk of misuse. It becomes easier for outsiders of the village to initiate a pseudo-WUA with other objectives than to organize collective action. It would reduce the burden of collecting a large number of signatures. Anyhow this formal requirement was relatively easy to bypass.

In general, in the amending bills to the WUA Act no measures could be found which would directly facilitate the emergence of cooperation, such as spreading information, strengthening the advisory system or increasing interpersonal trust among the rural community members. The latter, regarded as one basic prerequisite for collective action to emerge.

3 Public Choice Perspective on the National Irrigation Sector Policy

The political power play and the frequently changing governments in Bulgaria's transition period are extensively reviewed in the transformation literature (Dobrinsky

2000, 599; Swinnen 1997a; Davidova et al. 1997). Swinnen (1997b) suggests analyzing the reforms in Central and Eastern European countries at "analytical stages" representing varying governmental power holders. Swinnen (1997b, 371) and Hanisch (2003, 109ff.) use these stages to subordinate agricultural property reform. Each time the power shifted between the Reds[3] and the Blues, agricultural reform policy objectives were modified, for instance, by amendments to the Law for Ownership and Use of Agricultural Land (LOUAL). Aspects of the Public Choice Theory of Institutional Change provide a good basis to analyze formal institutional change at national level and of politically initiated local processes. In a narrow sense the Public Choice Theory of Institutional Change explains the drivers of institutional change as the competition for votes (Meyer 1996). The maximization of votes regarded as the core effort of the politicians can help to explain the actions of political parties and, in particular, the design and implementation of legislation. Such kind of mechanisms are indeed not so distinct in the irrigation sector reform as they can be found in the frequent amendments of the LOUAL, but politically motivated decisions are clearly perceptible in designing irrigation sector policy. Phased or delayed property rights assignments proved to be a new object in the political power play.

In general, the explanation of competition between parties assumes a long-term relationship between voters and politician. This does not hold true for Bulgaria – a country in transition. The fact that decision-makers loose their positions with each political power change increases the instability of formal structures and reduces the time horizon for the actors. Likewise, their short-term of holding office leads to a behavior known as the "grabbing hand" (Olson 2000). Self-interested actions are more important for politicians than implementation of their promised policy agenda. It encourages position holders to strive for individual profit maximization as long as they have the power to do so. Frequent position replacement applies to the ISC state firm as well. For instance, the Executive Director from 1996 was replaced in 1997. In the first 3 months of 1997 there have been three Executive Directors. The last held office until 2001 and was replaced by the government change in June 2001.

Considering several phases of the World Bank project supporting WUOs, and subsequently enforcing the WUA Act created frequent legal insecurities perceivable even at the local level. The evolving instabilities at the local level are either accepted or deliberately encouraged by the decision-makers. Legal instabilities and insecurities allow private profits of power holders and make what Rabinowicz and Swinnen (1997, 20) call "economic and political hold-to-power-strategies" feasible, where legislation is set up in a fuzzy way to help political adherents to manage and control irrigation systems. In the same way, Christophe (2005, 16) describes Georgia's lack of formal enforcement as a result of intended processes.

In the following, irrigation sector policy reform is analyzed with regard to subordinate political events offering a public choice perspective. Table 1 emphasizes three periods of political power holding – governmental terms – decisive for national

---

[3] The political system went from "red" to "blue", referring to the political organizations: the "Reds" were the Bulgarian Socialist Party – the former Communist party which had renamed itself – while the "Blues" were the Union of Democratic Forces, the reformers.

Table 1 Formal institutional change in post-socialist irrigation sector policy

Analytical periods	Political power holders	Actions	Political economy perspective
Period I 1995/1996	BSP government (Videnov), "grain crises"	Second World Bank project phase: Extensive foundation of WUOs on paper with the help of "facilitators"	Red WUOs to keep political influence in rural areas. Political and economic hold-to-power strategy.
Period II 1997–2001	UDF government (Kostov)	Third World Bank project phase: Foundation of WUOs by political adherents Adoption of Water Law Adoption of WUA Act	Blue WUOs to keep political influence in the rural areas. EU accession negotiations led to the Water Law. Superficial WUA Act solely to demonstrate caring for the rural population prior to elections.
Period III 2001–2004	SNM government (Tsar Simeon) Pre-election phase	Passing of by-laws Amending Bills to the WUA Act	Deliberately complicated process to establish WUAs. Intention of government and ISC to keep influence in the irrigation sector. Law amendments increase state power.

Source: own compilation

irrigation sector policy. In each period the political power holders influenced actions in order to facilitate their economic and political hold-to-power-strategies.

3.1 Period I – Establishing Water User Organizations on Paper

After several changes in governmental power holders from 1990 onwards, the Socialists came into power again with the parliamentary elections in December 1994. The Bulgarian Socialist Party (BSP) won with overwhelming majority and left minor influence to the reformers. With the new government under the lead of Videnov, impacts were recognized in the irrigation sector, marking the first analytical period. By the end of 1996, the country had suffered from food shortages, the so-called "grain crisis", as well as from a complete breakdown of the financial sector. As a result of public outrage, the socialist government had to step down in December 1996 and make way for new elections.

This period is to a large extent determined by the World Bank project which pursued the official aim to register at least 60 WUOs. The Executive Director of the ISC was dismissed and the new one strongly supported the establishment of WUOs. In summer 1996, the BSP government together with the World Bank pushed the foundations by employing facilitators, who received rewards for each register entry.

The BSP pursued two objectives: First to receive a World Bank loan for which the reforms and the establishment of WUOs were a precondition (Koubratova 2002), and second to keep strong influence on the rural electorate. The nomenklatura tried to enlarge its economic and political hold-to-power-strategy with the WUO establishments (Rabinowicz and Swinnen 1997, 20). Swinnen (1997a, 135) explains the BSP's strategy as "to cover up their crimes and use privatization to move into key positions in the emerging market economy." Swinnen's latter aspect applies here as controlling the scarce resource water and with this moving into key positions in the agricultural sector. In rural Bulgaria, people often refer to these WUOs as "red WUOs", which exemplifies their political character. During the BSP's governmental term, mainly socialist successor agricultural cooperatives established WUOs which were active in the above described sense. Their motivation was to control not only land and machines but also water as an additional resource and to strengthen their often monopolistic role with the expansion of their activities to the irrigation sector.

3.2 Period II – Third World Bank Project Phase and WUA Act

Prior to early re-elections, Sofianski's constituted a 90-day interim government from January 1997 onwards. In the subsequent parliamentary elections in April 1997, the UDF won the majority of seats and the Kustov government took over political power. In more detail, the faction of UDF formed a coalition with the Peasant Party and the Democratic Party.

Along with this change in government power, which marks the beginning of the second analytical period in post-socialist irrigation policy reform, the World Bank project went into its third phase. With the radical change in governmental power from socialist party to reformers party, a large number of new persons was appointed to leading positions in the economy. Besides the regional WUO facilitators who were dismissed, the management of the ISC was exchanged once more and the representatives of the World Bank in Bulgaria were replaced. Similar to the foundation of WUOs described in the previous period, the continuing foundation of WUOs during the UDF government was politically intended, too. This means that politicians were directly involved in their foundation using their access to information and networks to bring water resource management under their control or to help political adherents to do so. From 1997 until 1999, the UDF supported young politicians by sending them to World Bank seminars. In these seminars they were taught how to establish a WUO under the Cooperative Law, as described in the case above. These WUOs were frequently referred to as the "blue WUOs".

In contrast to the engagement of the socialist successor agricultural cooperative farms during the former BSP government, in this phase not so many cooperative farms tried to register as WUOs. They were mostly rather outsiders of the village but political adherents of the UDF who tried to gain prestige in their party by founding WUOs. The welcome side effect was the opportunity to gain profit from resource management. From the viewpoint of the UDF adherents, these WUOs created the opportunity to bring a scarce resource under their control and with this to exercise power and to gain influence in rural areas. Due to the maintenance of socialists organizational structures, the rural electorate continued to represent a support base for the BSP which the UDF politicians tried to break. The establishment of blue WUOs was a governmental strategy against its political opponents. Another observation supporting this reasoning is the fact that the non-accessibility of documents on the foundation of WUOs was often explained by the ISC with the argument that "its foundation was too political."

During the final years of the UDF government until June 2001, the actions of this government were increasingly influenced by EU accession negotiations which started in 2000. Thus, the enforcement of the Water Law clearly incorporated aspects of the European Water Framework Directive, such as the river basin management (European Commission 2000/60/EG).

In the light of local mismanagement of the irrigation sector by a variety of WUOs, the complaints of the local population increased, which led to the enforcement of the WUA Act just before the parliamentary elections in June 2001. The WUAs were announced in rural areas as a measure to improve rural life and increase agricultural output. In the rural election campaign, the idea was used to propagate the re-election of the UDF and to maximize votes. The design of the law clearly shows that the establishment of WUAs was not taken seriously with regard to long-enduring self-organizing local management. One indication is the fact that the law contains only an opportunity to transfer use rights and later ownership rights to the WUAs instead of an obligation to concede these rights to WUAs, as highlighted above. They are treated as an organizational form to be imposed from top-down. In transition countries, particularly in Bulgaria, there is a high level of interpersonal mistrust and skepticism in organizations based on cooperation. Therefore, the law does not provide enough time to let bottom-up processes grow. Moreover, the local processes are not accompanied by any other measures, for instance one that would rebuild the eroded social capital.

3.3 Period III – Restricting the Impact of Legislation

The UDF lost its political mandate in the national elections in June 2001 due to an ongoing economic crisis in combination with various corruption scandals. Tsar Simeon Sakskoburggotski, former King Simeon II, an outsider who had lived in exile in Spain since 1946, came back to Bulgaria and started his political campaign only a few weeks before the elections. The political grouping behind Simeon – the Simeon II National Movement (SNM) – won the election with 43 percent and has been ruling the country in coalition with the Movement for Rights and Freedom (MRF), a representative body of the Muslim minorities in Bulgaria. Simeon became Bulgaria's new Prime Minister.

During the SNM government, which represents the third analytical period, there have been indications for a strategy to keep the state influence and to protect the ISC's leading role in the irrigation sector. For instance, by the year 2002, decentralized organizational structures had still not been established, such as the Basin

Directorates and the Hydromelioration Agency, although they had been introduced by the new water legislation. Instead, the legislation has been implemented in such a way that a Temporary Committee, composed of high ranking officials from the Ministry of Agriculture and Forestry and the ISC, decides over the devolution of the irrigation infrastructure management on a case-by-case basis, as exemplified above. The unbalanced composition of that committee bears the risk of biased decisions in favor of the state management of the irrigation system. Scientists in Bulgaria estimated that around 25 percent of the irrigated area would be managed independently of the ISC provided that microdams could easily be transferred to local management organizations. Instead, a deliberately complicated process for WUA registration has been implemented. Due to the competition between the ISC state firm and the planned WUAs, whenever the infrastructure, microdams or distribution canals are considered as too important for the ISC, the establishment of a WUA is denied. If the Temporary Committee members consider the command area of a planned WUA as not suitable, boundaries are shifted. Thus, by means of implementation, the WUA Act's formal objective – bottom-up processes – is transferred into a top-down decision-making process.

In contrast to decentralization and privatization attempts during the UDF government in the previous period, the SNM government tries to establish closer relations again to the remaining state companies. This is illustrated by the Ministry of Agriculture and Forestry linking its decision-making power to the ISC state firm. The composition of the Temporary Committee is only one indication of this interweaving. Another evidence for the close connection between the ISC and the Ministry of Agriculture and Forestry is given by the fact that the Vice Minister of Agriculture has also held the position of Head of Board of Directors at the ISC. This personal union has led to the paradox that the Board of Directors and the Head of the Board had to sign the proposals for receiving subsidies, to be approved by the Ministry of Agriculture and Forestry, in particular by the Vice Minister of Agriculture. In the daily work the Ministry tries to prevent that the Vice Minister has to approve his own applications, but the role of the Ministry of Agriculture and Forestry as neutral approval body has been at stake. Furthermore, the state restricted the decentralization impact of the WUA Act by passing regulations and several amendments to these regulations contrasting the original Act, for instance, the Executive Hydromelioration Agency's Structural Rules. After the power change in government in June 2001 three legal amendments were issued to the Hydromelioration Agency's Structural Rules. With these amendments, especially the latest one dating from March 2003, the power of the Agency was weakened and its role was degraded to an advisory one for the Minister of Agriculture.

According to Koubratova (2002), a clear and targeted irrigation management transfer policy has been missing during the SNM government, as the registration procedure and the membership requirements do rather hinder the WUA establishment than support the management transfer. This observation is in line with the strong competition observed between the ISC and the planned WUAs. The ISC would loose power if WUAs based on local self-management could freely evolve.

4 Conclusion

In Bulgaria, the design and implementation of irrigation sector reform is strongly influenced by economic and political hold-to-power-strategies of political actors, including the competition for votes of the rural electorate. The implementation of the current legislation and the result of the political processes are contesting devolution in the water sector. A further concentration of decision-making power within the state authorities is the result. The WUA Act seems democratically and devolution-oriented, but its execution opens up ways for governmental representatives to control and hamper the reforms and to maintain the power of the state irrigation company.

During the last years, the Bulgarian government was increasingly influenced by the EU accession negotiation, which started in 2000. In line with Bulgaria's EU accession, foreseen for 2007, the country is confronted with numerous judicial reforms and approximation of legislation. With the pressure to implement the *acquis communautaire*, the enforcement of the Water Law in Bulgaria in 2000 already clearly incorporated aspects of the European Water Framework Directive.

Besides the EU expectations, the pushing of international donors for certain institutions, such as WUAs, to solve the mismanagement in the water sector may do more harm than good. Individual actors who are capable to achieve short-term access to and the authority for managing the resource system are able to take advantage of the way reforms are implemented and to gain private rents. This provides a ground for opportunistic behavior and destroys the already low level of social capital and trust of local population in such kind of organizations.

Driven by the World Bank, the transplantation of WUAs to Bulgaria as an institutional blueprint was not effective. Pistor (2000, 73) investigates the consequence of the law imported into transition countries, due to a belief in the transferability of another country's practices. She announces three premises for effective legal transplants. Besides the alignment of formal norms with underlying social norms and the provision of solutions for actual conflicts, she calls for a legal system that should respond to and foster demand. Disseminating organizational blueprints for WUAs worldwide is generally inadequate to change people's behavior and incentives. Central officials frequently design the basic structure of the farmer organization that is formally accepted. This design is conceived as a predetermined blueprint for farmers' self-organization. Frequently described, for instance by Tang (1992) for the irrigation sector, central authorities often direct the creation of farmers organizations without considering farmers' incentives and capabilities. This is already a severe problem. But, what is even more neglected are the incentives of the political actors. Pistor (2002) describes that legislation will not be effective, if there is no need among the population for it. This article has shown that interest and commitment of political leaders is just as important. With every new rule, also the distribution of benefits and duties among influential political and economic actors changes. Distributional aspects and power relations have to be taken into account as actors in the fear of loosing their powerful favorable positions will oppose the new rule. Besides technical and financial conditions, development cooperation has to take into account the political power conditions prevailing in a country. When the hidden political agenda of national actors interferes with the original objectives of the reform-blueprints, which donors or advisors try to enforce, the success of the reform will be at stake.

References

Agrawal A (2001) Common property institutions and sustainable governance of resources. In: World Development 29 (10), 1649–1672

Baland JM, Platteau JP (1998) Division of the commons: A partial assessment of the new institutional economics of land rights. In: American Journal of Agricultural Economics 80 (3), 644–650

Christophe B (2005) Metamorphosen des Leviathan in einer post-sozialistischen Gesellschaft. Georgiens Provinz zwischen Fassaden der Anarchie und regulativer Allmacht, Transcript Verlag, Bielefeld

Davidova S, Buckwell A, Kopeva D (1997) Bulgaria: Economics and politics of post-reform farm structures. In: Swinnen J, Buckwell A, Mathijs E (eds) Agricultural privatization, land reform and farm restructuring in Central and Eastern Europe, Ashgate Publishing, Aldershot, 23–62

Dobrinsky R (2000) The transition crisis in Bulgaria. In: Cambridge Journal of Economics (24), 581–602

European Commission (2000) EU Water Framework Directive (2000/60/ EG). Official Journal L 327, 22.12.2000

Hanisch M (2003) Property reform and social conflict: A multi-level analysis of the change of agricultural property rights in post-socialist Bulgaria. In: Institutional Change in Agriculture and Natural Resources, Vol. 15, Shaker, Aachen

Herrfahrdt E, Kipping M, Pickardt T, Polak M, Rohrer C, Wolff CF (2006) Water governance in the Kyrgyz agricultural sector. On its Way to Integrated Water Resource Management, Studies No. 14, German Development Institute, Bonn

Knox A, Meinzen-Dick R (2001) Workshop summary. In: Meinzen-Dick R, Knox A, Di Gregorio M (eds) Collective action, property rights and devolution of natural resource management: Exchange of knowledge and implications for policy, Zentralstelle für Ernährung und Landwirtschaft, Feldafing, Germany, 1–33

Koubratova HM (2002) Bulgarian water user associations: Situation and problems. Paper presented at the Central and Eastern European Sustainable Agriculture (CEESA) – Bulgarian Policy Learning Workshop, July 18–21, Plovdiv, Bulgaria

Meinzen-Dick R, Raju KV, Gulati A (2002) What affects organization and collective action for managing resources? Evidence from canal irrigation systems in India. In: World Development 30 (4), 649–666

Meyer S (1996) Ökonomische Theorie der Umweltpolitik: Der Erklärungswert der Neuen Politischen Ökonomie für umweltpolitische Entscheidungsprozesse, Eul., Bergisch-Gladbach, Köln

Olson M (2000) Power and prosperity: Outgrowing communist and capitalist dictatorships, Basic Books, New York

Pistor K (2002) The demand for constitutional law. In: Constitutional Political Economy 13, 73–87

Rabinowicz E, Swinnen J (1997) Political economy of privatization and decollectivization of Central and East European agriculture: Definitions, issues and methodology. In: Swinnen J (ed) Political economy of agrarian reform in Central and Eastern Europe, Ashgate, Aldershot, 129–158

State Gazette No. 67/27.07.1999. Bulgarian water law

State Gazette No. 34/06.04.2001. Water user association act

State Gazette No. 53/12.06.2001. Executive hydromelioration agency's structural rules

State Gazette No. 88/12.10.2001. Amending law of the executive hydromelioration agency's structural rules

State Gazette No. 4/14.01.2003. Amending law of the executive hydromelioration agency's structural rules

State Gazette No. 25/18.03.2003. Amending law of the executive hydromelioration agency's structural rules

Swinnen J (1997a) On liquidation councils, flying troikas and Orsov co-operatives: the political economy of agricultural reform in Bulgaria. In: Swinnen J (ed) Political economy of agrarian reform in Central and Eastern Europe, Ashgate, Aldershot, 129–158

Swinnen J (1997b) The choice of privatization and decollectivization policies in Central and Eastern European agriculture: observations and political economy hypotheses. In: Swinnen J (ed) Political economy of agrarian reform in Central and Eastern Europe, Ashgate, Aldershot, 363–398

Tang SY (1992) Institutions and collective action: Self-governance in irrigation, Institute for Contemporary Studies Press, San Francisco, CA

World Bank Office Sofia (1999) Irrigation rehabilitation project. Main report, Sofia

Two Steps Forward, One Step Back: Institutional Change in Kyrgyz Water Governance

Elke Herrfahrdt-Pähle

Abstract This article analyses water sector reforms in the Kyrgyz Republic since 2002. It is argued that objective reasons for change in the form of endogenous and exogenous factors provided a good point of departure for the reform process, but that later in the process other factors, namely informal institutions and vested interests of major actors, came to influence the course of reform, resulting in its delay and unsatisfactory outcome.

The article tries to give some answers concerning the role of the main institutions and organizations within Kyrgyz water governance and their influence and interests in the reform process. It concludes that exogenous and endogenous factors play a major role as push factors in "setting the stage" for the initiation of reforms. When it comes to the processes of formulating and implementing these reforms, however, additional factors come into play. These factors include the power wielded by, the incentives offered to and the resources of formal and informal stakeholders at all levels of water governance as well as the passivity of water users, combined with an autocratic style of leadership or the influence of (former) elites.

1 Introduction

The Kyrgyz Republic has been committed to reforming its water management sector since independence in 1991. Since 2002, the internationally recognized concept of Integrated Water Resources Management (IWRM) has shaped the reform process.[1] The government (with the support of several donors) has since then drafted, and the

Elke Herrfahrdt-Pähle
e-mail: Elke.Herrfahrdt@die-gdi.de

[1] IWRM recognizes the need to integrate the ecological, social and economic spheres into water management. The basic principles of IWRM are the recognition of water as a finite and vulnerable resource and as an economic good, the acknowledgement of the role of women and the general need for broad stakeholder participation in water management (so-called Dublin principles; GWP n.d.).

W. Scheumann et al. (eds.), *Water Politics and Development Cooperation*,

doi: 10.1007/978-3-540-76707-7_13,

Parliament has passed a number of laws on the reorganization of the water sector and on the restructuring of water governance.[2] Some progress has been made primarily in the economic sphere and in local water management. Water pricing has been introduced, and the transition to volumetric water fees has been launched. At local level water management has been decentralized and become participatory through the establishment of Water User Associations (WUAs). Nevertheless, serious deficiencies can be identified in the reform process. They relate mainly to ecological and social sustainability and to sectoral integration. Ecological issues such as minimum flow requirements and social concerns, e.g. gender equity, which are fundamental elements of IWRM, are mostly neglected. Besides these imbalances, the reform process is generally rather slow. Even in the spheres where good progress was initially made, the process slowed down considerably when it came to implementation.

The Kyrgyz Republic enjoys relatively good conditions for the reform of its water sector and has made good progress in adopting a comprehensive new legislative framework. The question now is why these good conditions cannot be used to ensure the proper implementation of the reforms. In the conclusions of this chapter it is recognized that institutional change is driven not only by good conditions but also by the interplay of formal and informal institutions and organizations. This has been shown in other contexts, by, among others, Wegerich (2002), who stresses the importance of elites for reform processes in Uzbekistan. To shed some light on this issue in Kyrgyzstan, which, despite the deficiencies referred to above, is making comparatively better progress in the reform of its water sector than its neighbors, the following provides an analysis of the various formal and informal institutions and organizations at the various levels of Kyrgyz water management, their interests and their room for maneuver in implementing water governance reforms. The analysis will focus on the agricultural sector, i.e. irrigated agriculture, since it is the country's main water user, accounting for 93 percent of total water consumption (FAO AQUASTAT 2004).

The article is structured as follows: it begins by defining the role of formal and informal institutions and organizations involved in institutional change in the water sector. It then outlines the conditions for reforming Kyrgyz water governance, using the framework of endogenous and exogenous factors supporting water governance reforms. In Chap. 3 water governance reform efforts and results will be scrutinized against the background of formal and informal institutions and organizations, with a view to explaining the discrepancy between good conditions for reforms and relatively poor implementation. The final chapter sets out the conclusions drawn on the influence of the various factors. It suggests that the subjective reasons for institutional change (such as incentive structures etc.) need to be taken into account when water governance reforms are designed and analyzed.

---

[2] In contrast to water *management*, which concerns the operational aspects of water development, distribution and delivery, water *governance* is defined here as being focused on the setting in which water management takes place and which influences its outcome: "Water governance refers to the range of political, social, economic and administrative systems that are in place to develop and manage water resources, and the delivery of water services, at different levels of society" (Solanes M, Gonzalez-Villarreal F 1999, cited in Rogers and Hall 2003, 7).

2 Institutional Change and Water Governance

Conventional water governance has undergone constant change and has been subject to comprehensive restructuring and reform, especially since the introduction of the new paradigm of IWRM in the 1990s. However, the speed of transition and the outcomes of these restructuring processes have differed widely, even among countries with similar points of departure. What, then, are the enabling factors, and what are the obstacles to water governance reform?

Several researchers have stressed the importance of formal and informal institutions and institutional arrangements[3] both in general and with regard to water governance in particular: in this volume Theesfeld argues that distributional aspects and political power constellations must be considered when it comes to transferring models from one part of the world to another and to enforcing legislation in general and water law in particular. Mollinga and Bolding (2004) dedicated a whole volume to the politics of irrigation reform, addressing policy formulation and implementation in Asia, Africa and Latin America. They reason that policy implementation must be seen as a process and that the interests, strategies and resources of the various actors must be taken into account. With a reference to Grindle, they identify the "politics of policy" as the crux of the problem, which "refers to the political nature of policy formulation and implementation processes, policy as contested by different interest groups at all stages of its existence, and with all interest groups trying to shape it in particular ways" (Mollinga and Bolding 2004, 6). Gatzweiler and Hagedorn point in the same direction when stressing that "institutional change depends on the characteristics and objectives of the actors involved" (Gatzweiler and Hagedorn 2001, 10). As mentioned above, Wegerich (2002) illustrated these considerations with regard to Uzbekistan, emphasizing the importance of elites and the need for their support for institutional change. The interplay of formal and informal institutions is identified as one of the main factors in explaining why some countries are not able to replicate others' more successful policies and strategies (Pejovich 1999, 164).

2.1 The Role of Formal and Informal Institutions and Organizations in Institutional Change

In accordance with new institutional economics, institutions are defined here as "the rules of the game in a society; more formally, they are the humanly devised constraints that shape human interaction. In consequence they structure incentives in exchange, whether political, social, or economic" (North 1997, 2). One of the main functions of institutions is to influence transaction costs und thus provide

[3] The term institutional arrangement refers to the organizational structure of the water sector or water administration as well as all types of organization relevant for the management of water resources (Saleth and Dinar 2004, 25).

incentives for related actions or behavior. In line with North and others, a distinction is made between formal and informal institutions. Formal institutions are all kinds of legally binding norms, such as constitutions and laws. They are defined as institutions which "determine the political system (i.e. the governance structure and individual rights), the economic system (i.e. property rights and contracts), and the enforcement system (i.e. the judiciary and the police)" (Pejovich 1999, 167). Informal institutions, on the other hand, cover such cultural norms as customs, traditions and moral values. They can be defined as "socially shared rules, usually unwritten, that are created, communicated, and enforced outside of officially sanctioned channels" (Helmke and Levitsky 2004, 727).

If institutions are characterized as the rules of the game, then organizations are usually depicted as the players in that game. On the one hand, they take a certain institutional environment as given and act accordingly; on the other hand, organizations try to reshape institutions according to their preferences and thus change the rules of the game they play (North 1992, 5). In the following a distinction is made between formal and informal organizations. Formal organizations include all players or organizations, which are legally entitled to engage in the relevant societal sphere – with defined roles. Applied to water management, formal organizations are those, which have a mandate, an area of responsibility within the water sector under Kyrgyz legislation or a formal status within the sector. Informal organizations, on the other hand, are not necessarily informal in terms of legal status (unlike, for example, more "traditional" institutions as the court of elders).[4] Rather they are termed "informal" here because they do not have formal, legal or other foundations with respect to the water sector. Applied to water management, this concerns those organizations, which do not, or no longer, have a legal mandate to act in water management within the water governance framework.

According to Köllner (2005, 22), both formal and informal institutions affect political actors in that they structure their interests and incentives, influence access to and the allocation of resources or define the range of possible actions. Reforms predominantly signify change to the formal institutions and organizations. Informal institutions or rules, which can be supposed to be in harmony with well-established formal institutions, usually do not change at the same pace, which makes that reform processes often result in an imbalance, if not contradiction, between the new formal and formerly existing informal rules. This imbalance may hinder institutional change since "... if that [formal] rule is out of harmony with informal institutions, people will view it with apprehension, uneasiness, and even outright hostility" (Pejovich 1999, 174).

Reform, i.e. institutional change, signifies a change in transaction costs and incentives for the organizations (and other actors) involved. Often the aim of reforms is to lower specific transaction costs and raise the efficiency of a system. At the same time, though, the implementation of reform itself induces transaction costs. Institutional change is undertaken if the costs of maintaining the status quo (e.g. mitigation of negative effects on the status quo, etc.) are higher than the cost of the

[4] Courts of elders (*aksakal sotu*) exist in most villages and are responsible for law enforcement.

implementation or transaction costs of reforms (political cost of implementing reforms etc.; Neubert and Herrfahrdt 2005, 240). North and Ostrom assume that institutional change at higher levels causes higher transaction costs than changing institutions at lower levels (Wegerich 2002, 19; Ostrom 1990, 52). "Changes in the rules change the way the game is played" (Pejovich 1999, 165). Reforms thus force organizations either to "play in a different way", i.e. according to different (new) rules (and transaction costs), or to find ways to circumvent the new set of formal institutions in order to adhere to the former (or other) informal rules, which seem more attractive to them, e.g. because of lower transaction costs. Thus "a great deal depends on the incentive structures under which rule-makers operate and the effect of the prevailing political order upon those incentives" (Pejovich 1999, 173). These incentive structures are called subjective reasons for change here. They can be geared to supporting reforms or obstructing change if no institutional or other corrective incentives are provided.

In Chap. 3 a closer look is taken at the formal and informal institutions and organizations and related subjective incentive structures in the Kyrgyz Republic with a view to better understanding and explaining the outcomes of water governance reform. First, however, the objective reasons for change in the form of exogenous and endogenous factors in the Kyrgyz case are considered.

2.2 Push Factors for Water Governance Reform

Saleth and Dinar (2000) analyzed institutional changes in the water sectors of 11 countries. On the basis of this research, they formulated push factors for institutional change which catalyze water management reform: "institutional changes within the water sector occur due to the role of both endogenous factors (...) as well as exogenous factors" (Saleth and Dinar 2000, 175). Endogenous factors are those determined within the water sector and comprise, among other things, water scarcity, water conflicts, financial constraints, the state of the water infrastructure and the efficiency of institutions. The exogenous factors for institutional change focus on influences from outside the water sector. They include economic and political reforms, macroeconomic crisis, commitment of international donors and natural disasters. Saleth and Dinar argue that these factors promote institutional change either by lowering the transaction costs or by increasing the opportunity costs of change (Saleth and Dinar 2000, 177). They are referred to here as the objective reasons for change. They play a role in actors' incentive structures by "setting the stage" in that they increase the demand and need for change. At this point the objective reasons for change are closely tied to and translate into subjective reasons for change. Nevertheless, it must also be acknowledged that actors are not free from additional incentives, independently from outside-induced changes: they not only react to objective incentives from their environment, but usually have rather strong internal or subjective motives and incentives, which also affect transaction costs. Among these subjective reasons for change are power play, gaining influence, etc. In the following the objective factors for change are applied to the situation in Kyrgyzstan in order to determine its conditions and point of departure for reforms at the beginning of the 21st century.

Endogenous factors:

- *Water scarcity:* Kyrgyzstan is located in the upstream region of many Central Asian rivers and can thus be characterized as water-abundant at national average.[5] Despite this, water scarcity (due not only to the quantity of water but also to poor water quality) is a serious and growing problem in several regions. Poor water quality is prevalent in many parts of the country, while water quantity problems occur especially in the south. The latter is especially true of water availability in the agricultural sector, which is heavily dependent on irrigation: 75 percent of the agricultural area is irrigated, accounting for more than 90 percent of all water consumed in the Kyrgyz Republic (FAO AQUASTAT 2004). The Kyrgyz economy's dependence on (irrigated) agriculture and water makes it vulnerable, and this will worsen since climate change is likely to have a negative impact on river flow regimes, to reduce the availability of water resources and so to add pressure to a growing water problem (IPCC 2001, 548). Nevertheless, the main causes of water scarcity in Kyrgyzstan are deficient irrigation infrastructure, poor water management and few incentives for water saving (Herrfahrdt et al. 2006, 16).
- *Water conflicts:* Conflicts over water are especially common in the southern parts of Kyrgyzstan and in irrigated agriculture. They regularly occur when the agricultural season has begun, but water from the melting glaciers is not yet available (Bichsel 2006, 80). The main causes of conflicts over water include the scarcity, "theft" and inequitable distribution of water (Herrfahrdt et al. 2006, 126–128).
- *Financial constraints and physical deterioration of water infrastructure:* After the collapse of the Soviet Union, Kyrgyzstan lost the budget support formerly provided from Soviet funds, which made up about 50 percent of its revenues. This led to a sharp decline in public spending in many sectors, including agriculture (World Bank 2004, 8–9). Furthermore, tax revenues generally decreased as the economy shrank. After independence almost no public funds were therefore available anymore for maintaining water infrastructure, which led to the sharp deterioration of irrigation infrastructure (Herrfahrdt et al. 2006, 101–102). This process has been accelerated by the relatively common practice of breaking holes in irrigation canals to withdraw water illegally (Herrfahrdt et al. 2006, 126).
- *Operational inefficiency of water institutions:* During Soviet times the structure of water management and allocation was extremely hierarchical. Insufficient communication and information made water management inflexible and inefficient. For example, divisions of the same ministry communicated with each other via Moscow, and water has been allocated to water users without prior notice (O'Hara 2000, 376). Poor sectoral integration, communication and dissemination

[5] The Water Poverty Index shows Kyrgyzstan in the upper third (Sullivan et al. 2002, 31), and total renewable water resources per capita amount to 4,182 m^3 per year (UNESCO World Water Assessment Programme 2003, 72), while the threshold for water scarcity lies at about 1,000 m^3.

of information still persist today. For instance, water resources are monitored by several research institutes in parallel, yet communication and the exchange of data are limited (Herrfahrdt et al. 2006, 79).

Exogenous factors:

- *Economic and political reforms:* Since independence the Kyrgyz Republic has undertaken a whole range of reforms relating not only to its political but also to its economic and social systems. Because it introduced democratic and economic reforms at a faster pace than its neighbors in the region, Kyrgyzstan became famous in the 1990s as an "island of democracy" in the Central Asian region (Anderson 1999, 23). Considerable progress was made, for example, in public participation and decentralization, especially in comparison with such neighboring countries as Uzbekistan. Kyrgyzstan was the first of the former Soviet Union successor states to gain access to the World Trade Organization (WTO), doing so in 1998. These comprehensive reform processes lowered the transaction costs of the comparatively minor task of reforming water governance. At the same time, opportunity costs were raised through land reform, which included the break-up of the collective and state farms into many small private units. This raised the number of water users at local level and increased the demand for water management reforms because of the mismatch between water and land rights, on the one hand, and irrigation infrastructure, on the other hand.
- *Macroeconomic crisis:* After the disintegration of the Soviet Union, which led to the collapse of the little industry in the country and to the return of subsistence agriculture, the Kyrgyz Republic underwent serious economic destabilization. Inflation climbed to over 800 percent annually, and by 1995 GDP had dropped to 50.7 percent of its 1990 value (EBRD 2004, 40; Koshoev et al. 2003, 5). Membership of the WTO required the lowering of tariffs and thus an additional reduction of public revenue (UNESCAP 2001), which in turn further reduced the room for maneuver of the government. It thus prepared the ground for institutional change, since the state was no longer able to finance much of the irrigation and water infrastructure.[6]
- *International commitment:* Today, IWRM is internationally recognized as the leading concept for water management reform. During the 1990s IWRM was increasingly acknowledged as the state-of-the-art concept for modern water management practice and was subsequently introduced in developed as well as developing countries. This development is reflected by the fact that several donors have attuned their conditionalities, program design and project setup to the principles of IWRM. International donors stepped in by providing the resource-scarce country with the funds formerly received from the Soviet Union. If Kyrgyzstan is more open to change and new concepts than its neighbors, this is not least because it is heavily dependent on donor funding. These factors provided good leverage for

[6] In recent years, however, Kyrgyzstan has made good progress in stabilizing its economy. By 2004 GDP had already risen to 80.4 percent of its level in 1990, and inflation was limited to about 3 percent in the same year (UNDP 2005, 59).

external influence and served as the main push factors for institutional change at national level.

- *Natural disasters:* Kyrgyzstan has had to contend with several natural disasters, such as drought, floods and landslides, in recent years. More than 1,210 disasters occurred between 1992 and 1999 alone (UNECE 2000, 27). Some of the larger incidents included huge landslides, which killed 115 people in 1994 and caused damage amounting to US$ 36 million, and large-scale flooding due to heavy rainfall, causing damage to the tune of US$ 240 million (UNECE 2000, 27–28). While such disasters cannot be entirely avoided, their impact can be mitigated by improved water management practices, such as monitoring, early warning systems, better catchment area management, and close cooperation among the relevant agencies.

All of these factors increased the demand for change in the water sector, in that they raised the opportunity costs of the status quo compared to conducting reforms. They show that before and during the introduction of the IWRM-inspired reform process in Kyrgyzstan there was a variety of push factors (or objective reasons for change) which, taken together, amounted to a major demand for change and which are expected to argue in favor of reforms. The main drivers of reform must be sought in the exogenous factors and especially in the pressure exerted by international donors, who have influenced not only the water management sphere but also the direction and content of the general economic and political reforms. Among the endogenous factors, financial constraints and deteriorating infrastructure appear to have led to the most vociferous demand for change.

However, these objective reasons for change have not been sufficient to sustain reform dynamics. Thus, the following focuses particularly on subjective reasons and the role of formal and informal institutions and organizations in order to explain the discrepancy. It should be mentioned here that Saleth and Dinar explicitly excluded informal institutions or factors from their analysis and focused on formal water institutions to determine the enabling environment for institutional change. They characterize formal water institutions as consisting of the three components water law, water policy and water administration (Saleth and Dinar 2000, 176). As this article centers on the implementation of reforms, which as a rule depends on the (participation and) support of various actors and stakeholders with different aims and agendas, the analysis is broadened here to include informal and subjective factors. In the next chapter a closer look will therefore be taken at the actual results of the reform process and at the subjective reasons for change, which influenced that process.

3 Kyrgyz Water Governance: Reform Process and Reform Results

Kyrgyz water governance has been subject to ongoing reforms for several years. The main results of this reform process can be summarized as follows: overall, the reform process has led to the introduction of comprehensive new water legislation, but its enforcement has so far been deficient. Most progress has been made in regulatory integration,[7] especially in the economic sphere and at local level. Water pricing has been introduced (though at a fairly low level), and decentralization has made some progress, Water User Associations (WUAs) having been set up in many parts of the country, for example. Progress in the ecological integration sphere has been slow, and almost no results are available with regard to sectoral integration.

In consequence, the question arises how a country can, on the one hand, have compelling incentives for reforming its water governance, which it subsequently begins, and yet, on the other hand, make rather slow and selective progress in the implementation of those reforms. In order to find some answers to this question, the role of the main formal and informal institutions and organizations involved in the reform process and their influence on it are analyzed in the following.

3.1 Two Steps Forward ...

3.1.1 Formal Institutions

In recent years a number of laws have been passed, resulting in comprehensive change to formal institutions in the Kyrgyz water sector. They include the Law on Unions (Associations) of Water Users (2002) and the new Water Code (2004). They cover, among other things, the alignment of water management structures with hydrological boundaries, the introduction of water-pricing and water use rights as well as the decentralization of water management and the enhancement of the participation of water users in decision-making processes. With the passing of this new legislation, formal institutions have undergone sweeping changes. Despite minor shortcomings, it lays sound foundations for water governance reform and institutional change.

3.1.2 Formal Organizations

As described above, formal organizations are defined as those, which are formally and legally involved in water governance. They include the parliament, various

[7] For a detailed overview of the reform process see Herrfahrdt et al. (2006), which establishes an analytical framework consisting of three pillars: ecological, sectoral and regulatory integration. Ecological integration signifies that IWRM is based on the eco-system approach and so systematically considers ecological interdependencies, such as water quality and quantity issues and water-land interaction. The second pillar, sectoral integration, requires that water be allocated in the way that benefits the whole of society most. Thus economic, ecological, and social externalities of water use should be taken into account when water management decisions are made. The final pillar, regulatory integration, scrutinizes the organizational and hierarchical structure of water management. Decisions should be taken at the lowest appropriate level (principle of subsidiarity), and they should also support the introduction of demand management as a complement to supply-side management.

ministries, the Department of Water Management – with divisions not only at national, but also at provincial (oblast) and district (rayon) levels –, WUAs, water users and also donors (cf. Fig. 1). The interests of these actors in and their contributions to the reform process are examined in the following.

Parliament

The Kyrgyz parliament is responsible for developing, approving and amending legislation on the management of water resources and for setting water prices and tariffs (Kyrgyz Republic 2004, Article 7). It has impeded the introduction of water pricing in the past. The introduction of water tariffs was achieved under the new Water Code on the insistence of and following pressure from various donors, who,

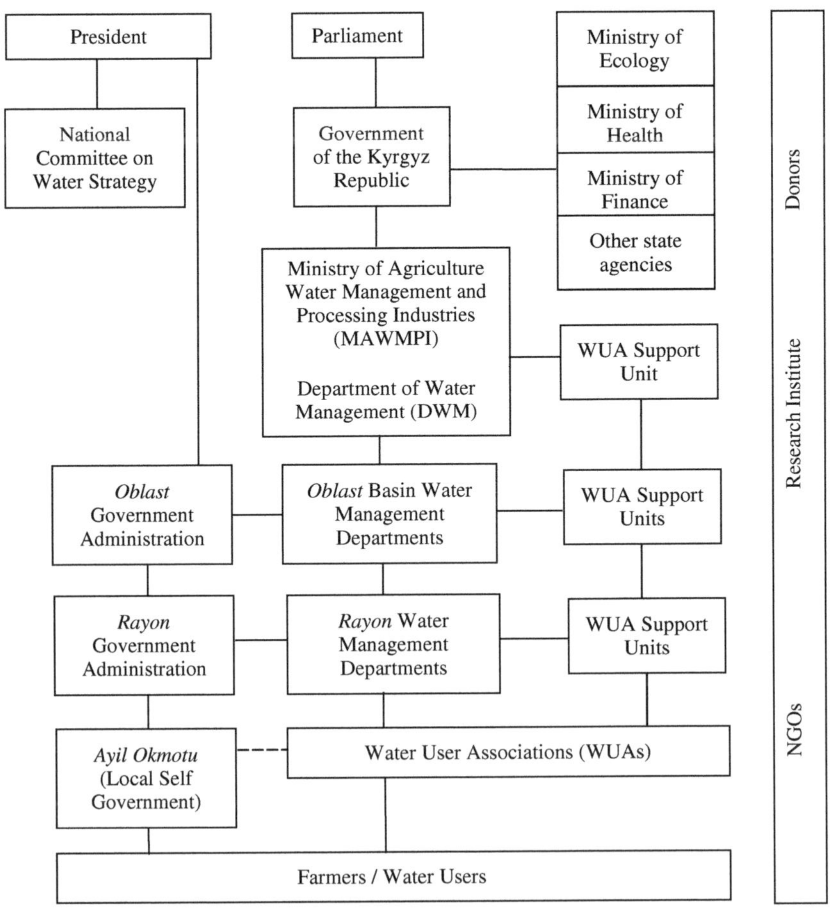

Fig. 1 Institutional arrangement of Kyrgyz water governance
Source: Herrfahrdt et al. (2006, 53)

through consultants and projects supporting the framing of laws, greatly influenced the drafting process. In fact, parliamentarians blocked the introduction of irrigation water tariffs for several years and, only reluctantly and because of increased donor pressure, approved the new Water Code (including the provisions for water fees) in 2004. In addition to their unwillingness to agree to the introduction of water prices in general, members of parliament were subsequently reluctant to increase water tariffs because of their obvious unpopularity, especially with their rural constituencies, which are largely dependent on irrigated agriculture. This attitude is likely to have continued after the "tulip revolution" in 2005, which was mainly supported by the rural population.

Ministries

Responsibility for water management is somewhat fragmented, being divided among several institutes and ministries. The Department of Water Management (DWM), which is part of the Ministry of Agriculture, Water Management and Processing Industry (MAWMPI), is, for example, responsible for irrigation and surface water, while the Ministry of Geology coordinates groundwater issues, and the Ministry of Ecology is responsible for water quality and monitoring. In addition, there is a National Water Council, which is responsible for policy-making and comprises ministers and heads of administrative agencies and other state bodies involved in water management (Kyrgyz Republic 2004, Article 9) but it is unclear in how far this body lives up to its tasks. Despite this variety of water management organizations and decision-makers and the resulting need for information exchange and coordination, communication among the various actors is poor.

The rivalry among the various ministries and their stakes in water governance reform is illustrated by the establishment of a new water governance body. The new Water Code envisages the merging of the various water-related ministerial divisions in a new body, the Water Administration (Kyrgyz Republic 2004, Article 11). The Code provides for the separation of the DWM (which had had the status of an independent ministry before being merged with the Ministry of Agriculture) from the MAWMPI, for it to assume the water management tasks currently performed by several other ministries and for the Water Administration to be established. This plan was strongly opposed by the Ministry of Agriculture, Water Management and Processing Industry (MAWMPI), which defines water management and the tasks of the DWM as its core responsibility. However, the other (potentially deprived) ministries, too, fought to retain their responsibilities (field notes). This dispute led to even less communication as distrust among the ministries involved grew. In the end, the DWM was too weak to insist on the creation of a new body as provided for in the Water Code, and the other actors succeeded in preventing the merging of all water responsibilities in one agency. Even though the DWM will become an independent agency, the new Water Administration, several core responsibilities (such as monitoring water quality) will remain where they are now. Thus, despite the wide range of water management and water policy organizations, which might suggest a fragmentation of power, the ministries involved retain major influence on water policy-making and find ways to circumvent provisions of the water legislation.

Department of Water Management (DWM)

The DWM is an important actor not only at the national level, where it is pushing for its independence and the establishment of the new Water Administration, but also at the lower administrative levels, where it is represented by province and district branches (see below).

Just like the other national stakeholders, the DWM at central level is opposed to handing over tasks to other (in this case, subordinate) units. It thus partly disapproves the decentralization of water management and the establishment of decentralized management structures outside its sphere of influence in particular. For example, it impeded the notion of WUA federations[8] in the new Water Code and is trying to obstruct donor initiatives to establish such entities in a bottom-up approach. In a Soviet management style, the DWM (frequently engaged in "micro-management") interferes in local water allocation and management, not leaving it sufficient resources for the performance of the more strategic policy-making tasks and the development of a long-term vision for water management and governance.

Thus the DWM is, on the one hand, a small and weak actor compared to others at national level, one who has not been in a position to force through, for example, the merging of all water-related responsibilities under its roof. On the other hand, it has done a great deal to help suppress the notion of WUA federations in the legislation and (through its local branches) has had a major influence on the enforcement of this legislation (e.g. bottom-up formation of WUA federations), as can be seen below.

Provincial/Basin DWMs (*OblVodKhoz*)

Provincial DWMs are branches of the national DWM. They are backed by donor-funded Water User Association Support Units (WSUs) in the performance of the task of establishing WUAs. Where their responsibilities and tasks are concerned, even the basin DWM/WSU staff find it hard to distinguish their work from that of the district DWM/WSU at a lower administrative level (Alymkulov and Kulatov 2001, 561; Herrfahrdt et al. 2006, 135). Both have the task of allocating water, providing data on water consumption and assisting with the formation of WUAs. This and the fact that most basin DWM tasks can be performed at district level have led to discord between the two levels of administration.

The provincial DWMs were recently renamed basin DWMs to enable the IWRM principle of management to be implemented along hydrological boundaries. The general political intention to abolish the provincial administrative level and the fact

[8] The establishment of WUA federations is primarily driven by donors and some WUAs. The aim of federating the WUAs in a region is to be able to introduce issues with relevance for WUAs at higher administrative levels.

that the former provincial water agencies were already more or less compatible with hydrological boundaries accelerated the transition to the basin system: this step also kept transaction costs to a minimum (causing no reassignment of responsibilities and only minor shifts in service areas) and simultaneously ensured the continued existence of these agencies. For the same reason, basin DWMs tend to support the establishment of WUAs and the formation of WUA federations: once established, the federations will compete directly with the district DWMs for their tasks and probably make them redundant.

District DWMs (*RayVodKhoz*)

As district DWMs used to supply the state farms with water in Soviet times, they formed the lowest level of water administration. With the land reform and the privatization of the state farms and their division into hundreds of small farms, the number of "clients" of the district DWMs increased dramatically. This was a major driving force or incentive for district DWMs to support the foundation of WUAs and thus help to create a new (non-state) level of water management, thus reducing the number of clients with whom they had to deal (and transaction costs) for the state administration.

For the same reasons that motivated basin DWMs to push for the formation of WUA federations, district DWMs might be expected to obstruct their establishment. It seems, however, that district DWM staff feels fairly safe even if faced with the possible closure of their departments, since there are few water management specialists in the provinces. Staff thus assumes the future federations will have to employ them. With relatively little to lose and probably even better income opportunities to look forward to in donor-supported federations (as compared to the chronically underfunded district DWMs), they also support the formation of WUA federations.

In general, the district DWMs wield some influence at local level. This is rooted in their higher level of know-how compared to that of the farmers and in the expertise of WSU staff and the donor funding of the WSUs. Nevertheless, the influence of district DWMs is weakened by farmers' reservations about state organizations in general.

Water User Associations (WUAs)

The implementation of reforms at local level depends mainly on the establishment of WUAs, i.e. it is influenced by the functioning of WUAs and such factors as the legitimacy of the WUA manager, his/her leadership and the reliability of water distribution. The starting conditions for WUAs were not good, since the irrigation infrastructure was usually in a poor state, thus inhibiting the proper functioning of WUAs. Furthermore, local water governance had been dominated by such informal practices as illegal water abstractions and nepotism for several years, making it difficult to re-establish and enforce formal institutions. The incentives for the foundation of WUAs are mostly of an economic nature. WUAs provide jobs and serve as a basis for the generation of finance from various donors for infrastructure rehabilitation. This is conducive to the formation of WUAs. The dependence on donor funds is also assumed to create incentives for the implementation of such IWRM principles as participation, which donors impose as conditions for the distribution of grants and credits. Nevertheless, IWRM principles are being implemented no more than half-heartedly: owing to information asymmetries donors are not able to monitor implementation of these principles properly. A list of WUA members (many of whom do not even know of the existence of their local WUA, let alone of belonging to such an organization), and a manager, a council and a charter often suffice to receive donor funds.[9] The result is the fairly comprehensive introduction of WUAs, although many exist only on paper and most have yet to perform their allotted tasks of water distribution, conflict resolution and infrastructure rehabilitation (see Theesfeld in this volume for a similar example).

Despite their deficient functioning, WUAs are now widespread and have the potential to become relevant water management organizations in the future. With the support of donors, many WUAs (i.e. their management) are also pressing for the unification of the WUAs in a catchment area in order to establish WUA federations. This would increase the WUAs' influence, enable them to replace the district DWM in the medium term by taking over the respective infrastructure and so increase their authority and access to funds.

Water Users[10]

Water users are mainly interested in the sufficient, reliable and free supply of water, as they were used to in Soviet times. Since water and water services were provided to them free of charge for decades, the motivation of farmers to participate in WUAs and to pay water fees is low. In most places they are only now beginning to understand their own new role and the role of WUAs in water management. Owing to the (in many places) continuing lack of interest and information, opportunities to participate in water management are seized no more than sporadically by the majority of farmers. Nevertheless, in regions where WUAs manage to improve water distribution and infrastructure maintenance and hence to improve reliability of water supply, water users' appreciation of the benefits of WUAs and the acceptance of water fees are growing.

[9] The World Bank has identified seven milestones in the assessment of the functioning of WUAs. They are (1) the establishment of the WUA, including legal registration and the opening of a bank account, (2) recruitment of WUA staff, (3) preparation and approval of an O&M plan, (4) payment of O&M costs and irrigation service fees (ISF), (5) identification of alternative rehabilitation measures, (6) elaboration of a rehabilitation plan, and (7) majority voting on rehabilitation measures and loan applications (Johnson et al. 2002). However, it is unclear how these are being monitored.

[10] Even without being a formal organization in the strict sense, water users are addressed here as major actors and stakeholders in water governance reform. In line with the focus on the irrigation sector, agricultural water users are of greatest relevance here.

Donors

International donors (the World Bank, the Asian Development Bank and others) play an important role in the introduction of IWRM in the political process and in institutional change in Kyrgyzstan in general. After it gained its independence in the early 1990s and budget support from Moscow ceased, the small (in terms of geographic area, economic performance and population) and industrially underdeveloped Kyrgyz Republic became heavily dependent on donor credits and grants.[11] In the water sector, this situation was exacerbated by the progressive deterioration of the irrigation infrastructure, which was in need of comprehensive rehabilitation, all the more so as the country was heavily dependent on (irrigated) agriculture. Through their financial support, donors have been able to exert considerable pressure on the Kyrgyz government with regard to the design of water sector reforms and the conditions attached to rehabilitation credits and grants for those reforms. This led to the introduction of comprehensive new legislation on water, in which – as mentioned above – donors had a considerable say. Nevertheless, the process was biased towards economic aspects. International development banks press mainly for such economic factors as water-pricing and cost recovery not least because they are interested in seeing their loans repaid. They have therefore concentrated on the economic aspects of IWRM and neglected most others (such as ecological integration and social aspects, including gender issues). This bias has influenced the implementation process not only at national level but at all levels of administration.

3.2 ... One Step Back?

3.2.1 Informal Institutions

There are various informal institutions interacting in one way or another with water governance reforms in Kyrgyzstan. Since the "legislation" governing informal institutions differs from that under which formal institutions operate, they cannot be changed deliberately and quickly. Informal institutions change more slowly than formal ones and, if they are not taken into consideration during the conceptualization of new formal rules, they may be inconsistent with new formal institutions. Accordingly, most informal institutions that are relevant in the water sector and influence the outcome of water sector reforms today stem from behavior patterns and practices that evolved in Soviet or even pre-Soviet times. Most of these behavioral patterns have acquired the status of informal rules: they guide human behavior though they are not rooted in formal institutions, but are nevertheless perpetuated and not questioned by a majority even though many of these patterns are inconsistent with formal rules. They include:

---

[11] For example, donors contribute 30 percent more to the DWM budget than the state does (Ministry of Finance, field notes).

- the influence of (former) elites (e.g. state farm directors, village elders) on local water management organizations and water allocation: through their status in society they have the ability to obstruct equitable water supply and access to management positions in favor of relatives;
- autocratic leadership and the passivity of water users: Both those who govern and those who are governed are used to hierarchical and centralized leadership, which inhibits new forms of governance, i.e. participatory and decentralized water management;
- the notion of water as a "gift of God": this pre-Soviet tradition implies free water access and use for all members of society. It was endorsed by the Soviet practice of providing water free of charge and hence still runs counter to the introduction and enforcement of water fees;
- illegal water abstraction: "Water theft is so common that it can be described as a local institution itself as it represents a widely not confronted rule of behavior" (Sehring 2005, 33). This obstructs the enforcement of formal water management rules and impairs equitable water allocation;
- the people's distrust of all forms of state organization: this distrust and skepticism of state organizations hamper the functioning and work of WUAs, which are widely perceived as state organs.

These informal rules are active at all levels of society, but their most direct influence on water governance is felt at local level, where they mainly obstruct the establishment and functioning of WUAs and the introduction and enforcement of water-pricing.

3.2.2 Informal Organizations

Informal actors are defined here as those who are not (or no longer) formally involved in the reform process, examples being local government bodies (*ayil okmotu*), the village heads (*ayil bashi*) and the courts of elders (*aksakal sotu*).

Local Government Bodies (*ayil okmotu*)

Ayil okmotu, a new form of local government body, were established in 1996 (Grävingholt et al. 2006, 47). During a transition period the *ayil okmotu* were made formal water management organizations: in the process of land reform and the fragmentation of the state farms, the tertiary water infrastructure was handed over to the local government body, which was however financially and technically overburdened by this task. The result was the further deterioration of infrastructure. Since 2004 it has been possible for irrigation networks to be transferred to the newly established WUAs, a development which changed the *ayil okmotu* into informal water governance organizations once this step was accomplished. The possibility of transferring irrigation infrastructure left the *ayil okmotu* with a huge economic incentive to promote the foundation of WUAs so that they might hand the infrastructure over to them and transfer this growing burden. Furthermore, the *ayil okmotu* have a considerable interest in seeing water management function well since their tax base largely depends on agricultural production and irrigated agriculture, the most important economic sector in rural areas. This explains the influence and activities of the *ayil okmotu* in the formation of WUAs as new forms of local water management.

Even though they have handed the irrigation infrastructure over to WUAs in most regions of the country, the *ayil okmotu* are still heavily involved in local water management. Local government bodies are, for example, still involved in conflict resolution (despite the existence of the WUAs' conflict resolution committees) and often have better irrigation and agricultural know-how than the infant WUAs themselves (making them the first point of contact for farmers when it comes to calculating water demand, etc.). The involvement of the *ayil okmotu* is ambiguous inasmuch as it stabilizes water management in the short term, but in the medium to long term obstructs WUAs in the independent development and performance of their tasks. This leads to WUAs being perceived by many farmers as irrelevant and further weakens many of them. On the one hand, the initial involvement of *ayil okmotu* in irrigation management had positive impacts because it gave them an incentive to support the formation of WUAs. On the other hand, *ayil okmotu* might in any case have had an incentive to support the formation of WUAs and thus the introduction of efficient water management, given the importance of agriculture as the main source of tax revenue. In addition, *ayil okmotu* now find it difficult to withdraw from water management tasks since local water management would then be left without guidance. Thus *ayil okmotu* fill a gap, which should have been considered when the reforms were being designed.

Courts of Elders (*aksakal sotu*) and Village Eads (*ayil bashi*)

"Traditional" informal institutions, such as courts of elders and village heads, generally wield considerable influence at local level. Their incentive to engage in water management can be explained by the relevance of water in society and their desire to consolidate social capital. Both *aksakal sotu* and *ayil bashi* are involved in the settlement of conflicts and jurisdiction on water conflicts under traditional law. They also disseminate information on WUAs and their work and help them to collect water fees and organize collective work (*ashar*) on the rehabilitation and clearance of irrigation canals. Thus (like the *ayil okmotu*) they compensate for the WUAs' weaknesses by serving as multipliers in the dissemination of information, etc., and they are taking over various tasks from the WUAs, thus depriving them in the medium term of the chance to develop these tasks.

4 Conclusion

Despite relatively good objective conditions for change in water governance, the Kyrgyz water sector is lagging behind in the implementation of reforms. The analysis of the various actors involved has revealed the discrepancy between the objective demand for change and the donor-driven formulation of reform through the drafting of the new Water Code, on the one hand, and the resistance of many actors to these changes and organizational restructuring, on the other hand. The analysis of institutions and organizations and their incentives to support or hinder institutional change explains much of this discrepancy. But the analysis has also shown that formal institutions and organizations are not necessarily "honest implementers" of institutional reforms. Nor can informal arrangements always be termed obstacles to institutional change and reform processes. A distinction must rather be made between the various levels of administration.

In Kyrgyzstan the role of formal organizations at national level is biased. On the one hand, the opportunity costs of maintaining the status quo at national level are very high because of the need for donor funds and the stringent requirements imposed and pressure exerted by donors on national agencies. This has done a great deal to commit the government to reforms and to agree to the framing of new legislation in the form of the Water Code. It gives rise to a situation in which the reform program does not necessarily reflect the need for reforms as perceived by Kyrgyz stakeholders, but is rather a reaction to the country's and government's financial constraints and so reflects the requirements of the international state of the art in water management strategies. When the opportunity costs (donor funds, infrastructure decay etc.) are considered, it is not surprising to find sweeping institutional change occurring primarily at national level (in the form of legislation at least) despite the fact that transaction costs tend to be higher at this level than at lower levels. When it comes to enforcing the new legislation, transaction costs are still perceived as very high at national level, since this would mean the renunciation of "micro-management" by the national DWM and thus the transfer of tasks and powers to lower-level agencies. Despite their finite influence on the design of water policy (due to donor involvement and pressure), actors at national level enjoy considerable room for maneuver in the implementation of these policies, of which they take ample advantage. Struggling to maintain their power, they defend their spheres of influence against the claims of other departments and Ministries and against subordinate administrative organizations, which are entitled to wield greater influence under the new legal provisions on decentralization and the principle of subsidiarity.

The same set of incentives leads the lower levels of administration to favor the implementation of the reforms, which give them greater influence and power and also underline their right to exist (especially with administrative reform pending).

Informal organizations seem to have little impact at national and regional level, but strongly influence the implementation of reforms at local level. Their influence mainly has a delaying effect on the establishment of new institutions and organizations. Although the involvement of informal organizations in the short term stabilizes the functioning of local water management, it contributes to weaken the new formal organizations (WUAs) in the longer term.

In sum, change agents, i.e. organizations supporting and smoothly implementing the reform agenda, can be found principally among international donors and at local level: both WUAs and district DWMs are faced with sufficient economic and political incentives to make the new institutions and organizations beneficial to them. Being among the main initiators of reforms, donors are pressing more or less successfully for their implementation by WSUs, but also struggling with the vested interests and power constellations referred to above. This supports the observation concerning the powerful effects which exogenous factors, mainly donors in this case, have on institutional change (Saleth and Dinar 2000, 197). But it also raises the question of the sense of excessive donor pressure, since it obviously has an impact only at national level, causing the formal rules to be changed, while the more powerful informal rules, which can be changed only from within, remain in place to thwart reforms. The question then is whether a less donor-driven process (focusing on a more country-specific approach to the implementation of the IWRM concept) might have achieved better results, all the more so as the donors themselves do not adopt a comprehensive approach, but focus rather on economic factors and so neglect some vital aspects of IWRM.

To summarize, the process of institutional change in Kyrgyz irrigation management is focused mainly on the establishment of formal institutions and organizations, without giving any consideration to informal rules. This results in the coexistence and inconsistency of (1) new and old water **institutions**, leading to partly inconsistent water management rules (such as the perception of water as a free gift from God, on the one hand, and the introduction of water pricing, on the other[12]), and of (2) new formal water management **organizations** such as WUAs and old informal water management organizations (such as the *ayil bashi*, etc.), which find it hard to withdraw from the scene. In Kyrgyzstan's case, some aspects of water governance reform are not only inconsistent with informal institutions but are also out of harmony with the interests and incentives of influential formal organizations. Thus, despite good objective conditions for reform, they have not been sufficient to ensure substantial institutional change. The required push has, in fact, originated from such exogenous factors as the international community in the form of donors and has not had the support of the relevant actors and organizations involved.

To avoid dual structures, informal rules need to be taken into account when new ones are being established: "[...] the design of efficient formal rules must take into consideration the interaction between new formal rules and existing informal ones" (Eggertsson 1996, 22). Old water management institutions and other forms of local government organizations should be taken into account when, for example, such new institutions as the Conflict Resolution Committees, WUA federations or water fees are being developed and introduced.

The framework established by Saleth and Dinar identified relevant objective factors able to generate demand for institutional change and water sector reform, such as water scarcity (endogenous factors) and donor commitment (exogenous factors). These factors are essential and necessary conditions for reforms. However, other conditions, too, must be satisfied if reforms are to succeed. They are subjective factors, such as incentives and power play among the actors or organizations involved in the process and the role of formal and informal institutions. In Kyrgyzstan's case,

[12] Water managers have been aware of this discrepancy and tried to circumvent it by referring to water prices as "tariffs for water services". However, this is something of a rhetorical trick, and it is still difficult to explain the difference to farmers and local water managers.

these other factors obstruct institutional change because they were not considered in the process and are thus in a position to "reverse" reform processes.

This gives rise to the situation observed in Kyrgyzstan, where initially good progress was made despite internal resistance due to the donor pressure which resulted in promising water legislation. In the process of enforcement, however, rivalry emerged between national water management organizations and informal institutions and organizations engaged in, if not reversing the process, then at least in slowing it down. This situation resulted in a "stop-go" approach to reform and in two steps first being taken forward, followed by one step back.

References

Alymkulov E, Kulatov M (2001) Local government in the Kyrgyz Republic. In: Munteanu I, Popa V (eds) Developing new rules in the old environment, Local Government and Public Service Reform Initiative (Local Governments in Eastern Europe, in the Caucasus and in Central Asia/ O. S. Institute) Budapest

Anderson J (1999) Kyrgyzstan: Central Asia's island of democracy? s.l.: Harwood Academic Publishers (Postcommunist States and Nations)

Bichsel C (2006) Dangerous divisions: irrigation disputes and conflict transformation in the Ferghana Valley, Inauguraldissertation der Philosophisch-naturwissenschaftlichen Fakultät der Universität Bern, mimeo

EBRD (2004) Transition report 2004: infrastructure, EBRD, London

Eggertsson T (1996) A note on the economics of institutions. In: Alston LJ, Eggertsson T, North DC (eds) Empirical studies on institutional change, Cambridge University Press, Cambridge, 6–24. http://www.ucab.edu.ve/economia/economia/eco_soc/archivos/guia1.doc. Cited 24 Aug 2006

FAO AQUASTAT (2004) FAO's information system on water and agriculture: AQUASTAT. http://www.fao.org/ag/agl/aglw/aquastat/dbase/index.htm Cited 16 Jun 2006

Gatzweiler F, Hagedorn K (2001) The evolution of institutions in transition, CEESA Discussion Paper 4, Berlin

Grävingholt J et al. (2006) Strengthening participation through decentralisation, German Development Institute (DIE Studies 16), Bonn http://www.die-gdi.de/die_homepage.nsf/6f3fa777ba64bd9ec12569cb00547f1b/81c64bcd2814f575c12570fb002cc37c/$FILE/DIE_study 16_ Graevingholt_et_al_decentralisation_Kirgistan.pdf. Cited 16 Jun 2006

Helmke G, Levitsky S (2004) Informal institutions and comparative politics: a research agenda. In: Perspectives on Politics 2 (4), 725–740

Herrfahrdt E et al. (2006) Water governance in the Kyrgyz agricultural sector: on its way to Integrated Water Resource Management? German Development Institute (DIE Studies 14), Bonn

IPCC (2001) climate change 2001: impacts, adaptation and vulnerability. http://www.grida.no/ climate/ipcc_tar/wg2/index.htm. Cited 22 May 2006

Johnson SHI, Stoutjesdijk J, Djailobayev N (2002) Irrigation reform in the Kyrgyz Republic, Sixth International Seminar on participatory irrigation management institutional options for user participation, Beijing, China, March 2002. http://www.fao.org/ag/agl/aglw/waterinstitutions/docs/CSkyrgyz.pdf Cited 22 May 2006

Köllner P (2005) Formale und informelle Politik aus institutioneller Perspektive: Ein Analyseansatz für die vergleichenden Area Studies, Deutsches Übersee-Institut (Working Paper 6) Hamburg. http://repec.giga-hamburg.de/pdf/giga_05_wp06_koellner.pdf. Cited 3 Sep 2006

Koshoev B et al. (2003) Ocenka ekonomicheskogo razvitija Kyrgyzstana, Bishkek

Kyrgyz Republic (2004) Water Code of the Kyrgyz Republic: #8 of January 12, 2005, mimeo

Mollinga PP, Bolding A (eds) (2004) The politics of irrigation reform: contested policy formulation and implementation in Asia, Africa and Latin America, Ashgate (Global Environmental Governance Series 04/2004), Aldershot, UK
Neubert S, Herrfahrdt E (2005) Integriertes Wasserressourcen-Management: Ein realistisches Konzept für Entwicklungs- und Transformationsländer? In: Messner D, Scholz I (eds), Zukunftsfragen der Entwicklungspolitik, Nomos, Baden-Baden, 237–252
North DC (1992) Institutionen, institutioneller Wandel und Wirtschaftsleistung, Mohr (Die Einheit der Gesellschaftswissenschaften 76), Tübingen
North DC (1997) The contribution of the new institutional economics to an understanding of the transition problem, UNU World Institute for Development Economics Research (UNU-WIDER) (WIDER Annual Lectures 1), Helsinki
O'Hara SL (2000) Lessons from the past: water management in Central Asia. In: Water Policy 2 (4–5), 365–384
Ostrom E (1990) Governing the commons: the evolution of institutions for collective action, Cambridge University Press, Cambridge
Pejovich S (1999) The effects of the interaction of formal and informal institutions on social stability and economic development. In: Journal of Markets & Morality 2 (2), 164–181. http://www.acton.org/publicat/m_and_m/1999_fall/pdf/mm-v2n2-pejovich.pdf. Cited 19 Mar 2005
Rogers P, Hall AW (2003) Effective water governance, GWP (TEC Background Papers 7), Stockholm. http://www.gwpforum.org/gwp/library/TEC%207.pdf. Cited 9 Jan 2005
Saleth RM, Dinar A (2000) Institutional changes in global water sector: trends, patterns, and implications. In: Water Policy 2 (3), 175–199
Saleth RM, Dinar A (2004) The institutional economics of water: A crosscountry analysis of institutions and performance, Washington, D.C.: World Bank
Sehring J (2005) Water User Associations (WUAs) in Kyrgyzstan: a case study on institutional reform in local irrigation management, ZEU (Zentrum für internationale Entwicklungs- und Umweltforschung) (Discussion Paper 24), Giessen. http://geb.uni-giessen.de/geb/volltexte/2005/2392/pdf/ZEUDiscPap24.pdf. Cited 12 Jun 2006
Solanes M, Gonzalez-Villarreal F (1999) The Dublin Principles for Water as reflected in a Comparative Assessment of Institutional and Legal Arrangements for Integrated Water Resources Management, GWP (TEC Background Papers 3), Stockholm
Sullivan CA, Meigh JR, Fediw TS (2002) Derivation and testing of the water poverty index phase 1, Final report May 2002, Vol. 1, Overview, Centre for Ecology and Hydrology Wallingford
UNDP (2005) Kyrgyzstan: at a new stage of development, Bishkek. http://www.undp.kg/english/publications/2005/2005112801.pdf. Cited 14 Oct 2006
UNECE (2000) Kyrgyzstan (UNECE Environmental Performance Reviews). http://www.unece.org/env/epr/studies/kyrgyzstan/welcome.htm. Cited 27 Mar 2006
UNESCAP (2001) Accession to the World Trade Organization: issues and recommendations for Central Asian and Caucasian economies in transition, s.l. (Studies in Trade and Investment 48). http://www.unescap.org/publications/detail.asp?id=539. Cited 2 Mar 2006
UNESCO World Water Assessment Programme (2003) Water for people, water for life: The United Nations World Water Development Report. Executive summary, Paris
Wegerich K (2002) The role of elites in changing water institutions at the local level in Uzbekistan. In: Neubert S, Scheumann W, van Edig A (eds), Reforming institutions for sustainable water management, German Development Institute, Bonn, 19–29
World Bank (2004) Kyrgyz Republic: Public expenditure review: fiscal policies for growth and poverty reduction, Vol. I, Main Report, Washington, DC

Part V
Development Cooperation

The World Bank's Water Sector Policy Reforms

Volkmar Hartje

Abstract Integrated Water Resources Management (IWRM) has been developed as a concept integrating the management of all water resources with the major water using sectors (e.g. water sup-ply and sanitation, irrigation agriculture, environmental protection) nearly two decades ago. It undertook three major efforts in the 1990s and between 2001 and 2004 in translating the concept into a set of policies guiding its own operations. The first effort consisted of the Water Resources Management Policy Paper of 1993 which was based on an internal review with limited involvement of NGOs. The second effort meant to come to a deal with its external critiques over dam and other water infrastructure building, leading to the set up of the World Commission on Dams. The latest efforts resulted in the 2004 Water Resources Sector Strategy, which was based on an internal process with the explicit inclusion of recipient countries, emphasizes a renewed inclusion of hydraulic infrastructure in the lending policy. The chapter assesses these policy developments in the framework of the principal–agent theory, focusing on the interest of the World Bank of maintaining its autonomy as an agent and the role of NGOs as third party to the principal–agent relationship between the owners and the bank.

1 Introduction

Integrated Water Resources Management (IWRM) has been developed nearly two decades ago as an encompassing new concept integrating the management of all water resources with the major water using sectors (water supply and sanitation, irrigated agriculture, environmental protection, flood protection, shipping and hydropower). Since then, it has been promoted as a concept for water policy reforms in most developing countries by a number of donor agencies and other international organizations dealing with water policy issues. The World Bank is clearly the most

Volkmar Hartje
e-mail: hartje@imup.tu-berlin.de

W. Scheumann et al. (eds.), *Water Politics and Development Cooperation*,
doi: 10.1007/978-3-540-76707-7_14,

prominent donor institution endorsing the concept analytically and providing substantial financial support for its implementation. It prides itself for its intellectual leadership role and it undertook two major efforts – in the early nineties and between 2001 and 2005 – in translating the concept into a set of policies guiding its own operations. The first round of reform was the result of the actions of external stakeholders, mainly Washington based environmental Non-Governmental Organizations (NGOs), pressuring the US Congress and the US Treasury to request from the World Bank various procedural changes which amounted to an upgraded environmental review process and an IWRM based sector policy. The second effort 10 years later originated from within the Bank and the shareholders emphasizing the implementation of IWRM principles, but reversing the retreat from dam to building that had occurred gradually over the 1990s. Here, the role of NGOs was minor and the stakeholders included the major dam–building countries which make up a large share of the non-IDA borrowers.

While not acting prominently in the process leading to and immediately following the United Nations Conference on Environment and Development (UNCED) in 1992[1] which helped to propagate the concept among the UN participants and their constituencies, the World Bank started its own process at the same time preparing its 1993 policy paper in *Water Resources Management* which contains most of the ideas of the IWRM concept, but has a Bank-specific emphasis on economic topics. A number of comparable processes in the water using sectors were pursued in the nineties, while creating analytical support internally with the water resources group and externally by engaging in supporting networks as the Global Water Partnership (GWP) and the World Water Council (WWC): The most prominent effort in dealing with a controversial topic was the participation in the creation of the World Commission of Dams (WCD). The internal review of the implementation of the 1993 policy paper by the Bank's own evaluation department led to its review and the development of the Water Resources Sector Strategy in 2004, explicitly referring to the Dublin principles.

The first level of the impact of the adoption of IWRM principles by the World Bank concerns the implementation of these policy guidelines on the operations of the Bank in water-related lending.[2] These have been subject to considerable controversies, but they were limited to two sub-areas of the water-using sectors, dam, building and privatization of urban water supply. For the evaluation of the 1993 concept's implementation in lending and the other water sub-sectors, the Bank's own assessment of the 1993 policy paper by the evaluation department is the major source and it identifies the progress, but also the shortcomings. The weaknesses consisted mainly of the incremental generation of projects and their marginal improvement without linking them to a coherent set of objectives and priorities. As a consequence, the Dublin principles are now, i.e. after the policy developments in the early 2000 being implemented with a "pragmatic but principled approach" (World Bank 2004, 2).

[1] For account of the role of the IWRM during this phase cf. Scheumann and Klaphake (2001).

[2] The second level concerns the implementation of the IWRM principles in the borrowing countries and the third level the effects of these water policy reforms for the actual water management. For these aspects cf. Saleth and Dinar (2004).

These problems in implementing the Bank's own policies at the level of project lending relate to its role as a bank and its competitive advantage as a lender for long-term, complex projects with reliable and competitive conditions. Here on the second level, the demand by the partner governments for the Bank's support originates which includes the demand for institutional support, but the demand for the concept of IWRM is limited and its uptake faces the potential opposition of the water using sectors which fear losing control or losing access to water at low cost. Thus, on one side analytical sector work and considerable efforts in the diffusion of the concept were expanded and on the other side the Bank adjusts by continued lending on a project level and by waiting, looking for opportunities and reacting more towards the demands of its partner, particularly theu larger ones.

The analytical literature about the aid delivery process, for both the bilateral and the multilateral aid agencies, and the NGOs commenting mainly on the multilateral agencies agree on an assessment that there is a considerable difference between stated objective and actual performance, particularly with respect to the inclusion of environmental objectives, but increasingly about social and human rights considerations as well. While NGOs emphasize the devastating effects of non-performing projects and suspect evil motives on the side of the multilateral financial institutions, economists and political scientists take the difference largely as given. Weaver (2006) calls it the "IO Hypocrisy" and Gutner (2005a) the "gap between mandate and performance", and focus on the question of explaining the reasons for the diagnosed performance gap. Considerable attention has gained the question related to environmental performance as a result of environmental policy changes in the World Bank, but the question has not been applied to the water resources sector of the Bank.

This article raises the question how the World Bank has developed and adopted the concept of Integrated Water Resource Management and the question whether and if yes to what extent the water policies of the World Bank have changed in the first half of this decade towards a more infrastructure-building orientation and how these changes can be explained. Here, the article employs the analytical concept of the principal–agent relationship to which the environmental NGOs play the role of a lobbying third party. The adoption of IWRM has to be seen in the context of the overall integration of environmental considerations in the lending activities of the World Bank. Because of the prominent role of dams as a focusing type of project for NGO campaigning against the environmentally damaging effects of World Bank lending, the position of the World Bank on lending for dams is an import indicator for policy changes. While the research on the environmental policies of the World Bank in general moves on the question of measuring and explaining the effectiveness of these policies, this article focuses on the content and process of various policy formulations which are important for water resource management: Two internally controlled processes, resulting in the *Water Resources Management* Policy Paper in 1993 and the 2004 Water Resources Sector Strategy are contrasted to the Bank assisted, but external process of the World Commission on Dams. Here, the question is pursued to what extent and by what means NGOs managed to influence the policy formulation process.

The article starts with the description of the analytical framework, the principal–agent model, and then it surveys the literature applying the model to international organizations in development cooperation. Here, the specific features of the delegation chain between the taxpayer in donor countries at the beginning and the beneficiaries in the borrowing countries at the end are analyzed and how these features and the attempts of the principals to control their agents influence aid delivery. This process of a delegating in multiple steps results in an unavoidable degree of *agent slippage* which is considered to be the price to be paid for the delegation by the principal. The analytical framework is then applied to the analysis of the governance of the World Bank. Here, the issue of the relative autonomy of the World Bank as an agent is raised based on the high share of refinancing by international capital markets and the specific mechanisms to control and influence the World Bank as an agent are analyzed. Next, the chapter summarizes the (empirical) research about the inclusion of general environmental objectives into the governance of the World Bank and the mechanisms employed to bring about this change of behavior of the World Bank.

Then, the framework is applied to the water policy of the World Bank. First, the multidimensionality of the water policy on a conceptual level and the need to transform it into an existing governance structure are discussed. Second, two highly visible policy changes – the 1993 *Water Resource Management* Policy Paper and the 2003 Water Resources Sector Strategy – are compared. They serve as contrasting cases of external actors influencing the World Bank's water policy as opposed to the self-controlled process by the agent. The summary places the 2003 water policy in the context of the increase of bargaining power of major borrowing countries.

2 The Analytical Frame of Analyzing Multilateral Banks

The literature on international organizations and on the operative international organizations in particular has continuously expanded the use of the principal–agent models to study the behavior of international organizations (Martens 2002; Ostrom et al. 2002; Nielson and Tierney 2003). Originally developed to explain management behavior and hierarchies in organizations (Furubotn and Richter 2005; Sappington 1991), it has been expanded to explain the incentive structure in delegation situations. The principal in a company, a non-profit organization or a public bureaucracy cannot perform all tasks himself/herself. They (he or she) need to delegate to agents. The principal defines the tasks and expects from the agent to perform them. While the principal covers the costs of performing the task (and receives the residual benefits in a for profit organization), the agent receives a reward for carrying out the task. Because of the delegation, the principal does not have full information about the activities (and properties) of the agent, resulting in two types of problems. First, when reaching an agreement the agent may have better access to information (or may manipulate the information) and use this to his advantage, resulting in adverse selection problems. Secondly, after the conclusion of the contract, the agent may not perform the tasks as agreed, either by reducing his efforts or by performing the task to further his own interest instead of those of the principal. This is called moral hazard, a problem observed in a number of market transactions as well. These incentive problems are considered unavoidable problems of delegation and need to be dealt with appropriate institutional devices for screening and monitoring and rewarding the agent to change their incentive structure.

The principal–agent theory has been applied to a number of organizations and contracts, mainly in the context of business transactions, but in the nineties increasingly to public (Tirole 1994) and non-profit organizations. The beginning of this decade saw then the application of the model to the aid delivery process (Martens 2002; Ostrom et al. 2002). There had been earlier applications to the study of conditionality between donor organizations and the recipients of aid (Killick 1997; Rodrik 1995).

Martens (2002, 11) views the informational characteristics of public administrations, including those involved in development cooperation, as central as opposed to private companies:

- Public administrations tend to have multiple objectives which mean for development agencies the pursuit of different sectoral and regional spending priorities.
- Public agencies have multiple principals – a feature they share with private enterprises – but their principals usually do not share the same objectives while the shareholders tend to agree on profit objectives.
- The measurement of the performance of public agencies is quite difficult since they have problems identifying the opportunity costs of alternative policies when pursuing multiple objectives.

The process of development cooperation differs from private business transactions and from public agencies working in the taxpaying countries by having no direct link between the benefits as seen by the beneficiaries in the recipient countries and the costs as seen by the paying taxpayers or by voluntary contributors in the donor countries. To organize the transfer between the taxpayers (or individual voluntary donors) and the beneficiaries, a process of multistage delegation takes place: First, between the voters and taxpayers the classical delegation of indirect democracies occurs. Second, the elected politicians are the principals to the ministries and other governmental agencies as their performing agent. Third, the management of these agencies is the principal to their employees delegating their task via employment contracts, quite often on the basis of civil services type contracts. Fourth, depending on the type of organizational structure, the management or politicians negotiate with partner countries' finance ministries or similar organizations the allocation of funds to be transferred (in terms of quantity and quality, grants vs. loans). Fifth, the staffs of the development agencies develop, assess and negotiate individual projects with sectoral ministries or agencies in the recipients' countries and become the principal to these ministries/agencies as they implement the projects. Sixth, in the process of this implementation (and often quite earlier in the planning and design phase of the projects) private commercial companies are contracted by the staff of the aid agencies in the donor countries and/or the implementing agencies in the recipient countries and serve as their agents. Once these projects are implemented, these resulting effects can create benefits for the targeted group in the recipient country. As the first three steps are common to all public organizations, the following phases and the involvement of ministries from other countries is specific to development cooperation. For each of the delegation steps, according to the principal–agent theory one expects to find adverse selection and moral hazard problems, resulting in a multi-stage *agent slippage* (Nielson and Tierney 2003).

Furthermore, development cooperation is characterized by a common device of retaining control by the principals in the donor countries by transferring the funds on a project basis with an individual assessment by the donor agencies. They tend to produce tangible outputs that are easier to monitor. The newer forms of transfer in terms of sectoral adjustment lending are the result of concerns about deteriorating conditions of developing countries in the early eighties and institutional reforms to promote the adjustment. These transfers do not allow for monitoring devices developed for projects and instead they are tied to a broad set of conditionality which have to be fulfilled before tranches of the transfer are paid out, but generating a new set of principal–agent problems as the institutional reforms "produce less tangible outputs that are much harder to verify" (Martens 2002, 17).[3] A second specific of development cooperation are the two types of funds employed – grants and loans. While the grants in most cases are refinanced by taxes in the donor countries and are subject to the budgetary appropriation procedures, and, thus to the control procedures of parliaments, the alternative form of financing, via the provision of public equity and guarantees for a bank and the use international capital markets for loans to developing countries, provides an aid agency with considerable more autonomy as some of their sources of income are independent of the budgetary control of parliaments. But, the role of these development banks as financial intermediaries between the international capital markets and the recipient countries forces them to focus more on the repayment qualities of the recipient countries and to present the bank-like qualities of its management to the capital markets. There are a number of donor organizations which use both sources of funds (the German KfW Development Bank as a bilateral agency and the World Bank as a multilateral organization) and those which use exclusively tax-based finances to provide mostly grants (bilateral agencies, e.g. US AID and UNDP as a multilateral agency). The development banks are attractive to their principals as they allow them to transfer money without taking it out of the domestic taxes once the equity is paid and as the risk of default is relatively low because of the focus on repayment quality. But this advantage comes at the price of relative autonomy of the development banks as aid agencies vis-à-vis the purely tax-financed aid agencies.

The third feature specific to development cooperation is the existence of international or multilateral agencies in addition to the national-bilateral agencies. The multilateral development banks add another element to the delegation chain as several donor countries (and to some extent recipient countries) join and delegate the transfer of their resources to an international organization, thus giving up control

[3] For a survey of the agency problems involved with conditionality cf. Kapur (2004); Killick (1997).

over the recipients and the forms of aid and adding to agency slippage. Rodrik (1995) explains the function of multilateral development banks (MDB) as being complementary to both, the private capital markets and bilateral aid programs. Compared to the operation of private capital markets which he characterizes as being at least partially inefficient due to their cyclical nature and geographical concentration despite the huge amounts of transfers, the multilateral development banks provide information gathering and monitoring services about the recipient countries which have public goods quality. The combination of lending with information gathering is based on the fact that the lending reduces the transaction costs of information gathering and support for information among the recipient countries (Martens 2005). Secondly, the lending increases the value of the information made public by the MDB as they commit their own resources based on the information assembled. Compared to bilateral aid agencies, the MDB have the advantage of a higher coverage and economies of scope in information gathering and monitoring. The second advantage lies in "achieving collective action in the presence of heterogeneous preferences among donors" (Martens 2005, 656) when they need to cooperate to achieve a common public good in the recipient country as an economic reform program or a sector reform. They are in a better position to exercise aid in a conditional way. Because of their lack of (national donor country) conflict of preferences they help avoiding the spill over of these conflicts to recipient countries. But as these advantages come at the price of losing control compared to bilateral aid agencies as an instrument of national foreign policy, multilateral aid covers roughly only a third of the total of foreign aid.[4]

The mechanisms of control for the principal tend to be similar for MDB as for the other governmental agencies (Hawkins et al. 2006). The major instrument potentially available for the principal is to define rules for the behavior of the agent. Here, the principal has the choice to define the rules of the behavior – leaving relatively little discretion to the agent – or to define the objectives and let the agent make the choices about the procedures in achieving the objectives. The actual contract between the principal and an agent will vary between the two approaches as they are considered to be ideal points along a dimension of rules versus discretion. The general assessment is that a rule focused contract fits better for a standardized task with a stable organizational environment while a discretion-based contract is best suited under circumstances of high uncertainty and thus when a high degree of flexibility is desirable and or when the task requires specialized knowledge possessed by only one agent (Hawkins et al. 2006). However, the higher the degree of discretion left to the agent, the greater the opportunities for opportunistic behavior of the agent.

The second mechanism for control is the reporting requirements of the agent and the monitoring effort of the principal. Hawkins et al. (2006) emphasize ex post reporting and distinguish here between regular control type reporting ("police patrol") and event based reporting ("fire alarms") which are specified in the delegation contract. These requirements are designed to monitor the routine operation, but there

[4] A further advantage of multilateral aid is that it provides the national governments in the donor countries with a shield against skeptic voters in their domestic principal–agent relationship (Milner 2006).

are ex ante reporting requirements as well for major changes, investments and other changes. An event based ex post reporting and high discretion in reporting ex ante changes leaves considerable room for opportunistic behavior.

The common characteristic of development aid – to manage the transfer process by allocating it to individually assessed projects – can thus be seen as a compromise between discretion and rules. The identification, evaluation and implementation of individual projects leaves considerable discretion to agent, but the rules for this process are defined by the principal. The major advantage for the principal is the ex ante reporting of planned aid projects which is the major interaction between aid ministries and aid executing organizations in bilateral aid. In the MBD, the reporting requirements to a Board serve to allow the control of the member countries as collective principals.

A third mechanism is to use screening and selection procedures to find agents – at the leadership and agency level – with preferences similar to those of the principal. This will allow the principal to grant more discretion to the agent and to lower his monitoring efforts as he can hope that the agent will pursue actions similar to those as if the principals would act on his own. In the case of MDB, the collective principals have already created agents according to their collective preferences once they set up the organization, leaving the selection device mostly for the leadership positions and the hiring practices. Here, however the choices of leaders are relatively limited if the hiring process is not competitive. This tends to be the case quite often and additionally the hiring is restricted according to national quota.

A fourth mechanism to control the agent is to build checks and balances to the organizational design of the agent. Within an organization, the principal can establish organization units with overlapping, competing or controlling functions that provide information to the principal about the behavior or performance of the agent which is not controlled by the agent, thus reducing the informational asymmetry. Another approach is to establish organizations with overlapping functions to provide to some extent competition between the organizations and thus endow the principal with additional information. In the context of the MDB, the creation of the regional development banks may be seen as overlapping agents which provide similar information to their collective principals.

The fifth mechanism is to apply sanctions for opportunistic behavior and to reward behavior conforming to the principal intentions. For governmental agencies, the most important sanctions and/or reward are changes in the size of the budget. Successful agents are usually rewarded with larger budgets allowing them to fulfill their function more comfortably and to reap tangible side benefits easier. The use of these sanctions and rewards is easier applicable in the context of bilateral development aid (and for the aid program of the European Union), as they depend on taxes as their source of revenues and are subject to the procedures of parliamentary budgetary control. These instruments are not available if the development bank is bank-like in its financing structure as it relies on capital markets and return flows of funds for its growth. This is the case for the European Investment Bank (EIB) for its EU wide lending and the European Bank for Reconstruction and Development (EBRD). It was also the original position of the World Bank until in 1960, when

IDA facilities were added which make the World Bank subject to a budgetary control procedure in the principal member countries on a tri-annual basis.[5] In addition, the principals can reward an agent by adding tasks which have a growth effect on the organization for which the expansion of regions outside the EU is an example for the EIB.

These control mechanisms are used in a combination depending on the circumstances of the principal–agent relationship and the balancing of the benefits of delegation by the principals against the costs of control and the degree of slack by the agent that can be attained by use of the control mechanisms. As the use of control mechanisms is resource consuming to the principal and the use of sanctions may be damaging to the reputation of the agent and thus to the delegation itself, resulting in costs, the principal will weigh these against the cost of the slack of the agent. As a result, agents will always have a degree of autonomy. The MDBs face collective principals who might disagree on the degree of control and the amount of resources they are willing to commit for control, thus increasing their autonomy further. The major sources of autonomy are, however, the complexity of the tasks and the degree of specialization required to perform these tasks, resulting in an informational advantage on behalf of the MDB as agents and their own control of budgetary resources via the refinancing via capital markets.

3 Looking with the Lenses of the Principal–Agent Model at the Governance of the World Bank

The research applying the principal–agent models of the World Bank focuses on the behavior of the World Bank as the agent and the owning donors countries as principals. The first point emphasized is the fact that the principals are of a collective nature and – as Nielson and Tierney (2003) argue – of a multiple nature. The collective principals are the member countries represented on the Board of Governors, mostly composed of finance ministers (or the Ministry for Economic Cooperation and Development in the case of Germany). The Board of Governors, which meets once annually, has the authority to alter Bank policies, approve the annual budget with a simple majority rule and amend the articles of the agreement with a supermajority of 85 percent. As probably well known, the votes are distributed according to the shares of the individual countries with the highest shareholder being the US with 16.4 percent and the Group of seven nearly controlling 50 percent. The US thus has a blocking minority for amendments, but the power of the US as a shareholder relies on its importance for forming majority coalitions.

The second tier of the collective principals is represented by the Board of Executive Directors to which the member countries delegated the supervision of operational policy and the approval of individual projects. This Board consists of 24

[5] Formally, the World Bank consists of two separate organizations, the IBRD (International Bank for Reconstruction and Development) for capital market based loans and IDA (International Development Association) for the tax-based financing.

persons who work on-site delegated by the member countries. While the five largest shareholders appoint one Executive Director on their own, the remaining 19 seats represent groups of countries with the remaining donor countries and recipient countries on a regional basis. According to the articles of the agreement, they select the President of the World Bank who is also the Chairman of the Board. But the right to propose a candidate has been informally reserved to the US and their proposed candidates so far have been accepted by the Board of Executive Directors. The role of the Executive Directors as guide to the World Bank is rather limited compared to the President and senior Bank management (Kapur 2002), as their stay is usually limited to 3 years and they tend to take a parochial view.

Because of the relatively well-known high US voting share but for other reasons as well (cf. below), there is a consensus that the World Banks' governance is dominated by the United States, despite the fact that the US Executive Director has also limited tenure. Although the voting shares declined from 35 percent in 1945 to 16.5 percent in 1999, the US dominance is supported by the willingness of the US to exercise its power, by the lack of countervailing pressures from other shareholders and the *soft power* of the USA as manifested in academia, the NGOs and the location in Washington, DC. (Kapur 2002). The gentlemen's agreement mentioned above on the United States government proposing a US national as the candidate for the president of the World Bank should be added as an additional source of the dominance of the US among the principals.

The concept of multiple principals, introduced to the governance of the World Bank by Nielson and Tierney (2003), tries to capture the special relationship between the World Bank and US institutions. They see the US President and Congress as additional principals (in addition to the collective principal represented by the two Boards), creating a situation for the agent where he has to deal with the collective principal and other principals. The US President and Congress interact more directly and not only via the Board of Executive Directors with the World Bank as they can influence the World Bank separately via the nomination process of the President and via the budgetary competence, having "numerous formal sources of authority that allow it to unilaterally recontract with the World Bank" (Nielson and Tierney 2003, 256).

As an established MDB, the World Bank has been granted considerable discretion. The use of rules as an approach to control the World Bank by the principals has been limited as an instrument. Kapur (2002) classifies the autonomy of the World Bank management in day to day operations as a central quality during its foundation when it was a lending institution to governments for infrastructure projects. What he categorizes as major changes, the inclusion of the International Development Association (IDA) in 1960 and the introduction of sectoral adjustment lending in 1980 can be classified as additional tasks. Here, the inclusion of IDA changed the control mechanism by establishing the budgetary control mechanism of tax financing although in a milder form by making the decision only every 3 years. This task was added by the principals against the Bank's management while the introduction of structural adjustment lending was initiated by the President against a reluctant Board of Executive Directors (Mosley et al. 1991). The introduction of policy-based lending did not change the rules of control by the collective principal, but the relationship between the World Bank as the principal and the borrowing countries as its agents. The use of conditionality as the controlling instrument of policy-based lending created new incentives problems.

Adding more tasks, like adding products to the portfolio of a private corporation changes the internal structure of the organization, the necessary skills of its personnel and the incentive structure. In addition to the three major products, providing capital markets based loans for physical infrastructure projects, giving tax based subsidized loans and grants to poorer countries and the policy based lending for sectors and structural adjustment based on a broader evaluation process than private banks, the World Bank provides research and monitoring to its shareholders and the borrowing countries and provides policy advice to the borrowing countries. While the research and monitoring are quite well received by the beneficiaries, the advice is not always welcome. As it is tight to the attractive loans and grants, the borrowing countries accept it as a condition as long as the loans and grants are attractive.

Similarly, the inclusion of the environment can be seen as adding new products by supporting projects with environmental benefits and changing old products by avoiding or mitigating negative environmental impacts up to the point of avoiding these types of loans altogether. Gutner (2002) emphasizes that the peculiarities of environmental behavior of MDB is that their definitions are not always clear cut and that they vary over time, particularly when projects consist of a number of components of which there might be those with the intentions of providing environmental benefits and at the same time having negative environmental impacts.

The use of reporting to the principals and their monitoring was restricted to conventional ex post reporting and the presentation of individual projects to the Board of Executive Directors and their approval. Kapur rates the degree of control with respect to project approval and reporting as being rather low as the informational advantage is high of the side of the World Bank vis-à-vis the Board of Executive Directors which have seldom equivalent resources to master the details. A better equipped monitoring instrument is the evaluation department, now called Independent Evaluation Group, which reports to the Executive Directors Board. Interestingly, this is a newer development as the department was created by the Bank's President in 1973 (World Bank 2003d) to inform the President and the senior management about project performance ex post. The construction of the evaluation department is not independent as an external auditor with interrogative rights, but relies on the cooperation of the project managers. Although the World Bank and the OED emphasize its independence and its objectivity, these claims were not fully believed as the separate one-time valuation commissions (Wapenhans-Report: World Bank 1992; Morse and Berger 1992) and the establishment of the Inspection panel in 1993 show. The creation of these control mechanism was, however, not meant to improve the degree of information of the principals, but to external third parties (Fox 2002).

The use of screening and selection devices is applied differently for the President and the management and the staff. With 5 year tenure and the potential for renewal, the selection of the President of the World Bank has the potential to select a person with similar preferences as the collective principal and provide an incentive to maintain these preferences. The practiced process of selection, based on the prerogative of the US, has been criticized for its limited range of candidates, closed door procedures and its bias for US preferences based on domestic political compromises, and not on experience in development policy (Kapur 2000). The staff screening and selection seems to be working better with competitive procedures, but there is a focus on US and Anglo academic institutions sharing the same preferences in development policy. The autonomy delegated to the President grants him considerable discretion in setting up the internal structure, organizational set up, budgetary procedures, promotional and personnel policies (Kapur 2002). The Presidents have used this discretion to change the organizational structure, particularly the relationship between the country departments and the technical/sectoral departments, and to change personnel.[6]

The use of checks and balances as an instrument for control by the principals has been limited to the creation of the regional development banks. The potential role of creating internal organizations with competing and/or overlapping tasks was not used by the principals. Kapur (2002) assesses that the relatively high autonomy the World Bank attained by reliance on US and international capital markets has been gradually undermined by introducing the IDA which increased leverage of parliaments and the NGOs, particularly in the US.

4 The Inclusion of Environmental Objectives in the Governance of the World Bank

The inclusion of environmental goals, the introduction of rules supporting these goals and the implementation of these rules is then subject to the general principal–agent constellation. There has been research on both steps of the process of changing the agent's behavior, but with an emphasis on the efforts of mitigating negative environmental effects via the introduction of corresponding policy changes and their implementation into the mainstream of project and sectoral lending. The changes in the other direction, designing environmentally beneficial projects and their development and structural composition over time have been the focus of presentations of the World Bank and its environmental staff and only recently the result of academic research (Nielson and Tierney 2003; Gutner 2002).

The first topic concerns the process of introducing environmental objectives into the activities of the World Bank. Here, one of the important observations made in the literature is that the environmental objectives received relatively low priority among the borrowing countries and in the beginning among the MDB principals as well. The establishment of the Office of Environmental Affairs has been granted to initiative of the President of the World Bank in 1970 with the support of the US Executive Director on the Board (Le Prestre 1989). The various changes of the rules/reorganizations of the World Bank to avoid and/or mitigate the negative

[6] Cf. Rich (2002) on the *Strategic Compact* in 1996/97.

environmental impact are to a large extent the result of actions of the principals, mainly in the US, reacting towards pressure of environmental lobby organizations which were successful in leveraging the US Congress on several occasions during the period of authorizing IDA replenishments:

- The establishment of a central environmental department and of the four regional environmental divisions by President Barber Conable in 1987 was a reaction to a NGO campaign which had started 4 years earlier in 1983, based on a small number of environmentally damaging projects supported by the World Bank,[7] and it was ultimately designed and timed to avoid a negative vote of the US Congress on IDA-8 and, as a rare occasion, on an increase in the US contribution to the World Bank's capital base. The capital increase had become necessary to implement the Baker-Plan formulated in 1985 which was designed to give the World Bank a larger role in lending to debt-ridden borrowing countries. The US Treasury changed its course towards pressing the World Bank for an environmental reform in 1986 when it realized the threat the environmental problems posed to the vote of Congress (Wade 1997, 667f.).
- The Environmental Assessment procedures of the World Bank were made more specific than the 1984 Directive, based on Bank's staff, but during the negotiations to replenish IDA-9 in 1988, US Congress demanded systematic environmental assessment procedures from the World Bank (Gutner 2002, 57; Wade 1997, 681).
- The Inspection Panel, as a three member commission to investigate complaints against the World Bank, is not an environmental institutional device per se, but it was intended to improve the accountability of the World Bank the lack of which was seen by NGOs as a major hindrance towards environmental improvement. But its establishment was linked to the appropriation of IDA-10 in 1993 (Gutner 2002).

The US principals and other countries are credited with supporting the Bank with further measures, namely the creation of the Global Environmental Facility in 1989 and of the Montreal Protocol's Multilateral Fund (Gutner 2002).

The literature identifies major reorganizations which were relevant for the inclusion of environmental rules and their implementation which were mostly credited to the presidents who have undertaken them (Wade 1997; Rich 2002; Gutner 2002). These reorganizations were important as they introduced larger departments with environmental functions and influenced the incentive structure for mitigating negative environmental effects and pursuing environmentally beneficial projects. Here, the role of the principals can be best described as pressing for reorganization to cut costs.

- The first reorganization in the context of introducing environmental objectives and rules was the creation of the matrix organization in 1987 by President

[7] These were most prominently the Polonoroeste project in Brazil, the Transmigration Program in Indonesia, a livestock project in Botswana and the irrigation and hydropower projects on India's Narmada river (for the campaign strategy cf. Wade 1997, 656f.).

Conable with operating departments on a country basis and policy and planning departments on a sector basis. The country departments were relatively large, included sector divisions and hosted the task managers responsible for managing the project cycle with a broad sectoral focus. Additional sector experts were located in the four regional technical departments. These technical departments were enlarged by the regional environmental divisions where the operational environmental staff was positioned to support the design of environmentally beneficial projects and to review the adherence to the Bank's environmental assessment procedures. Environmental research, policy advice and the integration of environmental evaluation in the procedures of the Bank and its further development was undertaken in the central environmental department. These regional technical departments had a core budget for administrative expenditures of their own, but they had to acquire about 50 percent of their costs from the country departments by selling services (Rich 2002). The chiefs of the regional environmental divisions were in the position to stop project proposals to continue in the project cycle for environmental reasons. These environmental divisions in the Bank were extended to a network to include similar units in the International Finance Corporation (IFC) and the World Bank Institute.

- A second reorganization took place under the Presidency of Lewis Preston in 1993–94 when the structure of vice presidencies was changed to include thematic vice presidencies and mergers of divisions were installed to cut costs.
- A major impact on the incentive structure of staff had the third reorganization in 1997 under President James Wolfensohn called the Strategic Compact which was set to reverse the trend towards cost-cutting. The decision to reorganize required additional financial resources as it included a move to update the Bank electronically, increase outside training for staff to a considerable extent and it was taken against the intentions of the Board of the Executive Directors. It moved a large share of country staff into the recipient countries and reorganized the country departments by downsizing their regular staff and creating a market type system for hiring sector experts from a central pool or the outside. The country directors and their small size of country staff were assembled in country management units, focusing on country strategy and controlling the country budgets with which they hired most of the personnel to perform the evaluation functions in the project cycle. The remaining country personnel and the personnel from the technical departments were moved to regional sector groups with one environmental sector group (Gutner 2002) and they had to rely for their administrative expenses more on the income they gained from being contracted by country managers. The environmental professionals in the various units are linked by an Environment Board consisting of the heads of these units and they are part of a network, called Environmentally and Socially Sustainable Development (ESSD) which contains professionals dealing with rural and social development topics under a sectoral vice-presidency with the same name. In 2006, the Vice Presidency of ESSD, however, was merged with the Vice Presidency for Infrastructure.

The size, the location and resources controlled by the environmental staff is one important factor for the incentives and resources to monitor and control potentially environmentally adverse project designs and to create and pursue environmentally beneficial projects. The other set of factors is the relationship between the environmental staff and the task managers responsible for the design and operation of the projects. If one follows the literature (Wade 1997; Rich 2002; Gutner 2002), three phases of the organization of environmental management, close to those outlined above, can be distinguished which created a specific institutional set up with a different incentive structure and expectations about the environmental quality of the World Bank's lending. Nielson and Tierney (2003) identify changes in 1994 as an additional cut off point, thus, introducing a fourth phase. These four phases created different incentives for environmental assessment and monitoring and for lending stand-alone environmentally beneficial projects:

- The first phase between 1970 and 1987, basically established by President Robert McNamara, led to the creation of the Office of Environmental Affairs which was understaffed for its screening function (3–5 professionals for more than 200 projects per year), was involved late in the planning cycle, lacked veto power and had to rely on relatively unspecific guidelines issued late in 1984. "These were easily ignored, since they left compliance to the discretion of the task managers" (Gutner 2002, 53). The understanding of environmental management was largely restricted to the control of industrial pollution and pesticides use. There was no structure for separate environmental lending.
- With the 1987 reorganization, separate environmental units were created, both as a central unit and within the regional divisions and within a year the number of environmental staff increased by 50 within a year. Although it took until 1989 before the specific Environmental Assessment procedures were in effect, the reorganization in 1987 created better resource conditions for the Environmental Assessment process, the external literature agrees that remaining conflicts between environmental staff and country managers about the adequate procedure for the integration of environmental criteria into the design, implementation and operation of project were not adequately resolved.[8] The main explanations center on the incentive structure of the project managers.[9]
- The 1987 reorganization is considered more successful in creating incentives for separate environmental lending. Environmental professionals in the regional divisions were free and had the resources to identify potential environmentally beneficial projects in borrowing countries and find supporters within the governmental agencies. An additional incentive to identify and promote environmentally beneficial projects is the result of the establishment of the Global Environment Facility (GEF) in 1991 for which the World Bank is one of the implementing agencies and for which it hosts the independent secretariat. The GEF has a completely different principal structure and decision rules about projects, so that the GEF

[8] NGOs claimed that their cases proved the overall picture. But even World Bank internal reports raised a number of weaknesses in the appraisal and monitoring process (World Bank 1992, 1996b).

[9] Rich (1994) focuses on the incentives to get approval for projects, later (Rich 2002) adds staff turnover and lack of institutional continuity; Wade (1997) diagnoses a bargaining situation between task managers and environmental staff; Gutner (2002, 93) sees the restriction in the pursuit of keeping the transaction costs of projects low as a major point.

projects cannot simply be added to the portfolio of the Environment Department of the World Bank, but they increase the leverage of the environmental staff of the Bank in promoting environmental projects.

- Nielson and Tierney (2003, 2005) see the creation of the Inspection Panel in 1994, which was based on the same influencing mechanism of an NGO demand leveraged by the US Congress who had to agree to an IDA replenishment, as the organizational "reform that sticks" (Nielson and Tierney 2003, 260). But, they add that this change was supported by the Board which reflects changes of the preferences of the major principals around the UNCED process and complemented by changes in the reporting requirements: One was to make project documents publicly available including the environmental assessments and to get the Board involved earlier in the planning cycle of the project and not only for approval.
- The 1997 reorganization changed the incentive structure from a supply oriented to a more demand (by the recipient governments) driven approach (Gutner 2002). The country managers controlled the operating and administrative budgets and the Bank wide technical networks which included the environmental staff had their administrative budgets lowered so that they had to find country or task (project) managers who would contract for their services, imitating a market for consultancy services. Furthermore, the country managers formulated the lending priorities with the recipient country finance ministries which had lower priorities for environmental projects. One of the previous sources of environmental lending, the environmental staff with staff time and own resources, had dried up. Environmental assessment now relied on an established set of guidelines, but the incentive structure of environmental staff changed as they became dependent on task managers for work assignments. In the words of the author of the 2001 Operations Evaluation Department (OED) report on the Bank's environmental performance "the quality of the EA [Environmental Assessment] process deteriorated" (2001a 2001a, 21).

The empirical research on the effects of these changes progressed considerably in the last 5 years (Gutner 2002; Nielson and Tierney 2003, 2005), but remained somewhat inconclusive, mostly for the difficulties of operationalizing the variables: First, the dependent variable "environmental behavior" and second, "environmentally relevant changes in the institutional structure of the World Bank" as the independent variables are subject to an intensive debate. True, for the dependent variable, there are the cases of damaging projects which were the basis for the criticism and the NGO-campaigning against the World Bank since 1983 (Rich 1994), the results of internal evaluations of the World Bank (World Bank 1996b, 2002a), the results of the evaluations by the Wapenhans Commission and the Morse Commission 1992 and the Inspection Panel evaluations (World Bank 2003c). This evidence is sufficient to identify the problem, but for analytical purposes it suffers from the small numbers problem.

The most comprehensive effort so far to solve this problem has been by categorizing Board approved project commitments according to their environmental quality in time series between 1980 and 2000 (Nielson and Tierney 2003). The authors first classified 5300 World Bank (and Global Environmental Facility) projects (numbers and volume of committed spending) into free-standing environmental, traditional, agricultural, economic and social. In a later chapter (Nielson and Tierney 2005), they classified the projects on an ordinal five point scale from strictly dirty to strictly environmental independently from the World Bank classification (see Fig. 1): Here, the major changes are the continuous decline of broadly "dirty" projects over the whole period, the decline of the strictly "dirty" project only in the first decade and the increase of environmentally neutral projects. Environmental projects increased only slightly until the mid-nineties and then declined (Nielson and Tierney 2005, 795).

This measure of environmental behavior has been debated as problematic since it does not measure behavior in terms of implemented projects and their performance (Gutner 2005b). Nielson and Tierney (2005) counter claim that the commitments in the projects approved by the Board is the best available performance measure of the agent as the actual project implementation is the result of joint actions of the World Bank, the borrowing governments, the national implementing agencies and the project contractors. Their actions cannot be described by the principal–agent relationship as they are mostly voluntary contracts.

The measures of institutional reform are subject to a similar debate. While Nielson and Tierney (2003) emphasize institutional changes in 1994 (creation of the Inspection Panel, increased reporting requirements, changes in the environmental assessment process) that stuck because it increased the role of the Board, Gutner (2005b, 778) described the institutional change as a continuous, slow, ongoing

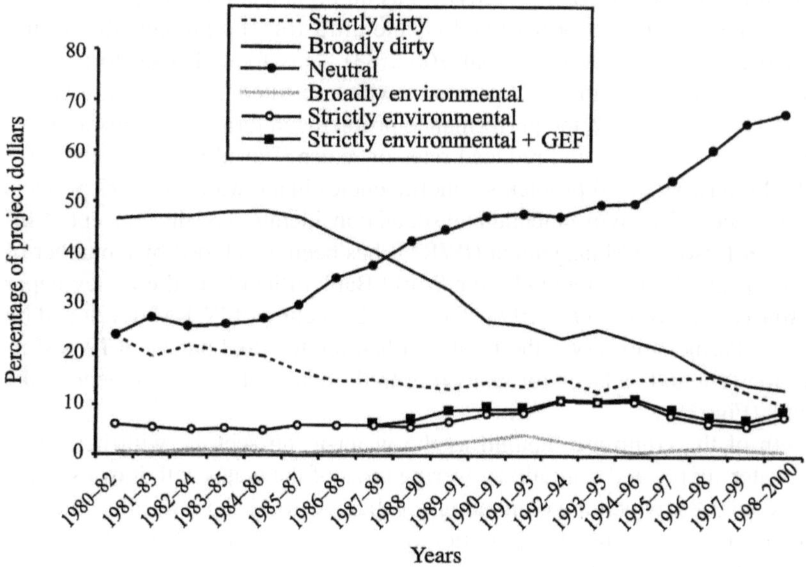

Note: GEF = Global Environment Facility.

Fig. 1 Percent of World Bank project dollars by category (3 year rolling average)
Source: Nielson and Tierney (2005)

process with advances (the 1987 organizational change, the 1989 introduction of Environmental Assessment, the 1994 changes), but also with backward or sideways steps such as the 1997 *Strategic Compact.* Not included in the above analysis are the effects of highly visible cancellations of advanced projects proposals (Narmada dam 1992; Arun dam 1995; Qinghai project 2000) on the incentive structure of task managers (Wade 1997; Mallaby 2004).

If one wants to apply a similar analysis to water related projects, a similar problem has to be resolved, i.e. the question which projects can be classified as being supportive to IWRM has to be answered. Here, an independent measurement is not possible as this is mostly a desk study and the corresponding data for a separate analysis are not available yet. Thus, only policy changes of World Bank in the water sector can be interpreted within the framework of the principal–agent model.

5 The Multidimensionality of Water Policy of the World Bank

The analysis of the development of a water policy or water policies in Multilateral Development Banks needs an understanding of what this policy or these policies cover. Traditionally, in the 1950s towards the end of the 1970s, the understanding was restricted to lending for water projects which consisted of a number of sub-sectors, i.e. irrigation, water supply, hydropower, ports, flood protection and multipurpose. This emphasis on the infrastructure side of water policy has been an important part of the World Bank's activities as an organization which specialized on lending "to governments for hard and visible infrastructure projects that would ensure healthy income streams" (Kapur 2002, 83). This focus helped to create a number of environmental and social problems which resulted in the wave of NGO-criticism and it did not avoid the performance problems in terms of technical water inefficiencies, cost overruns and reduced rates of return (World Bank 2002b). As a reaction to these two general problems – inefficiencies in the water services projects and lack of concern for environmental and social considerations – the concept of Integrated Water Resource Management (IWRM) has been developed by a number of authors (Hartje 2002) and adopted by the World Bank officially in the Policy Paper *Water Resources Management* in 1993. The "comb" picture of IWRM developed by Global Water Partnership serves the Bank to illustrate the dual nature of IWRM as an integration tool and as basis for analyzing traditional water management of the World Bank (Fig. 2).

The teeth of the comb are thought as the water-using sectors, with irrigation, drainage, water supply and sanitation covering most of the water withdrawals in the borrowing countries, which are competing for their sectoral share of allocable water and both causing considerable pollution of the surface waters. The third sector, here presented according to the OED paper 2002 as the sum of energy, navigation and flood management, consists of activities changing rivers and their natural water regime for in-stream uses, causing changes in the river ecosystem and in riverine land uses. All three sectors have traditionally been the core of World Bank

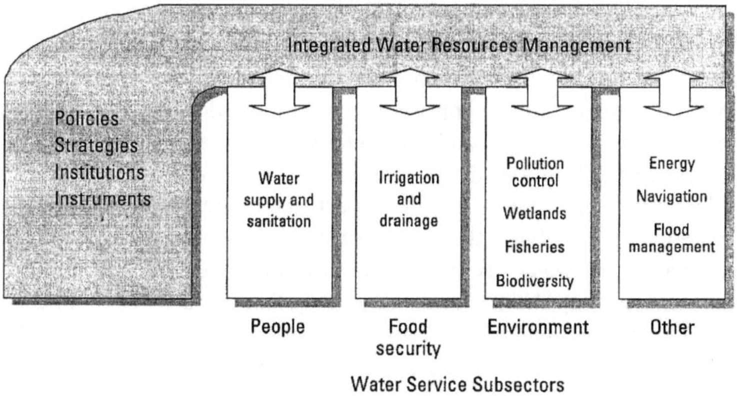

Fig. 2 A comprehensive framework for water management
Source: World Bank (2002b) based on Global Water Partnership

lending for physical infrastructure projects. A fourth tooth is called environment and includes pollution control, the protection of wetlands (presumably freshwater), biodiversity and (presumably inland) fisheries, which has been added to the project portfolio of MDBs later, mainly in the nineties. The "handle" of the comb consists of the management of the resource itself, the integration of the components of the resource – surface and groundwater –, the informational basis in terms of measuring and monitoring, the design and implementation of the general user rights and restrictions, the intersectoral allocation of water consumption, the planning procedure for projects and the organization structure for the administration of the resource and its coordination with the water-using sectoral administrations. Projects or project components supporting these functions have been added to the portfolio in the nineties as a new category of the activities of MDBs, mainly in terms of capacity building and analytical and advisory work. The "handle" basically covers the content of the Water Resources Sector Strategy (World Bank 2004) while the policy issues of the teeth are dealt with in separate sector strategies. Usually, they tend to be part of wider topical/sectoral area, such as irrigation and drainage is part of the rural development strategy (World Bank 2003a) or hydropower is part of the energy strategy (World Bank 2001b). Again, they are often connected to broader perspectives, e.g. the general institutional set up of providing infrastructure services and their financing or the Infrastructure Action Plan (World Bank 2003b).

One of the first concerns raised was the environmental impact of these projects which caused the development operational policies with respect to environmental impact assessment. While there was a minor public concern outside the professional communities for the environmental impact of water supply, sanitation and irrigation projects, the construction of dams led to the well-known controversy between the World Bank and other development banks, on the one hand, and environmental NGOs, on the other, which is still unresolved despite the efforts of the World Commission on Dams in the late nineties. For these water using sectors, a number of efficiency and sustainability problems arose and have not been completely solved which are a main concern of the professional community, but attract less attention in the environmental NGO community with again one exception, the efforts to promote private sector participation in water supply and sanitation (McCully 2002). This concern with environmental and subsequently social impact was complemented by a growing awareness of the limits of the available water resources in quantity and increasingly by declining water quality and the experience of increasing costs of the development strategies for supplying the water consuming sectors. In the eighties, this raised early concerns of water use efficiencies within the technical domain of the distribution systems among the engineers to be complemented by economic arguments about the incentive structure of pricing, water use rights, usually within the sectors concerned, and the allocation of water among sectors.

Water policies are multidimensional policies with a relatively high degree of complexity as they cover the use, the allocation among uses and the protection of the different water resources (surface freshwater in rivers and lakes, groundwater and coastal waters) and the management of the infrastructure for its various uses, for its storage, conveyance and distribution to the users, its collection, treatment and discharge. This involves two different general sets of problems within the water sector, i.e. first the use and protection of the resource in the context of the hydrological cycle, its function in the biosphere and the institutional structure for allocating the resource and second, the financing, building and maintaining of an infrastructure with a high level of networks. For a development bank, the second problem structure of the infrastructure side is similar to the other infrastructure networks, as transport, electricity, and telecommunications while the other problem set creates similarities with other natural resource sectors, e.g. forests, fisheries and mining and oil (extractive industries in the terminology of the World Bank).

6 The Development of the Water Policies at the World Bank

Water policies at the Bank were traditionally defined from the various users' point of view, i.e. they were sector-specific and concerned with technical questions, economic evaluation procedures and institutional topics, but related to each other, as the first crosscutting perspective came basically from the Environmental Assessment Procedures, introduced in 1984 for all type of projects, but which remained relatively vague about procedures and criteria (Wade 1997, 633) to be followed by the 1989 Operational Directive on environmental assessment.[10] They were backed by Sourcebook documentation filling the details for a number of questions in 1991 with sector specific guidelines (World Bank 1991).

The implementation of these procedures along with the establishment of the corresponding institutional backup in form the environmental departments and regional divisions constitutes the battleground between the Bank and the NGOs about the

[10] The process of establishing these procedures is documented in Wade (1997, 681–687).

adequacies of this process. Wade (1997) portrays the implementation as a bargaining situation between the environmental specialists in the regional environmental divisions and the task managers in the country departments. These actors bargain over a number of issues, i.e. the classification of the project proposal, the comprehensiveness of the studies, the degree and stringency of modification of project design until the environmental specialists have signed off the requirements. But the bargaining does not include control procedures on the post bargain phases, i.e. the implementation of the agreed modification of the project design and implementation of accompanying procedures, resulting in considerable discretion for the task managers "to cut corners". These procedures cover water-using projects in irrigation, hydropower and water supply and sanitation as well.

A number of water related projects, especially dams, featured prominently in the fight about the adequacy of their implementation in the late eighties and early nineties with the Narmada campaign as a peak resulting in a cancellation of a well advanced project in 1993. This process and the additional cancellation of the Arun dam in Nepal in 1994 were the results of campaigns by US-based NGOs using an additional change in environmental governance, the introduction of the Inspection Panel in the early 1990s. The analysis of these cases (Wade 1997, 687ff.) shows the mechanism of the reluctance on the part of the task managers and their superiors. Concurrently with the cancellation of the lending operations for the Indian dams and partly as a consequence of it, Wade diagnoses a change in the bargaining balance of the two parties involved in the early 1990s with the preparation of the 1992 UNCED summit and the 1992 World Bank report, Development and the Environment, an established Environmental Department and a changing public opinion in the developing countries.

7 The Policy Change in the 1993 *Water Resource Management* Policy Paper

The first policy change relevant for the above conceptualization was the preparation and acceptance of the World Bank Policy Paper on *Water Resources Management* of 1993, which was the result of a 3 year effort. This effort can be classified as an internally initiated and decided process with a relatively high degree of participation, but with limited influence of US-based NGOs. The second process of water policy is the establishment and the support by the World Bank of the World Commission on Dams, an external process jointly developed with the World Conservation Union (IUCN), but contrary to the development of the Water Policy Paper with considerable discretion of turning it into an internally binding document, but providing the external stakeholders with a policy reference. These two policy changes with a strong role of NGOs in its formulation were followed by two policy changes, internally initiated and influenced by stakeholders from borrowing countries and their representatives among the Executive directors. The Water Resources Sector Strategy, published in 2004, can be seen as a reaction towards the implementation of the earlier Water Policy Paper, focusing on the management of the resource itself and providing general principles for the water using sectors. The water using sectors have policy papers of their own, but the latest, most debated policy documents with relevance to the water sector is the Infrastructure Action Plan of 2003. Here, the process of its formulation was similar to that of the Water Resources Sector Strategy and left out the environmental NGOs.

The policy process leading to the Policy Paper on *Water Resources Management* was initiated by a number of evaluations of the water relevant projects by the World Bank itself (World Bank 1993, Appendix E) and accompanied by a general report on the performance of the Bank on a project level (Wapenhans Report: World Bank 1992). While the internal evaluation of the Bank focused on problems of the sustainability of water supply and sanitation projects and on irrigation and drainage activities (World Bank 1995), the Wapenhans Report focused on the implementation problems of projects in all sectors and found those in water supply to have a higher degree of major problems. The internal assessments were to a large extent the results of the work of the Environment Department, created in 1987.

Last, but not least, there was the criticism of individual projects in the water sector, mainly those involving larger dams, especially the report of the Morse Commission investigating at the request of the Board of Directors the compliance of the Bank with its own policies with respect to the Narmada dam in India (Morse and Berger 1992). The negative assessment of the compliance and the recommendation of the Commission to the Bank to withdraw support finally resulted in the cancellation of the World Bank funding by the Indian government. As one consequence, the Bank's management initiated a larger internal review process at the end of 1992 to survey the compliance with resettlement policies comprehensively which was undertaken by a task force led by the environment department and included not only a review of staff reports, but consistent documentation of the resettlement portfolio and review of the numbers of displaced persons (Fox 2000; World Bank 1996a).

The process to formulate a water policy took place earlier and had considerably lower visibility. Officially, the process was announced in early 1991, but Moore and Sklar in their account attribute the beginning of the policy formulation to a year earlier in 1990 (Moore and Sklar 2000, 354). The process was largely conceived as an internal process, led by a team from the Agriculture and Rural Development Department. The consultative process of Bank staff started with representatives from borrowing governments and the NGOs, restricted to one workshop in July 1991 (Moore and Sklar 2000). Two of the major US anti-dam NGOs organized a protest mailing which caused a new workshop in 1992 at which the NGOs demanded a policy with binding guidelines, based on a set of objectives which centered on giving priority to alternatives to large-scale projects, to the restoration of rivers and relying on public participation. After an initial search for a common ground, the process of consultation stopped, presumably due to internal differences within the Bank. The NGOs relocated their efforts to the US Congress and executive directors. After two reviews of drafts by the Board, the final draft was approved in May 1993 and released in September.

The policy change in the final Policy Paper *Water Resources Management* is mostly conceptual, adopting the Dublin principles and linking it to the Bank's 1992 World Development Report which mainstreamed environment among the non-environmental economists in the Bank (Wade 1997, 712). It combines the call for a comprehensive framework, especially on a river basin level and its institutional reform with a broadening of the options in the water using sectors from the traditional supply augmentation to efficient service delivery and demand side options and includes a higher priority for environmental objectives without calling it IWRM (World Bank 1993, 10f.). It includes the concerns of economists (pricing, financial accountability, decentralization and private sector participation) and the call for a comprehensive approach, based on river basin management which was important to the water engineers. It does not follow the demands of the NGOs for binding guidelines and for explicit anti-large projects priorities. The final chapter on the implications is rather short, but the strategic orientation is that water policy changes in the borrowing countries have to designed and implemented and project lending has to be linked to these processes. This implies a higher share of analytical and advisory work for the Bank.

After the Water Resource Management Policy Paper was approved by the Board, another water-related policy formulation process of a different nature took place which dealt separately with dams, the major physical link between water resource management and the water using sectors, namely the World Commission on Dams. Large dams had played a major role in the NGOs' campaigns criticizing the World Bank for the environmental and social impacts of its lending, but up to the early nineties mostly around individual projects. In 1994, a large number of NGOs demanded a moratorium on dams which should only be lifted, if the World Bank set up an independent review (Dubash et al. 2001, 28). Partially as a reaction to these demands, the OED undertook an internal desk review of 50 dams supported by the World Bank as a first phase and found the performance to be mixed, but argued for continued support for large dams with stronger emphasis on compliance (World Bank 1996c). Based on leaked copies, NGOs reviewed the report, found it inadequate, and asked for its rejection by the World Bank and an independent review instead (McCully 1997). The World Bank reacted by convening with IUCN (which had little experience with social impacts and human rights violations but was considered as an legitimate organization to the critiques) the well known workshop in Gland in 1997 to review the report, to develop evaluation criteria and define an acceptable evaluation process which resulted in the World Commission on Dams.

The trilateral structure of the World Commission on Dams, its ability to include all NGOs active in the area and to produce a common report is commonly seen as a positive example of a global civil society (Dubash et al. 2001; Witte et al. 2000). The objectives of the World Bank in participating have not been examined from a strategic point of view of an agent. The WCD constitutes a very special review process of its behavior (as part of the global review of dam planning, effects and impacts) in comparison to the other review processes in the nineties which were mostly controlled by the Bank with the exception of the Inspection Panel, which was forced on the Bank with the IDA-Congress leverage. The participation of the World Bank in the WCD process was voluntary, restricted in terms of activities performed (set up, seed financing, thematic contribution), but it was without any guiding functions. The expectations of the World Bank about its results are usually described by the motive of breaking the stalemate of dam construction (WCD 2000, 26; Witte et al. 2000, McCully 2001, 1468). The potential deal between the World Bank and the critical NGOs of accepting stricter standards and granting/receiving higher acceptance did not materialize as it became clear already during the selection process of the commissioners (Dubash et al. 2001, 39; McCully 2001, 1470). After the WCD had published its report in 2000, the World Bank did not adopt its conclusions and the NGOs remained as opposed to the financing of large dams as before.

8 Evaluating the Impact of the 1993 *Water Resources Management* Policy Paper in 2002

There is no independent analysis of the impact of the policy change that was introduced by the 1993 *Water Resource Management* Policy Paper. To get an impression one has to rely on the internal evaluation of the World Bank. The OED evaluation of the implementation of the 1993 *Water Resource Management* Policy Paper was a multi-year effort, approved by the Board of Directors Committee on Development Effectiveness (CODE). Starting in 1998, it is based on a comparison of a 6-year period prior (1988–1993) and 6 years after the beginning of the *Water Resources Management* Policy (1994–1999) by covering all and completed water projects and projects with water components. It was based on the desk evaluation of 410 projects and numerable analytical activities, the in-depth review of four countries, thematic studies (later cancelled) and a survey of World Bank staff. In addition workshops were held within the Bank and with stakeholders in borrowing countries to discuss methods and progress (World Bank 2002b). The draft report was discussed with stakeholders in a forum organized by the Water Resources Management Group. The main result has been that the 1993 *Water Resource Management* Policy Paper has been implemented broadly, but only partially and unevenly. The implementation varied widely across regions, countries and water sub-sectors. In water supply and environment, there has been considerable progress, but less so in irrigation: There have been changes on a project level, but there the impetus for policy changes has been limited, as well as in country analytic work.

The portfolio of the Bank has been responsive to the impetus of the 1993 *Water Resource Management* Policy Paper: Economic and sector work has increased considerably and dealt more with policy issues and less with the traditional technical topics (World Bank 2002b, 8). There has been an increase in commitments for water lending after 1993, but it lasted only until 1997 when it declined. Its sub-sectoral composition shifted from the traditional service sectors, water supply and sanitation and irrigation and drainage, to environment and "other" sectors, which means water components in projects with an agricultural or social sector classification.

The diagnosis in the OED report of the causes of the incomplete implementation focuses first on the complexity of the comprehensiveness of introducing IWRM and its inherent difficulty and the need to find an opportunity to install the necessary reform in the borrowing country. It identifies the long time horizon necessary to sustain the reform process and the need to sequence a number of steps on multiple governance levels. It sees the World Bank as being comparatively well qualified in identifying opportunities of reform, generated by fiscal problems or reforms in other sectors; it has had a number of problems: It sent mixed signals about the desirability and directions of reform as there were limited interactions among the water sub-sectors. The long time horizon necessary for the reform can in principle be addressed by series of projects, but the continuity was often missing. With respect to large water infrastructure, it points out that the cost of compliance to the World Bank guidelines could be significant for the borrowing country and identifies the need for the harmonization between domestic legislation and the Bank's policies prior to project processing. It sees the reduction of the number of persons to be resettled down to 40 percent in water projects as an improvement, but it sees a considerable room for the Bank to improve the institutional basis for safeguard policies in the borrowing countries as the report emphasizes the need for additional storage for water (World Bank 2002b).

With respect to the internal incentives of the Bank, it sees the lack of an implementation plan of the Bank's management in 1993 as a major deficit so that the adoption remained voluntary for the regional management which was differently prepared after the 1997 internal reorganization. Regional water teams with an integrated perspective existed only in two regions in the first phase. Often on a regional level and on the management level, the water sector was organizationally fragmented with the irrigation and the environment sub-sector located in rural units and the water supply side in infrastructure units with an emphasis on private sector participation. A measure to deal with the organizational fragmentation was the establishment of the Water Resources Management Group in 2000. Two additional problems do not seem to have been addressed: One is the low level of staffing and the limited extent for integration for the operational staff and the other is the development of better, meaningful guidelines helpful for operational activities. A World Bank Technical Paper on the formulation of a Water Resources Strategy for policy makers in developing countries in 1994 (LeMoigne et al. 1994) and its republication as a FAO Report in 1995 can be seen as one-time efforts. A larger effort constitutes the foundation of water policy related networks, the World Water Council and the Global Water Partnership[11] in 1996 which support the concept of IWRM and which were supported financially with a minor share by the World Bank.

[11] Both organizations were set up with the support of bilateral donor organizations and the World Bank and consist mostly of organizations responsible for water management and related suppliers. Both expanded in the last 10 years and organized a number of activities supporting the IWRM concept. There are only a few green organizations members although they participate in conferences and meetings.

9 A Partially New Direction in the 2004 Water Resources Sector Strategy

The development of the Water Resources Sector Strategy (WRSS) is presented as the result of dual process: The review process on the 1993 *Water Resources Management* Policy Paper which as been initiated by the Operations Evaluation Department (World Bank 2002b) and the preparation of the Sector Strategy as a result of the request of the Board of Directors in 1999. Both processes were coordinated, but had their own structure: The OED review process presented the first results in 2001 and the final version was made public in 2002. The development of the sector strategy was prepared by the Water Resources Management Group chaired by the Bank's senior water advisor. Here the process started in 1999 and resulted in the document approved by the Board in 2003. Partially parallel was the process of developing the Infrastructure Action Plan which was presented to the Board of Directors in 2003. It has been designed to reverse the declining trends in infrastructure lending in the 1990s and it encompasses seven sectors, including the water resources sector. For these seven sectors, sector strategy papers were developed. Thus, The WRSS is part of a larger strategy for the infrastructure as a whole.

The Water Resources Sector Strategy has been developed at the request of the Board by a group led by the Bank's senior water policy advisor, guided by the (informal) Water Resources Sector Board in cooperation with the OED Review of the implementation of the 1993 *Water Resources Management* Policy Paper. It relied on an additional assessment of the Bank's portfolio and extensive in-country consultations in 2000 as a brainstorming exercise with water stakeholders in lead countries. As another input, the chapter mentions an internal, high level panel which examined the options for the World Bank's engagement in hydraulic infrastructure, chaired by the Vice President for South Asia.[12] After the group produced a draft, it was widely discussed, put on the Bank's external Web for comments, and a second consultation process took place in borrowing countries and with donors, the private sector and NGOs. After considerable debate within the Bank and its Board, it was approved in 2003 and published in 2004.

As a strategy for water resources, the strategy covers the resource itself and the concept of Integrated Water Resource Management, but not the individual water sectors in detail as they have strategy papers on their own (e.g. World Bank 2001b for energy; World Bank 2003a for rural development). It explicitly covers water storage as a problem under the conditions of high precipitation and runoff variability and emphasizes the need for additional hydraulic infrastructure. It starts with an introduction that puts water resources in the context of development and here emphasizes its contribution to the overall objectives of sustainable economic growth and poverty reduction. It then advances the concept of IWRM by relying on the water management comb, but without using the term "integrated" and then provides

[12] There are two references to the panel in the strategy text (World Bank 2004, V and 28), but it is not documented further. It seems to be the basis for the introduction of the concept of high-reward-high-risk hydraulic infrastructure.

a survey of the linkages between the concept of water resources management and four of the five main water services (irrigation and drainage, energy, water supply and sanitation, environmental services; not for industry and navigation) and the main strategic issues in these sectors. The linkages are seen in the institutional framework, the management instruments, the development and management of the hydraulic infrastructure and in the political economy of reform.

The next topic – stocktaking – relies on the OED evaluation of the selective adoption of IWRM to a large extent, adding the dimension of regional variation in the adoption of the IWRM principles and derives a need for a customized approach to problems and solutions. The stocktaking of the lending portfolio by OED and a separate effort by the Water Resources Management Group (WRMG) yielded the following results: The World Bank has historically invested US$ 3 billion on average annually, which accounted for 5 percent of the US$ 70 billion total investment in water-related projects in developing countries of which 90 percent is financed from domestic sources (World Bank 2004, 32). Over the decade between 1993 and 2002, the lending for water was increased to 16 percent of total World Bank lending. The Middle East and North Africa, South Asia and East Asia and the Pacific are the most important regions. There has been a shift towards investment in the environment and resource management accompanied by a sharp decline in hydraulic infrastructure, a reduction of 90 percent for hydropower, resulting in a share of the World Bank in new dam financing of 0.5 percent (World Bank 2004, 35). The next points are the identification and explanation of the Bank with the recommendations of the World Commission of Dams (WCD 2001). Here agreement with the core values and the seven strategic priorities is clarified, but also the differences between three of the 26 guidelines emphasized. The first important is the guidance on "prior informed consent" of affected and indigenous people which would amount in the view of the authors in a virtual veto right. The second concerns the continued "proactive engagement" of the World Bank in countries that are not "not already negotiating with their neighbors on international waters" (World Bank 2004, 38) instead of the recommended disengagement. The third is the WCD's suggested "multi-stage, negotiated approach to project preparation" which is considered impractical and the equivalent to the preclusion of dam construction (World Bank 2004, 38).

The World Bank's role is described as having comparative advantages in its combination of knowledge, financial resources and engagements in all water sub-sectors and scale and its ability to integrate them. For middle income countries, the Bank's financing is less preferred, compared to all other sources because of the complexity of the procedures of the Bank, the resulting high transaction costs and the risk aversion of the Bank.

The identified strategic options deal with the improved adoption of the IWRM approach, the role of the Bank in attracting additional sources of financing for water resource infrastructure and the decision-making about re-entering hydraulic infrastructure (the business model dealing with high risk). The engagement in water resources management is seen as an objective with a broad consensus; the major problem is the identification of opportunity, the sequencing of steps in widely varying country circumstances. Here the major answer is the improvement of country analytic work, by including the political economy of reform and the integration of the Country Water Resource Assistance Strategy in the overall country strategy. The financing section analyses the development of private investment in the water infrastructure, particularly the shortcomings for small countries and the resulting collaborative public–private partnership approach in which the Bank can play a major knowledge-based role. The final discussion of the more effective business model deals generally with the approach to risky decisions and the implication of risk aversion on the Bank's side for low and middle income borrowing countries, i.e. the interest of these countries in having a less risk averse approach of the Bank to complex water infrastructure. Then, the implications for the management of these projects are debated with a proposal of treating them as corporate projects with responsibility on a higher management level, i.e. country directors and regional vice presidents, being able to command the Bank's internal resources to improve the implementation of environmental and social safeguards. This elevation of management responsibility will be accompanied by the establishment of water resource management units and an Human Resources Strategy for water resources.

The general perspective of the WRSS was supported by an external Commission on the Financing of Water Infrastructure, established by the Global Water Partnership and the World Water Council at the 3rd World Water Forum in 2001 in Kyoto, chaired by Michel Camdessus which delivered a report "Financing Water for All" in 2003 in view of the Millennium Development Goals (WWC 2003). It focuses on the financing problems and potential solutions to water service delivery, but makes a supporting reference to the WRSS of the World Bank in its section on major hydraulic works (WWC, GWP 2003, 26).

The observation of the reduction infrastructure investment lending, especially in IBRD countries, by 50 percent between 1993 and 2002, led to a request for a policy change in a Board meeting in early 2003. This was supported by the report of the Camdessus panel, the New Partnership for Africa's Development (NEPAD) and the World Summit on Sustainable Development in Johannesburg (WSSD). The Infrastructure Structure Action Plan is, thus, a concerted action of the Bank's management to fill (at least partially) the financing gap between the MDGs and the decline of private investment and World Bank lending in water supply and sanitation, among other infrastructure sectors (World Bank 2003b).

10 Assessment of the Changes in World Bank Water Resources Policy

The 2004 WRSS constitutes continuity and considerable change at the same time: The continuity applies to the adherence to the principles and for the implementation of the IWRM concept, but it implies a "realistic" pragmatic approach to the implementation in the borrowing countries giving the heavy political opposition it faces as it often means the redistribution of power in the water sector. The WRSS focuses on identifying the opportunities for reform, tailoring the proposals to the country specific circumstances and providing the Bank with a better organizational structure and more, better human resources to implement the concept. This has been relatively uncontentious in the Bank and among its owners, the other donors and the various stakeholders.

The change has occurred in the renewed inclusion of dams or hydraulic infrastructure in the lending activities of the Bank. The reduction in dam lending has been the result of a combination of causes: The inclusion of environmental evaluation procedures in operational policies which made the planning of these projects more time-consuming and cost-intensive for the Bank and the staff, potentially revealing the lack of economic feasibility of proposals is certainly a contributing factor. Another more direct effect can be attributed to the international campaigning against the dams in general which resulted in the Inspection Panel and in the cancellation of two dams which were relatively well advanced in their internal process: Sardar Sarovar and Arun II. The inability of staff to answer the questions of the Inspection Panel added to the reduced incentive of operational staff to support dams.

The reduction of lending for hydropower and for other hydraulic infrastructure has implication for the level for IBRD lending, which constitutes the core of lending of the World Bank. These activities form the basis of the relative autonomy of the World Bank as an agent. As Kapur pointed out, the expansion of IDA-based loans reduces the autonomy of an MDB and it needs to maintain the loan-based share of its activities (Kapur 2002). The major customers for these loans are middle-income countries which have access to other sources of financing nationally and internationally. This diagnosis is repeated again in the WRSS, its accompanying documents and in the Infrastructure Action Plan (IAP). As the process of developing the WRSS and IAP have been designed by the higher management in cooperation with the Board of Executive Directors to focus on these borrowers and to give international NGOs a reduced role, compared to the WCD process, it seems clear that the Bank changed it course to placate the major borrowers and it accepted the potential increase in conflicts with the international NGOs.

The Bank's management seems to be aware that resumption for hydraulic lending risks a resumption of NGO-campaigning against new dam projects and an effective use of the leveraging in the US Congress and in the US Treasury. The proposed "new business model" of elevating the responsibility these projects to "corporate" projects is designed to deal explicitly with the complexity of these projects and their applicable criteria. The NGOs are not convinced that their concerns will be adequately addressed (Bosshard et al. 2003).

Another interesting observation concerns the process leading to the water policy document. The 1993 Water Resource Management Policy Paper was based on a comparatively low key internal process which reflected probably changes in the preferences among the major principals and which served as a basis for integrating policy views within the Bank's management and staff. It seems as if the NGOs were largely excluded from it despite the smaller workshops; although they managed to reach selective Executive Directors, but a real impact is not discernable. With these characteristics, the influence of the NGOs was considerably lower compared to all the other campaigns which resulted in reorganizations and new procedures. The process leading the 2004 WRSS was controlled by the Bank's staff with considerable external involvement emphasizing borrowing country representatives who could be expected to be supportive of the planned changes.

In a wider research perspective, the consultation process and the presentations at the World Bank's water week can be seen as selecting another set of third parties supporting the view of the agent as opposed to NGOs as self selected third parties trying to convince the principals about their view of the agent. The view of NGOs as third parties to the principal–agent relationship can be seen as a source of information for the principal when self-reporting of the agent is problematic. This function of providing agent-independent information can be seen as a new research direction to shed additional light on the role of NGOs in the context of global water policy (Lake and McCubbins 2006).

References

Bosshard P et al. (2003) Gambling with people's lives – What the World Bank's new "high-risk/high reward" strategy means for the poor and the environment, published by Environmental Defense, Friends of the Earth and International Rivers Network, Washington, DC

Dubash NK, Dupar M, Kothari S, Lissu T (2001) A watershed in global governance? An independent assessment of the World Commission on Dams, World Resources Institute, Washington, DC

Fox, JA (2000) When does reform policy influence practice? Lessons from the bankwide resettlement review. In: Fox JA, Brown LD (eds) The struggle for accountability. The World Bank, NGOs, and the grassroots movements. MIT Press, Cambridge, MA, 303–344

Fox J (2002) The World Bank inspection panel and the limits of accountability. In: Pincus JR, Winters JA (eds) Reinventing the World Bank, Cornell University Press, Ithaca, NY, 131–163

Furubotn EG, Richter R (2005) Institutions and economic theory: the contribution of the New Institutional Economics, 2nd edn, University of Michigan Press, Ann Arbor, MI

Gutner T (2002) Banking on the environment – multilateral development banks and their environmental performance in Central and Eastern Europe, MIT Press, Cambridge, MA

Gutner T (2005a) Explaining the gaps between mandate and performance: agency theory and World Bank environmental reform. In: Global Environmental Politics, Vol. 5, 2, 10–37

Gutner T (2005b) World Bank environmental reform: revisiting lessons from agency theory. In: International Organization, Vol. 59, 773–783

Hartje V (2002) International dimensions of integrated water management. In: Al Baz I, Hartje V, Scheumann W (eds) Co-operation on transboundary rivers, Nomos, Baden-Baden, 7–34

Hawkins DG, Lake DA, Nielson DL, Tierney, MJ (2006) Delegation under anarchy: states, international organizations, and the principal–agent theory. In: Hawkins DG, Lake DA, Nielson DL, Tierney, MJ (eds) Delegation and agency in International Organizations, Cambridge University Press, Cambridge, MA, 3–38

Kapur, D (2000) Who gets to run the world? In: Foreign Policy, Vol. 121, November–December, 44–50

Kapur, D (2002) From shareholders to stakeholders: the changing anatomy of governance of the World Bank. In: Pincus J, Winters J (eds) Reinventing the World Bank, Cornell University Press, Ithaca, NY, 80–114

Kapur, D (2004) Conditionality and its alternatives, Harvard University, Cambridge, MA, unpubl. manuscript

Killick, T (1997) Principals, agents and the failings of conditionality. In: Journal of International Development, Vol. 9, 4, 483–495

Lake DA, McCubbins MD (2006) The logic of delegation to international organizations. In: Hawkins DG, Lake DA, Nielson DL, Tierney, MJ (eds) Delegation and agency in International Organizations, Cambridge University Press, Cambridge, 341–368

LeMoigne G, Subramanian A, Xie M, Giltner S (eds) (1994) A guide to the formulation of water resources strategy, World Bank Technical Paper No 263, Washington, DC

Le Prestre P (1989) The World Bank and the environmental challenge, Associated University Press, London, Toronto

Mallaby S (2004) The world's banker, a story of failed states, financial crises, and the wealth and poverty of nations, Penguin Press, New York, NY

Martens B (2002) Introduction. In: Martens B, Mummert U, Murrell, P (eds) The institutional economics of foreign aid, Cambridge University Press, Cambridge, 1–33

Martens B (2005) Why do aid agencies exist? In: Development Policy Review, Vol. 23, 6, 643–663

McCully P (1997) A critique of "The World Bank's experience with large dams: a preliminary review of impacts", International Rivers Network. http://www.irn.org/programs/finance/index.php?id=critique.html. Cited 19 Feb 2007

McCully P (2001) The use of a trilateral network: an activist's perspective on the formation of the World Commission on Dams. In: American University International Law Review, Vol. 16, 6, 1453–1475

McCully P (2002) Avoiding solutions, worsening problems. A critique of World Bank water resources sector strategy: strategic directions for World Bank engagement, draft, March 25, 2002, International Rivers Network. http://www.irn.org/new/0205/wrsscritiquea.pdf. Cited 19 Feb 2007

Milner HV (2006) Why multilateralism? Foreign aid and domestic principal–agent problems. In: Hawkins DG, Lake DA, Nielson DL, Tierney, MJ (eds) Delegation and agency in International Organizations, Cambridge University Press, Cambridge, 107–139

Moore D, Sklar L (2000) Reforming the World Bank's lending for water: the process and outcome of developing a water resource management policy. In: Fox JA, Brown LD (eds) The struggle for accountability. The World Bank, NGOs, and grassroots movements, MIT Press, Cambridge, MA, 345–390

Morse B, Berger T (1992) Sardar Sarovar: report of the independent review, Resources Future International, Ottawa, Canada

Mosley P, Harrigan J, Toye J (1991) Aid and power. The World Bank & policy-based lending, Vol. 1: Analysis and policy proposals, Routledge, London, New York, NY

Nielson DL, Tierney MJ (2003) Delegation to International Organizations: agency theory and World Bank environmental reform. In: International Organization, Vol. 57, 241–276

Nielson DL, Tierney MJ (2005) Theory, data, and hypothesis testing: World Bank environmental reform redux. In: International Organization, Vol. 59, 785–800

Ostrom E et al. (2002) Aid, incentives and sustainability: an institutional analysis of development cooperation, Sida Studies in Evaluation 02/01: 1, Swedish International Development Cooperation Agency, Stockholm

Rich B (1994) Mortgaging the earth: the World Bank, environmental impoverishment, and the crisis of development, Boston, MA

Rich B (2002) The World Bank under James Wolfensohn. In: Pincus JR, Winters JA (eds) Reinventing the World Bank, Cornell University Press, Ithaca, NY, 26–53

Rodrik D (1995) Why is there multilateral lending? Centre for Economic Policy Research, Discussion Paper Series, No 1207, London

Saleth RM, Dinar A (2004) The institutional economics of water. A cross-country analysis of institutions and performance, Edward Elgar, Cheltenham

Sappington DEM (1991) Incentives in principal–agent relationships. In: Journal of Economic Perspectives, Vol. 5, 2, 45–66

Scheumann W, Klaphake A (2001) Freshwater resources and transboundary rivers on the international agenda: from UNCED to Rio +10, report for the Ministry of Economic Cooperation and Development (Bundesministerium für wirtschaftliche Zusammenarbeit und Entwicklung), German Development Institute (DIE), Bonn

Tirole J (1994): The internal organization of government. In: Oxford Economic Papers, Vol. 46, 1–29
Wade R (1997) Greening the Bank: the struggle over the environment, 1970–1995. In: Kapur D, Lewis JP, Webb R (eds) The World Bank, its first half century, Vol. 2: Perspectives, Brookings Institution Press, Washington, DC, 611–734
Weaver, C (2006) The IO hypocrisy trap: explaining the rhetoric and the reality of the World Bank, paper presented at the annual meeting of the International Studies Association, March, San Diego
Witte, JM, Reinicke WH, Benner T (2000) Beyond multilateralism: global public policy networks, in: Politik und Gesellschaft. http://www.fes.de/IPG/ipg2_2000/artwitte.html. Cited 16 Feb 2007
World Bank (1991) Environment Department: environmental assessment sourcebook, Vol. I: Policies, procedures, and cross-sectoral issues, World Bank Technical Paper No 139, Washington, DC
World Bank (1992) Effective implementation: key to development impact, Portfolio Management Task Force, Washington, DC (Wapenhans Report) (mimeo)
World Bank (1993) Water resources management, A World Bank Policy Paper, Washington, DC
World Bank (1995) Operations Evaluation Department. The World Bank and irrigation, Report No 14908, Washington, DC
World Bank (1996a) Resettlement and development. The bankwide review of projects involving involuntary resettlement 1986–1993, Environment Department Papers No 032, Washington, DC
World Bank (1996b) Operations Evaluation Department. Effectiveness of environmental assessment and national environmental actions plans, Report 15835, Washington, DC
World Bank (1996c) Operations Evaluation Department. World Bank lending for large dams: a preliminary review of impacts, OED Précis, Washington, DC
World Bank (2001a) Operations Evaluation Department. OED review of the Bank's performance on the Environment 16, Washington, DC
World Bank (2001b) Energy and mining sector board. The World Bank Group's energy program: poverty reduction, sustainability and selectivity, Washington, DC
World Bank (2002a) Operations Evaluation Department. Promoting environmental sustainability in Development. An Evaluation of the World Bank's Performance, Washington, DC
World Bank (2002b) Operations Evaluation Department. Bridging troubled waters, assessing the World Bank water strategy (lead author: G K Pitman), Washington, DC
World Bank (2003a) Reaching the rural poor, a renewed strategy for rural development, Washington, DC
World Bank (2003b) Infrastructure action plan, note on informal board meeting, July 8, 2003, http://siteresources.worldbank.org/INTTRM/Resources/InfrastructureActionPlan.pdf. Cited 2 Feb 2007
World Bank (2003c) Accountability at the World Bank. The inspection panel 10 years on, Washington, DC, 16
World Bank (2003d) Operations Evaluation Department. World Bank's Operations Evaluation Department: the first 30 years, Washington, DC
World Bank (2004) Water resources sector strategy. Strategic directions for World Bank engagement, Washington, DC
WCD (2000) Dams and development, a new framework for decision-Making, Earthscan, London
WCD (2001) A New Framework for Decision-Making, Earthscan Publications
WWC, GWP (2003) Financing water for all, report of the panel on financing water infrastructure. http://www.unsgab.org/Recommended%20Paper/II-1.1.pdf. Cited 5 Feb 2007

Challenges for German Development Cooperation in the Water Sector

Martin Kipping

Abstract Water is key to poverty reduction and to achieving the Millennium Development Goals (MDGs), in particular those related to health and environmental sustainability. However, 1.1 billion people continue to lack access to safe drinking water, 2.6 billion remain without access to sanitation facilities. Germany is the third biggest bilateral donors in the water sector with regional foci in the Middle East and Africa. The German Federal Ministry for Economic Cooperation and Development (BMZ) and its implementing agencies are tackling six main challenges to further improve their water sector activities: (1) mainstreaming the concept of Integrated Water Resources Management (IWRM); (2) addressing water management in agriculture; (3) taking seriously questions of water governance and water politics; (4) prioritizing waste water management, sanitation and hygiene; (5) ensuring the sustainability of large-scale hydro-infrastructure; as well as (6) increasing funds for development cooperation in the water sector through innovative funding mechanisms. The probable re-organisation of BMZ's implementing agencies would significantly contribute to these efforts.

1 Introduction: German Development Cooperation in the Water Sector

"Water for life" is the heading of the UN Water Decade 2005–2015. This not a void slogan, but expresses a simple truth: Water is key to poverty reduction and to achieving the Millennium Development Goals (MDGs). Halving the percentage of people without access to safe drinking water and of those without access to sanitation between 1990 and 2015 has been defined as indicators of target 10 under MDG 7 ("Ensure environmental sustainability"). While this target is formulated at outcome-level, progress on target 10 will contribute significantly to achieving most of the

Martin Kipping
e-mail: martin.kipping@gmx.net

W. Scheumann et al. (eds.), *Water Politics and Development Cooperation*,
doi: 10.1007/978-3-540-76707-7_15,

other MDGs, such as MDG 1 ("Eradicate extreme poverty and hunger"), MDG 2 ("Achieve universal primary education"), MDG 3 ("Promote gender equality and empower women"), MDG 4 ("Reduce child mortality"), MDG 5 ("Improve maternal health") and MDG 6 ("Combat HIV, Aids, malaria and other diseases"), which are formulated at impact-level. Financial, time and health burdens of inadequate water and sanitation services inhibit peoples' economic and educational opportunities. Women and girls are especially disadvantaged, as households' provision with drinking water and hygiene services frequently is their responsibility. About 5,000 people die every day from water-induced diseases, diarrhoeal diseases in particular, among them 4,000 children (cf. UNICEF 2005). However, 1.1 billion people continue to lack access to safe drinking water, 2.6 billion remain without access to sanitation facilities. While the MDG target on drinking water is likely to be achieved on a global scale, the sanitation target will be missed unless much more efforts are made in this area. Sub-Saharan Africa will probably miss both targets by far (cf. WHO and UNICEF 2004, 12).

In addition to drinking water and sanitation, improvements in other fields of the water sector are prerequisites for achieving MDG 1 in particular: Halving the global proportion of people suffering from hunger – about 800 million people in absolute terms (cf. GTZ ed. 2004, 4) – is supposed to mean doubling the amount of water used for irrigation (Hoff and Kundzewicz 2006, 14). This will be impossible without major improvements in agricultural water use efficiency and water resources management. Water also figures as a main ingredient of most production processes and thus is a key factor for economic growth. In addition, better (waste) water management contributes significantly to reducing the incidence of malaria and other diseases (MDG 6) by eliminating breeding grounds for disease vectors. All in all, the importance of the water sector for poverty reduction and sustainable development is hard to overestimate, and much more progress is needed if the water sector is to accelerate – instead of slowing down – the achievement of the MDGs.

These challenges will not be solved by development assistance provided by bilateral donors and multilateral financing institutions.[1] Overall commitments of official development assistance (ODA) on drinking water, sanitation, waste water and water resources management stuck at about US$ 4 billion on average in 2003/04 (OECD 2006b), while estimations of financing needs range between US$ 25 billion and 49 billion annually (World Panel on Financing Water Infrastructure 2003, 3) for reaching MDG target 10 alone. In countries like India and China, where water problems are enormous however, development assistance represents only a minimal proportion of funding available, let alone needed. Nevertheless, development cooperation can contribute modestly to narrowing funding gaps and to improving institutional, legal and regulatory frameworks for investments, service provision and sustainable water resources management.

[1] Lena Partzsch seems to suggest the opposite when critically noting that "classic intergovernmental development assistance has failed up to now in solving the global water crisis" (2006, 20; author's translation).

Germany is the third biggest bilateral donors in the water sector[2] with regional foci in Africa (35 percent bilateral ODA in 2005) and in the Middle East (27 percent of bilateral ODA in 2005). The water sector represents approximately 10 percent of German ODA and constitutes its second biggest sector. Water has been agreed upon as a focal area of cooperation with 28 partner countries, among them 11 in sub-Saharan Africa (cf. BMZ 2006a). German development cooperation is under the overall responsibility and guidance of the Federal Ministry for Economic Cooperation and Development (BMZ). Its main implementing agencies are KfW Development Bank, Deutsche Gesellschaft für technische Zusammenarbeit (GTZ), InWEnt – Capacity Building International, and German Development Service (ded).

The water sector is generally considered as a particular strength of German development cooperation, especially in the Middle East and Africa. However, the Ministry and its implementing agencies need to steadily work on further improving approaches and strategies in response to new challenges within and demands from partner countries, while taking up new solutions developed by the academic and professional spheres. This article discusses six challenges that deserve particular attention by German development cooperation in the water sector. It aims at providing insights into the discussion process within German development cooperation in order to complement the more external and analytical perspectives taken by most of the other articles in this volume.

2 Challenge 1: Mainstreaming IWRM

In the course of the last decades, criticism has grown vis-à-vis the formerly predominant "engineering" or "supply-side" approach to water management. Not least due to increasing environmental problems induced by the over-exploitation and pollution of water bodies, demands for a more holistic perspective on the water sector have progressively gained ground. In consequence, the concept of Integrated Water Resources Management (IWRM) has developed as the new state-of-the-art model in water management since the early 1990s (GWP 2005). IWRM introduces interdisciplinary thinking into a field traditionally dominated by engineers and technicians, as it aims at

1. integrating economical, environmental and social externalities;
2. promoting an ecosystem approach, i.e. taking into account ecological interdependencies; and
3. moving water governance from traditional top-down hierarchies towards a more open, participatory decision-making structure in line with the principle of subsidiarity.

These principles of IWRM have gained wide acceptance within the international water discourse, and they are actively promoted by sector organizations such as the

[2] The biggest is Japan. The United States have recently replaced Germany as the second biggest donor because of their large portfolio in Iraq (OECD 2006b).

Global Water Partnership (GWP). The GWP defines IWRM as "a process which promotes the co-ordinated development and management of water, land and related resources, in order to maximize the resultant economic and social welfare in an equitable manner without compromising the sustainability of vital ecosystems" (Argawal et al. 1999, 22, cf. also WWAP 2003, 299). Although IWRM cannot provide blueprints for water policies, the concept sets a general framework of basic principles and guidelines for good water management (Argawal et al. 1999, 44–45; GWP 2005, 1). In this sense, IWRM is not something that can be "implemented" or "put into practice" once and forever, but IWRM is a set of basic principles and guidelines that need to be continuously mainstreamed into water (and related) policies in order to maximize the overall societal benefit of water use.

If IWRM is to be mainstreamed into national and regional water (and related) policies, the same applies to development cooperation in the water sector (and neighboring sectors). What does this mean in practice? First, a strategic analysis guided by the holistic perspective of IWRM should indicate the sub-sectors and regions where assistance is most urgently needed. Many partner countries have already started so-called "IWRM Processes", conducting strategic assessments of national water sectors (GWP 2006). The adoption of national and regional IWRM and water efficiency plans resulting from these processes has actually been set as an objective (to be accomplished by 2005) in the Johannesburg Plan of Implementation (Art. 26), adopted at the World Summit on Sustainable Development in 2002. Such assessments identify key current and future water sector problems, and spell out and prioritize possible solutions. The assessments may be undertaken under the label of IWRM or not, what counts is that policy-makers and projects managers use the key principles of IWRM for screening where the problems are or will be, and what should be done to resolve them. This knowledge and prioritization in turn should guide partner countries and donors when considering development cooperation in the water sector. Of course, this does not mean that every donor needs to engage in the top priority area or region identified in the respective screening. Donors need to coordinate and harmonize, as stated in the 2005 Paris Declaration on Aid Effectiveness.[3] However, donors should make sure that, in sum, their scarce funds are directed to solving the key problems. Thus, IWRM means getting the priorities right when deciding where to engage.

Second, when intervention areas have been agreed upon between donors and partner countries, the design of the concrete development projects and programs needs to conform to the principles of IWRM. This means that also interventions that take place in a narrowly defined sub-sector or region are expected to take into consideration their impact on neighboring sub-sectors or regions. For example, upstream–downstream effects of water use need to be analyzed in order to avoid water quantity or quality conflicts with competing users.

In short, mainstreaming IWRM into development cooperation in the water and neighboring sectors means drawing on IWRM principles both for the selection of areas in priority need of assistance and for the design of specific interventions. In

[3] http://www1.worldbank.org/harmonization/Paris/FINALPARISDECLARATION.pdf, Cited 29 Jun 2006.

the past, German development cooperation has most often followed IWRM principles within specific projects and programs, in particular regarding environmental aspects, even before the concept was known as such and enshrined as explicit BMZ sector policy. German development cooperation is also very much engaged in overall water sector reforms and national as well as regional water resources management, where comprehensive, IWRM-like reforms are main project and program objectives. However, there is still room for improvement in mainstreaming IWRM into German water-related development assistance.

Consequently, the new BMZ water sector strategy (to be officially adopted in autumn 2006) defines IWRM as the general framework for German development cooperation in the water sector (BMZ in prep.).[4] It also states that every German development project or program – where it touches the water sector – needs to conform to IWRM. This move towards explicitly adopting IWRM as a general guideline triggered some skepticism, in particular within KfW Development Bank. In subsequent discussions with the implementing agencies' water sector professionals, the main reasons for this skepticism got apparent: On the one hand, IWRM was frequently understood as "water resources management" alone, not as a concept covering the whole water sector, including drinking water, sanitation, waste water management, water for industrial purposes and power generation, water for food, and – as one segment among others – water resources management. Thus, IWRM was perceived as the domain of technical cooperation, as German financial cooperation in the water sector focuses on drinking water supply and waste water management. On the other hand, IWRM was equaled with the above-mentioned IWRM-Processes supposed to "implement" IWRM. Thus, the professionals from KfW feared having to wait until the IWRM-Process of a partner country has led to a corresponding and explicit "IWRM-Plan", before they would be allowed to finance investments. However, as we have seen, this is not what IWRM is about. IWRM is about adopting a holistic view on the water sector, independently of the concrete form this can take.

These misunderstandings have been eliminated in an intensive dialogue between the Ministry and its implementing agencies. However, the current challenge is to assist the implementing agencies in translating the Ministry's more ambitious standards on IWRM into practice. The new water sector strategy gives some broad guidelines and principles in this regard. For operationalizing these principles, more practical tools and reference papers are being developed. They will serve for guiding both the identification of key intervention areas and the design of projects within these areas. In the end, IWRM will have to be anchored not only in water sector strategy papers, but in the minds of project managers as well. Only if IWRM is perceived as adequate and useful also at the operational level, it will be effectively mainstreamed into concrete activities on the ground.

[4] The need for an explicit reference to IWRM in BMZ's water sector strategy was also highlighted in an evaluation of German development cooperation in the water sector carried out by the German Development Institute (Neubert and Horlemann 2005).

3 Challenge 2: Addressing Water Use in Agriculture

One challenge where the need of IWRM is most obvious is the consequences of agricultural water use on water availability for other uses, household consumption in particular. Agriculture accounts for 70 percent of global water consumption, rising to more than 90 percent in many partner countries located in arid and semi-arid regions (World Bank 2006a, 146ff.). At the same time, the water use efficiency in irrigated agriculture is frequently low, evaporation and infiltration losses of more than 50 percent being rather the norm than the exception (WWAP 2006, 255). Water demands of agriculture increase accordingly, and a lack of adequate regulation and limitation of water abstraction frequently leads to stressed water balances and quickly dropping water tables. Expected population growth and corresponding demands for food will further boost water demand for agricultural purposes. This is accelerated by changing diets and consumption patterns induced by modernization and economic development: While a kilogram of wheat necessitates 1,500 liters of water, a kilogram of beef takes more than 5,000 liters to produce in intensive cattle raising (WasserStiftung 2006).

Water use in agriculture is also a major source of water pollution caused by return flows contaminated with residues of fertilizer and pesticides as well as with rising concentrations of salt washed out of the soils. Over-abstraction of freshwater for agricultural purposes can also lead to the contamination of aquifers by subsequent intrusion of brackish or salt water. While water quality standards for cooling and processing purposes in industry tend to be modest, the growing contamination of surface and groundwater caused by agriculture particularly threatens drinking water supply. For instance, growing numbers of kidney diseases have already been reported from the Gaza strip where salt concentrations have risen far above international standards (Wimmen 2000), mainly due to the overexploitation of the Coastal Aquifer for agricultural water use, giving way to sea water intrusion into the aquifer.

At the same time, inadequate water management in agriculture (together with deforestation) does not only affect other water uses, but puts into question the sustainability of agriculture itself. Key problems are erosion and the salinisation of soils. The latter alone affects about 25 percent of area under irrigation in arid and semi-arid zones and spreads to another 250,000–500,000 hectares of agricultural land every year (WWAP 2006, 264). In addition, salinisation increases water needs of agriculture even more, as significant amounts of water are needed to leach fields before cropping seasons, which however only postpones the decline of soil productivity, as the vicious circle of over-watering and soil degradation continues.

In consequence, water management in agriculture needs to be urgently addressed.[5] The inter-sectoral re-allocation of water resources, the incorporation of environmental principles into irrigation and drainage policies, and increased water use efficiency in irrigation are imperative if ecological sustainability is to be achieved in the water sector as a whole. As the potential for increasing water use efficiency in agriculture is enormous, progress in this area can already be sufficient to

[5] This has also been a main recommendation of the OECD InfraPoor Report (OECD 2006c).

relax water balances. This potential is however difficult to realize, not least because distorted agricultural markets induce downward pressure on output prices. Possible measures to increase agricultural water use efficiency include better regulation via the definition and enforcement of water rights, technological innovations (e.g. sprinkler and drip irrigation instead of furrow irrigation) and economic instruments (e.g. service charges, tradable water permits) in order to create incentives for water savings, a key element of demand management. If this is not sufficient to assure sustainability of water use, a re-allocation of water resources from agriculture to household water services is needed from an economic and environmental point of view. This holds particularly true for countries where agriculture represents only a small fraction of GDP but consumes disproportionately much water, which is the case in many countries in the Mediterranean and the Middle East.[6] Many of these countries already rely on the import of "Virtual Water" incorporated in grain as a means to cover national food demand (Horlemann and Neubert 2006). However, a proactive policy of re-allocating water away from agriculture needs to be aware of possible social and economic consequences: Farmers might lack economic alternatives and risk bearing the social consequences of such a policy. In addition, export revenues need to be sufficient for simultaneously increasing food imports.

While it is obvious that agricultural water use is a key issue for water policy in general and thus for development cooperation in the water sector, the question arises how German development cooperation can contribute in this area. There is only limited irrigation in Germany,[7] and the irrigation portfolio of German development assistance has steadily decreased in the course of the last decades, also reacting to a perceived lack of success of past projects. However, as a main donor in the water sector, Germany cannot ignore water use in agriculture, which would also contradict its commitment to IWRM. In many places, disproportionate water use in agriculture already casts doubts on the sustainability of German development programs in drinking water supply: Wells are simply running dry. In addition, there are significant potential synergies with existing foci of German development cooperation in rural development and economic reforms, where they touch agriculture and agro-industry. Furthermore, German development cooperation is strong in water resources management, where impacts of agricultural water use are major issues of concern, and has developed institutional approaches for agricultural water management.

In consequence, it seems appropriate for German development cooperation to focus on the regulatory side of water use in agriculture: water rights and allocation mechanisms, economic instruments, abstraction and emission rules, environmental protection, institutional setting (catchment authorities, water parliaments, water user associations etc.). Increasing engagement in these areas – where appropriate and in line with partner demand and donor coordination – might complement and

[6] In Jordan, for example, 75 percent of total freshwater withdrawals are consumed by agriculture (World Bank 2006a, 147), while agriculture represents no more than 3.5 percent of Jordan GDP (CIA 2006).

[7] However, Germany has broad experience with water and soil associations managing water in agriculture.

widen on-going water resources management activities of German development co-operation and thus tackle more directly the key factors of sustainability or non-sustainability of water use.

Better integrating water use in agriculture into German development cooperation in the water sector also poses challenges to its own institutional structures: while the water sector as such is handled by the infrastructure departments of BMZ and its implementing agencies, agricultural water use resides in rural development or agriculture departments. This is indeed the same challenge of institutional fragmentation German technical assistance tries to alleviate in partner countries. What is even more important, differing views on agricultural water use prevail in water and agriculture departments: While water sector professionals tend to regard agriculture as a threat to sustainability in the water sector and thus as something to be controlled and limited, agriculturalists tend to see water as a production factor whose supply is to be secured at low cost or even augmented in case of increasing demand. Thus, more cross-department communication and coordination will be needed to effectively address the challenge of better integrating agricultural water use into water sector development cooperation.

4 Challenge 3: Dealing with the Political Economy of Water Governance

Water use in agriculture also is object of vital economic and political interests, in particular in situations of water scarcity. If water tables drop and drinking water supply is interrupted due to over-abstraction for irrigation, this is not necessarily due to a lack of knowledge and competence on the side of national governments and regulatory authorities. In many cases, the commercial farming sector has a strong lobby, frequently linked to a politically influential land aristocracy or to a strong export-oriented agricultural business community. Thus, politically well established groups would suffer from the introduction of water pricing or from a re-allocation of water resources from agriculture to household water services. Together with concerns for food autarky and ideological reasons ("let the desert bloom"), this frequently constitutes a main stumbling block for sustainability-oriented water sector reforms.

But not only agriculture is riddled with vested interests; the water sector in general tends to be a highly politicized field. Most illustrative in this regard is the swinging pendulum of the privatization debate. After the private sector euphoria of the 1990s, when unrealistic expectations saw the private sector closing both the funding and the capacity gaps in developing countries' water sectors, private sector participation in general has got under severe criticism since. In addition to absolute ethical arguments ("Water is a human right and should not be source of private profit"),[8]

[8] Although access to a minimum amount of drinking water and to basic sanitation is indeed a human right (deduced from Articles 11 and 12 of the International Covenant on Economic, Social and Cultural Rights), state parties are free to decide how they ensure this service delivery, including through private operators.

economic interests are rarely absent in this debate, as private sector participation is normally only one component of overall sector reforms, aiming at increased efficiency and improved services: Overstaffed and hence expensive bureaucracies fear job losses, middle-income households oppose cuts in factual subsidies to their drinking water supply, and alternative water vendors selling water to the poor in areas unserved by public utilities resist the extension of formal water services to their claims. Interestingly, the groups who most aggressively resisted reforms in urban water services in Bolivia were the Neighborhood Committees that had developed a business out of selling water in yet unserved areas (Fritz 2006).

In short, vested interests can severely hamper water sector reforms that aim at rational water use, economic sustainability and improved and affordable water services to the poor. Consequently, the water sector and its governance need to be understood and handled as a field of political economy: particularistic interests, resource conflicts and corruption are key factors and need to be addressed, also by development cooperation. However, translating into practice this general insight is more challenging than its mere formulation, and this holds true for all sectors of development cooperation.

The first challenge is to figure out what measures would really serve the overall objectives of poverty reduction and sustainable development. The Paris Declaration demands alignment to partner country's priorities and policies in order to ensure ownership and to put the partner countries into the "driver's seat". But what if these priorities and policies lack a poverty orientation or contradict sustainability criteria? A key question is whom the partner country's government does represent. Many partner countries display young or defect democracies with considerable regional, social or ethnical fragmentation. This leads to structural majorities and minorities, and governments cannot be expected to act in the interest of the society as a whole,[9] but rather in the interest of a particular fraction of society. Thus, sector policies and investment plans are not necessarily directed at alleviating the most pressing problems, correcting regional disparities or improving services to marginalized groups. They can indeed contradict donors' sector policies and thus lead to trade-offs between the ownership of partner governments, on the one hand, and donors' principles and objectives, on the other hand. In consequence, donors still need to cross-check partner government priorities and possibly identify alternative, more suitable ways for sustainable poverty reduction in line with IWRM.

Second, even if partner government and donor priorities diverge, possible fields of cooperation can still be found and agreed upon. In any case, partner governments are not homogenous, and reform-minded officials can most often be identified in specific ministries or departments (e.g. in the Ministry of Finance but less in the Ministry of Agriculture) and strengthened as "agents of change". Thus, development cooperation may act as a catalyst for change by convincing and helping partner governments to focus water sector policies on the poor. However, development cooperation never is a "deus ex machina" – a parachuting almighty savior – and a

[9] To avoid misunderstandings: This also holds true for industrialised countries; no government can rightfully be expected to be impartial. However, mature democracies are much more inclusive, majorities are rather political than regional or ethnical and change more frequently.

main trap is to expect donors to be omnipotent. Development cooperation in the water sector – just as development cooperation in general – also depends on its (mostly governmental) counterparts and needs their consent. Sometimes a delicate balance needs to be struck between donors' and partner governments' interests, between conditionalities and flexibility.

In order to further improve German development cooperation with so-called "fragile states", where particularly serious shortcomings in governance hamper development, BMZ is participating in a corresponding process hosted by the OECD Development Assistance Committee (DAC). The overall objective is to identify and to agree on "Principles for good international engagement in fragile states" (OECD 2005). As inputs into the OECD-DAC discussion process, BMZ has commissioned two case studies on its development cooperation with fragile states in the water sector (cf. Lindemann in this volume). The case studies' results have illustrated the difficulty of distilling general conclusions on how to deal with the political economy of water in partner countries. Political contexts are very heterogeneous but key for success or failure, and need to be taken into account when identifying opportunities for development cooperation. At the same time, the water sector can constitute a suitable entry door for cooperation with fragile states, as – ironically – water is frequently perceived as a rather technical, less sensitive field for cooperation. Development cooperation in the water sector can thus open the way for broader governance reforms in partner countries (GTZ 2005).

5 Challenge 4: Accelerating Progress in Sanitation Coverage

An area where lacking political prioritization significantly slows down progress is sanitation – the safe evacuation, treatment, eventual recycling or disposal of human excreta. In spite of numerous appeals[10] and goodwill declarations, waste water management, sanitation and hygiene frequently remain the poor cousins of drinking water supply: While investments into waste water treatment plans are on the rise in emerging economies, both partner countries' and donors' investments into sanitation and hygiene represent only a fraction of overall disbursements in the water sector,[11] although increased engagement in these areas would reap enormous benefits at comparatively low costs: Econometric calculations show "on the whole, benefits from sanitation investments being greater than those from water interventions" (WWAP 2006, 232, cf. also WHO 2004). Positive benefits mainly arise out of improved health, as contamination of drinking water with human excreta is the main cause of water-borne diseases. Thus, drinking water supply should not be handled in isolation from sanitation and waste water management. Also not to be forgotten are the contributions of improved sanitation services to human privacy and dignity.

[10] The UN Secretary General's Advisory Board on Water and Sanitation (UNSGAB) as well has recently urged for accelerated progress in sanitation (UNSGAB 2006).

[11] Exact overall numbers for donor engagement in sanitation are not available, as the statistical categories of the OECD common accounting system do not distinguish between water supply and sanitation.

The comparatively low engagement of development cooperation in sanitation parallels most partner countries' lacking demand: The urgency of progress in this area is underestimated, and taboos on human excreta can further complicate discussions on the issue. In addition, the urban poor suffer most from missing sanitation facilities and waste water management, as their living space in informal or slum settlements is increasingly contaminated, also by waste water from better-off areas. However, the poor frequently lack political voice and municipalities or their utilities hesitate to factually legalize informal settlements by providing services to these areas. In addition, public toilets or private sanitation facilities are bad tools for public communication: Few politicians will perceive it as rewarding to publicly inaugurate a new toilet block.

In addition to these problems of political prioritization by partner countries, implementing agencies also face particular practical difficulties in sanitation. These include decentralized systems, greater difficulties in costs recovery and the challenge to orchestrate behavioral changes of users. In contrast to water supply, centralized waste water systems are frequently inappropriate in developing countries: Settlement patterns, income levels and water availability often preclude the construction of water-based sewage systems. Settlements are too dispersed, canal systems too expensive and water too costly to be used for the transport of waste. Consequently, decentralized systems, like improved pit latrines, represent appropriate solutions, at least in peri-urban and rural areas, where coverage rates are worst. However, most instruments of financial cooperation are badly suited for promoting the installation of private sanitation facilities by households. Loan and grant procedures rely on public borrowers and counterparts, being tailored for bigger, more easily to handle, clearly identifiable construction projects.

Cost recovery is also particularly difficult to achieve in sanitation and waste water management. While cost recovery is already challenging in drinking water supply, people (and governments) are even less willing to pay for waste water disposal.[12] Thus, the economic sustainability of waste water and sanitation projects is in doubt, what makes them problematic for implementing agencies as well as for national governments and municipal authorities.

Finally, progress in sanitation and waste water management is seldom a question of hardware alone. Frequently, awareness needs to be raised on the health effects of lacking hygiene, and behavioral patterns need to change for achieving expected health outcomes. The WASH (Water, Sanitation and Hygiene) campaign of the Water Supply and Sanitation Collaborative Council (WSSCC) is the most prominent example for efforts in this regard. However, there are justified doubts concerning "social engineering" attempts for changing deeply-rooted everyday behavior. In any case, social change is a long-term process, and development projects and programs rely on rather quick, measurable results.

Vis-à-vis these practical difficulties, implementing agencies need to improve their conceptual approaches and instruments to facilitate increased engagement in sanitation. Micro-finance schemes might offer solutions for enabling households to

[12] Even in industrialised countries like Germany, waste water management is frequently cross-subsidised, e.g. by drinking water tariffs and land taxes (Boss and Rosenschon 2006).

fund and maintain their own latrines. Frequently, sanitation challenges are directly linked to issues of urban development in general, to housing rights and property structures for example (regarding the ownership of land for facilities and ways of access), and need to be integrated into these contexts. The World Bank has started to actively address these difficulties by setting up a "Sanitation & Wastewater Advisory Team" (SWAT), a sort of "in-house capacity building unit", which provides practical support to portfolio managers (Kolsky 2006). The SWAT assists them in improving sanitation components of planned water sector programs in such a way that they will actually get financed and implemented and not cancelled due to insufficient design as has frequently been the case before.

Another way of thinking forward is the concept ecological sanitation (ecosan). Ecosan as a general approach to sanitation and waste water management aims at closing material flow circles in order to re-use nutrient and energy components of waste water as well as the water itself. Theoretically, ecosan can provide additional incentives for prioritizing sanitation, as human excreta are not seen as simple waste to be safely evacuated, but as resources that can generate income, for instance as fertilizers in agriculture or as source of methane gas. In many arid and semi-arid countries, waste water is already used because of sheer water scarcity, however often under unsafe conditions. The promotion of treatment technologies that are close to nature (e.g. reed ponds) can facilitate an adequate re-utilization of wastewater in agriculture (cf. e.g. Neubert and Benabdallah 2003). However, both conceptual and perceptional difficulties still need to be resolved, before ecosan and related concepts will find wide-spread application.

BMZ's old and new water sector strategies draw a compulsory link between drinking water supply, on the one hand, and sanitation and waste water management, on the other hand (BMZ 1996; BMZ in prep.). As a general rule, all drinking water projects need to include a sanitation and waste water component. In case the sanitation and waste water component is postponed or skipped completely, the implementing agencies need to provide specific reasons for this. In past practice, however, many drinking water projects included only minimal sanitation or waste water components. In the future, both BMZ's oversight and implementing agencies' efforts need to make sure that sanitation and waste water are adequately represented in water sector portfolios.

6 Challenge 5: Large-Scale Hydro-Infrastructure in Africa

While latrines constitute the discrete lower end of water infrastructure, large dams in contrast represent the "concrete-and-steel" images the term "water infrastructure" frequently evokes at first. There are more than 45,000 large dams[13] around the world (WCD 2000, 11). Dams can have multiple uses; hydropower generation,

[13] The International Commission on Large Dams (ICOLD) defines large dams as dams (1) with a height of more than 15 m or (2) with a height between 5 and 15 m and a storage capacity of more than three million cubic meters.

irrigation and drinking water storage, flood protection and navigation being the most important. Dams are pillars of economic development in many regions of the world and will remain important sources of renewable energy, contributing to limit global greenhouse gas emissions. However, dams have come under severe criticism from ecological and social movements, pointing at the negative consequences many large dams have had on the environment and resettled communities. Many dam projects have also turned out to be more expensive and economically less rewarding than originally projected. In response to this criticism, IUCN, an environmental NGO and think tank, and the World Bank had set up a "World Commission on Dams" (WCD) in 1998, which included all relevant stakeholder groups – governments, the dam building industry, NGOs, affected community groups and academics. The WCD has come out with a report in 2000, including a framework for decision-making (WCD 2000). This framework of "strategic priorities", "policy principles" and "guidelines" is supposed to enable dam-planners to make sure that appropriate criteria and procedures are used for deciding

1. if a dam is needed, taking into account all alternative options, and
2. how the dam and associated measures (resettlement, compensation etc.) are designed, ensuring their sustainability and acceptability.

These recommendations have however not been universally accepted, not even by the World Bank as a co-sponsor of the WCD. Regarding development cooperation, Sweden, Switzerland and Germany are the only donors that have made conformity with the WCD recommendations compulsory for their own projects and programs.

Reacting to growing criticism on large dams in the 1980s and 1990s, most major donors have significantly reduced their dam portfolio since. However, a new trend towards large hydro-infrastructure is emerging, particularly in sub-Saharan Africa, where countries rely most heavily on donor funding. African governments are actively pushing the dam agenda, as illustrated by the South African Minister of Water Affairs and Forestry's cry for dams at the Stockholm World Water Week 2005 (Sonjica 2005). The most prominent project under discussion possibly is Grand Inga on the Congo River, which is supposed to provide the whole of Africa (plus southern Europe) with electric energy (Hütz-Adams and Gecks 2004). The World Bank in particular is re-engaging in large dams, the first example being Nam Theun II in Laos (World Bank 2006b). The European Commission comes on board with the Africa-EU Infrastructure Partnership: Large, mainly transboundary infrastructure projects in energy, transport, information and telecommunication technologies and water will be financed by a trust fund hosted by the European Investment Bank.

In part, this re-engagement of donors in large-scale water infrastructure responds to a clear need: per capita storage capacity in sub-Saharan Africa is only a fraction of US or Australian storage capacities.[14] Large dams also constitute significant economic opportunities, as only seven percent of African hydropower potential is already exploited (WWAP 2006, 316). Climate change makes infrastructure

[14] While there are more than 6,500 large dams in the United States alone, there are only 1,269 on the whole African continent (WCD 2000, 374).

development even more urgent, as increasing variability of rains and regionally decreasing rainfall patterns will necessitate compensating storage capacity. At the same time, climate change also poses additional risks to the sustainability of existing water infrastructure: reservoirs might be inadequately sized, and an exclusive reliance on hydropower could lead to power shortages if reservoirs run dry, as has happened in Eastern Africa in spring 2006.

Overall, particular efforts needs to be directed at avoiding new "white elephants" – unsustainable, over-dimensioned and unprofitable infrastructure projects – by applying adequate sustainability criteria, as proposed by the WCD report. Distributional effects of large-scale infrastructure projects are of particular importance: Frequently, urban populations, industry and export-oriented agriculture make the biggest profit out of dam projects, while marginalized communities in the project areas bear the negative consequences.[15] Corruption as well is a key problem with large dams, as with all big infrastructure projects.[16] Bribes increase construction costs and decrease construction quality. And corrupt practices – just like prestige considerations – impede rational decision-making on dams and their alternatives.

Alternatives to large dams in particular deserve much more attention: More often than not, energy efficiency increases, water demand management, rainwater harvesting and small dams provide more sustainable and less costly solutions to the problems large dams are supposed to resolve. Out of prestige and practical reasons – alternatives are often small-scale and decentralized – the latter may be less attractive to national governments and financing institutions alike. They may also pose particular political problems, such as water demand management, as it might necessitate water pricing or abstraction limits. However, both partner countries and donors need to make sure that their decision-making is guided by rational and comprehensive option assessments in the overall interest of poverty reduction and sustainable development.

In general, donors need to choose between actively taking up the dam debate or ignoring it and depriving themselves of opportunities to influence dam policies. If western donors do not get engaged, non-traditional donors like the People's Republic of China or Arab countries might provide the necessary funds with less demanding sustainability guidelines. In order to avoid race-to-the-bottom style negotiations on single large dam projects, a general acceptance and implementation of adequate sustainability criteria by partner governments would best be able to ensure that the new conjuncture on large dams in Africa will be beneficial, not detrimental, to peoples' well-being. BMZ is willing to stay engaged in this debate both with partner countries and in governing councils of financing institutions in order to promote the WCD recommendations as the relevant framework for decision making.

---

[15] The ENCOP research project has labelled these areas "national sacrifice zones" (Baechler et al. 1996, 324).

[16] The role of Lahmeyer International in the Lesotho Waters Highland Projects might be the best known example of recent years (BBC 2003).

7 Challenge 6: Mobilizing Finance

Influence in international water policies never is a simple function of the ODA a country disburses in the sector. The Netherlands, Switzerland or Sweden, for example, shape the international water discourse through their widely recognized competence to an extend far beyond the financial importance of their development cooperation in the water sector. However, significant financial engagement and extended experience with managing an own water portfolio certainly increase a country's or institution's weight in the international water discourse, also regarding the definition of sustainability criteria for large-scale infrastructure.

More importantly, the achievement of the water-related international development goals will at least partly depend on the amount of ODA channeled into the water sector, as it requires a substantial increase in overall investments. Although progress on the MDGs will never be accomplished by ODA alone, development cooperation can improve framework conditions and investment opportunities in the water sector and insert seed money triggering additional investments. In many countries, required investments are unlikely to be provided by private concessionaires. Exchange rate risks and political risks are frequently too important for private water companies, as recent years' experience has clearly shown. However, ODA-financed investments into improved public water supply, waste water management or irrigation systems as well as capacity building for ministries, regulatory agencies and utilities can enable public bodies and service providers to raise funds on the national and international capital market for subsequent service extensions and improvements. In this regard, capital market access of sub-sovereign entities like municipalities, which operate most water infrastructure, is of critical importance (Task Force on Financing Water for All 2006). Thus, development cooperation can have considerable multiplier effects, further reducing funding gaps.

Aware of the financial needs for reaching the MDGs, many donor countries have committed to increase their ODA. The European Union member countries have agreed to reach average ODA levels of 0.51 percent of GNI in 2010 and 0.7 percent of GNI in 2015 (EU 2005). It can be expected that corresponding increases in BMZ's budget will at least partly be devoted to the water sector. However, this effect is not automatic: In recent years, renewable energies and energy efficiency have attracted much more attention in the political arena. In consequence, BMZ's spending in the water sector – which however represents only a portion of overall German water ODA – has decreased significantly from € 397 million in 2002 to less than € 290 million in 2005. This trend has not been the outcome of a deliberate decision, as German development policy keeps basic needs and services at the top of its agenda. It has rather resulted from generally weakening public and political attention directed at the water sector, after a certain "water conjuncture" at the turn of the century (World Water Fora in Marrakech 1997, The Hague 2000 and Kyoto 2003; World Summit on Sustainable Development 2002 in Johannesburg; G8 Water Action Plan adopted 2003 in Evian), in addition to statistical[17] and administrative[18] reasons. In order to stop this unintended downward trend of its water portfolio, BMZ has recently introduced a new minimum benchmark for German ODA in the water sector, set at € 350 million to be committed each year.

However, tax-funded development assistance is not the only option for mobilizing additional resources for the water sector. Several European countries have already made positive experiences with forms of decentralized cooperation, frequently funded by "water pennies", which are collected in conjunction with household water bills. These cooperation mechanisms range from private actors such as WaterAid – a British development NGO founded by water utility employees – or the "Water for Life" Foundation owned by a Dutch private water company, to public schemes, the French *Loi Oudin* being probably the best known. This law, adopted in 2005, authorizes public water utilities to spend up to one percent of their annual turn-over on "international solidarity projects" and "emergency programs". Building on preexisting semi-legal decentralized water partnerships, this mechanism raises more than € 120 million per year. BMZ is currently preparing a similar initiative for a German framework for decentralized North–South water partnerships. However, the federal structure of the German State, attributing competence for drinking water supply and waste water management to the Länder, leaves the outcome uncertain up to now.

In general, however, the debate on development cooperation in the water sector should not solely focus on the "quantity side" of the topic. The quality of development assistance is at least as important as the amount of ODA spent, as large funds do not equal big impact. The debate on aid effectiveness, which has culminated in the Rome and Paris Declarations of 2003 and 2005, also influences development cooperation in the water sector. Standards are rising regarding the results-orientation of development assistance and regarding the monitoring of impacts. Donors are also expected to increase the efficiency and significance of their contributions by agreeing on a division of labor according to comparative advantages and on program-based or sector-wide approaches in order to reduce transaction costs for partners and to reach economies of scale. Donor harmonization and coordination are the leitmotifs in this debate. However, the water sector is lagging behind compared to other sectors. Reasons for this include the fact that investment projects in the water sector can relatively simple be set up as "stand-alone" and separately implemented. This is less the case with other sectors, health or education for instance, which are more prone to program and budget aid approaches. In addition, institutional interests and egoisms constantly need to be overcome, administrative procedures and organizational structures to be adapted in order to bundle donor interventions to a joint effort towards reaching the MDGs.

[17] Development cooperation in the water sector is more and more integrated into multi-sector programmes, e.g. in urban development, and thus less frequently reported as "water ODA" as such.

[18] German ODA commitments have switched from 2-year sequences to 3-year sequences.

8 Conclusion: Perspectives of German Development Cooperation in the Water Sector

Without being exhaustive, this article has discussed some of the main challenges German (and other donors') development cooperation in the water sector is confronted with. Taking these challenges seriously, BMZ and its implementing agencies are committed to further improve their water sector activities. At the conceptual level, BMZ's new water sector strategy and a Position Paper on *Transboundary Water Cooperation* (BMZ 2006a) move forward its overall approach and strategic orientation. This conceptual development relies on both implementing agencies' expertise and on specific evaluations of their water portfolios, most recently carried out by the German Development Institute (Scheumann and Neubert 2005; Neubert and Horlemann 2005) and by the BMZ evaluation program (Kaiser and Rothenberger 2005). This is complemented by analyses of regional water portfolios (already done for sub-Saharan Africa and the Mediterranean), which guide the sharpening of regional cooperation strategies in this sector. The implementing agencies in turn constantly work on improving their internal mechanisms and operational criteria, feeding in their and others' lessons learned.

A home-made challenge for German development cooperation – also in the water sector – is its institutional fragmentation, however (OECD 2006a, 14). Germany is probably the donor with the most disturbing multiplicity of implementing agencies – ranging from KfW Development Bank and GTZ to InWEnt, ded, the Federal Institute for Geosciences and Natural Resources (BGR), the Centre for International Migration and Development (CIM) and several smaller organizations. While all these institutions work professionally in their respective areas of competence, corresponding institutional rivalries increase coordination costs, slow down the pace of policy progress and complicate the articulation of a coherent single voice of German development cooperation in the water sector.

This institutional multiplicity also hampers an increased visibility of German development cooperation in the water sector within the international debate. While the German implementing agencies are well-known for their high-quality, professional work on the ground, they are less present at international fora, like the Stockholm World Water Week, or within international sector institutions, like the World Water Council or the Global Water Partnership. If German development cooperation puts emphasis more on results on the ground than on public relations – thus avoiding replacing the "projectitis" of the past by a sort of "conferencitis" of today – this certainly is a lesser evil. Nevertheless, German development cooperation has some lessons learned to tell and successful approaches to share, which can help both developing countries and other donors to tackle the above-mentioned challenges. There is certainly room for improvement in this regard, i.e. in developing a well-articulated, recognizable voice in the international water debate.

For the current election period (2005–2009) – also resulting from the OECD-DAC peer review (OECD 2006a) – the governing coalition of Christian Democrats and Social Democrats has announced to reform the institutional set-up of German development cooperation (Federal Government of Germany 2005, 134).

The expected result is the merger of KfW Development Bank and GTZ, giving birth to one major "GermanAid-style" implementing agency. It can be expected that an improved, more rational institutional structure of German development cooperation will also improve chances to take up the water sector challenges discussed above. In any case, Germany will remain a major donor in the water sector for the foreseeable future and thus contribute to progress in this crucial area for achieving the MDGs.

References

Argawal A, de los Angeles MS, Bhatia R et al. (1999) Integrated Water Resources Management, GWP Technical Advisory Committee, Stockholm

Baechler G, Böge V, Klötzli S, Libiszewski S, Spillmann, KR (1996) Kriegsursache Umweltzerstörung – Ökologische Konflikte in der Dritten Welt und Wege ihrer friedlichen Bearbeitung, ENCOP Abschlußbericht, Vol. 1. Rüegger, Chur, Zürich

BBC (2003) Lesotho fines second dam firm, news from 27 August 2003, http://news.bbc.co.uk/1/hi/business/3185145.stm. Cited 13 Aug 2006

BMZ (1996) Sector concept "drinking water supply and sanitation", Bonn

BMZ (2006a) The water sector in German development cooperation, Topics no. 153. Bonn

BMZ (2006b) Transboundary water cooperation – A BMZ Position Paper, Special no. 136. Bonn

BMZ (in preparation) Sector concept "Water", Bonn

Boss A, Rosenschon A (2006) Subventionen in Deutschland: Eine Bestandsaufnahme, Kieler Arbeitspapier Nr. 1267. Institut für Weltwirtschaft, Kiel http://www.uni-kiel.de/ifw/pub/kap/2006/kap1267.pdf. Cited 29 Jun 2006

CIA (2006) The World Factbook – Jordan. http://www.cia.gov/cia/publications/factbook/geos/jo.html. Cited 29 Jun 2006

EU (2005) The European Consensus on Development. http://ec.europa.eu/comm/development/body/development_policy_statement/docs/edp_declaration_signed_20_12_2005_en.pdf#zoom=125. Cited 29 Jun 2006

Federal Government of Germany (2005) Coalition Agreement between the CDU, CSU and SPD, http://www.bundesregierung.de/Anlage951220/Coalition+Agreement.pdf. Cited 29 Jun 2006

Fritz T (2006) Development aid and water privatisation – The example of German development cooperation in Bolivia, http://www.fdcl-berlin.de/fileadmin/fdcl/Publikationen/FDCL-Bolivia-Water-Privatisation.pdf. Cited 29 Jun 2006

GTZ (2004) Das Magazin zum Aktionsprogramm 2015, Ausgabe 2, Ernährung und Hunger. Bonn

GTZ (2005) Förderung von Good Governance in arabischen Staaten, Diskussionspapier. Eschborn

GWP (2005) Policy choices and challenges. Why and how IWRM? http://gwpforum.netmasters05.netmasters.nl/zappengine/objects/Policy_Choices_and_Challenges.doc. Cited 18 Jun 2005

GWP (2006) Setting the stage for change, Second informal survey by the GWP network giving the status of the 2005 WSSD target on national integrated water resources management and water efficiency plans http://www.gwpforum.org/gwp/library/IWRMSurvey-final.pdf. Cited 29 Jun 2006

Hoff H, Kundzewicz ZW (2006) Süßwasservorräte und Klimawandel. In: Aus Politik und Zeitgeschichte, no. 25/2006, 14–19

Horlemann L, Neubert S (2006) Virtueller Wasserhandel – Ein realistisches Konzept zum Umgang mit Wasserarmut in Entwicklungsländern? German Development Institute, Bonn

Hütz-Adams F, Gecks S (2004) Wasserkraft aus den Ingafällen: Große Potenziale für die Entwicklung der DRK und Afrikas, Südwind e.V. http://www.suedwind-institut.de/downloads/Wasserkraft-Inga_deutsch.pdf. Cited 29 Jun 2006

Kaiser M, Rothenberger D (2005) Evaluierung zur Strategie der deutschen Entwicklungszusammenarbeit im Bereich Siedlungswasserwirtschaft, Kurzfassung. Bonn http://www.bmz.de/

de/service/infothek/evaluierung/Projektuebergreifend/grundbildung09/siedlungswasserwirt schaft_pdf.pdf. Cited 29 Jun 2006

Kolsky P (2006) SWAT: A donor partnership for meeting basic sanitation needs of the poor, PowerPoint-Presentation at the informal consultative donor meeting on water supply and sanitation, Brussels, May 18th 2006

Neubert S, Benabdallah S (2003) La réutilisation des eaux usées traitées en Tunisie, Reports and Working Papers no. 11/2003, German Development Institute, Bonn

Neubert S, Horlemann L (2005) Empfehlungen zur zukünftigen strategischen Orientierung der deutschen EZ im Wasser- und Bewässerungssektor, German Development Institute, Discussion Paper 4/2005, Bonn

OECD (2005) Principles for good international engagement in fagile States, learning and advisory process on difficult partnerships (LAP), draft, DCD(2005)8/REV2 http://www.oecd.org/dataoecd/59/55/34700989.pdf. Cited 29 Jun 2006

OECD (2006a) DAC peer review of Germany, Paris. http://www.oecd.org/dataoecd/54/0/36058447.pdf. Cited 29 Jun 2006

OECD (2006b) Has the downward trend in aid for water reversed...? Measuring aid for water. http://www.oecd.org/dataoecd/53/4/36365514.pdf. Cited 29 Jun 2006

OECD (2006c) Guiding principles on using infrastructure to reduce poverty, by the task team on infrastructure for poverty reduction (InfraPoor). http://www.oecd.org/dataoecd/16/46/36301078.pdf. Cited 29 Jun 2006

Partzsch L (2006) Partnerschaften – Lösung der globalen Wasserkrise? In: Aus Politik und Zeitgeschichte, no. 25/2006, 20–25

Scheumann W, Neubert S (2005) Empfehlungen zur strategischen Orientierung der EZ im grenzüberschreitenden Gewässermanagement in Afrika, German Development Institute, Discussion Paper 10/2005, Bonn

Sonjica B P (2005) Speech at the 5th Stockholm Water Symposium, Stockholm, 22 Aug 2005, http://www.dwaf.gov.za/Communications/MinisterSpeeches/2005/StockholmWaterSym22 Aug05. doc.Cited 29 Jun 2006

Task Force on Financing Water for All (2006) Report 1: Enhancing access to finance for local governments, financing water for agriculture ("Gurría Report"), http://www.financingwaterforall.org/fileadmin/Financing_water_for_all/Reports/Financing_FinalText_Cover.pdf. Cited 29 Jun 2006

UNICEF (2005) World Water Day 2005: 4,000 children die each day from a lack of safe water. http://www.unicef.org/wes/index_25637.html. Cited 29 Jun 2006

UNSGAB (2006) Compendium of actions. http://www.unsgab.org/Compendium_of_Actions_en.pdf. Cited 29 Jun 2006

Wasserstiftung (2006) Wasserfakten – Zahlen & Fakten. http://www.wasserstiftung.de/ wasserfakten.html. Cited 29 Jun 2006

WCD (2000) Dams and development – a new framework for decision-making, Earthscan, London, Sterling

WHO (2004) Evaluation of the costs and benefits of water and sanitation improvements at the global level – Executive summary. http://www.who.int/water_sanitation_health/en/execsummary.pdf. Cited 29 Jun 2006

WHO, UNICEF (2004) Meeting the MDG Drinking Water and Sanitation Target – A mid-term Assessment of Progress, Joint Monitoring Programme (JMP) Report 2004. http://www.wssinfo.org/pdf/JMP_04_text.pdf. Cited 29 Jun 2006

Wimmen H (2000) Kampf um Wasser. http://www.uni-kassel.de/fb5/frieden/regionen/Nahost/wasser.html. Cited 29 Jun 2006

World Bank (2006a) World Development Indicators 2006, Washington DC

World Bank (2006b) The Nam Theun 2 Hydroelectric Project (NT2) – An Overview and Update, Washington, DC

World Panel on Financing Water Infrastructure (2003): Financing water for all ("Camdessus Report"). http://www.worldwatercouncil.org/fileadmin/wwc/Library/Publications_and_reports/CamdessusReport.pdf. Cited 29 Jun 2006

WWAP (2003) Water for people, water for life. The United Nations World Water Development Report 1, UNESCO/Berghahn Books, Paris
WWAP (2006) Water – a shared responsibility. The United Nations World Water Development Report 2, UNESCO/Berghahn Books, Paris

Addressing the Need for Water Service Delivery in Fragile States

The Case of German Donor Involvement in Yemen

Stefan Lindemann

Abstract Western donors recognize that the challenge of achieving adequate water service delivery at worldwide scale is closely linked to the problem of state fragility and seek guidance on how to deliver services more effectively. This study takes the example of German donor support to the water sector of Yemen and identifies lessons for good international engagement in fragile states. While Yemen has recently made important progress in the institutional and organizational consolidation of the water sector, its performance in terms of water policy development and implementation is still weak and the overall structures remain largely unsustainable. German development cooperation has taken up this challenge by devising a "multilevel strategy" that combines support to sectoral reform at the macro and meso level with the creation of decentralized and commercialized service utilities at the micro level. Specific "lessons learned" from German donor involvement in the Yemenite water sector include the need for (1) context sensitivity, (2) state building through intervention at different levels, (3) dialogue and participation, (4) conflict prevention, (5) alignment with local priorities, and (6) donor coordination.

1 Yemen as a Fragile State

1.1 (Water) Service Delivery in Fragile States

International donor agencies continue to struggle with the difficulties of their engagement in fragile states.[1] While the notion of "state fragility" remains somewhat elusive, the OECD-DAC Fragile States Group employs a concept of fragile states as suffering "deficits in governance" creating "conditions that make development

Stefan Lindemann
e-mail: S.Lindemann@lse.ac.uk

[1] For a useful summary of the debate on donor involvement in fragile states cf. Meagher (2005).

W. Scheumann et al. (eds.), *Water Politics and Development Cooperation*, 353
doi: 10.1007/978-3-540-76707-7_16,

difficult": In fragile states, there is generally a lack of ability or willingness to "establish pre-conditions for long-term development", including the delivery of basic services (OECD 2005a, 4; for a similar approach cf. Berry et al. 2004). As a consequence, fragile states pose a very significant challenge to meeting the Millennium Development Goals (MDGs) (cf. Branchflower 2004). This can be illustrated at the example of water service delivery: Fragile states now contain about a third of the number of people globally living without sustainable access to safe drinking water (ibid., 20). Western donors increasingly recognize the specific challenge of inadequate (water) service delivery in fragile states and seek guidance on how to deliver services more effectively. Against this background, this contribution takes the example of German donor involvement in the water sector of Yemen and identifies key lessons for good international engagement in fragile states.

1.2 Yemen

In 2003, Yemen was one of the least developed countries in the world, with a Gross Domestic Product (GDP) per capita of US$ 889 (Purchase Power Parity) and a Human Development Index ranking (HDI) of 151 out of 177 countries – the worst performance among Arab states (UNDP 2005). The country has a population of 19.7 million (74.3 percent in rural areas), of which 42 percent live below the national poverty line. Among the major problems are limited access to basic services, malnourishment (close to 50 percent), high illiteracy rates especially among females (71 percent), extreme water scarcity and a very high population growth rate that is estimated at 3 percent per year. The prospects for the Yemenite economy remain weak, with the country's level of economic growth (2.4 percent in 2005) likely to fall far short of the World Bank target of 8 percent needed to achieve sustainable development (The Economist Intelligence Unit 2006, 4ff.). Inflation is likely to increase to 15.8 percent in 2006, from an estimated 10.5 percent in 2005. While Yemen remains highly dependent on oil revenues, its oil reserves are gradually dwindling. With oil production in terminal decline, the government will be forced to introduce economic reform. However, fears of angering an already restive population are likely to keep reform progress slow and haphazard – as illustrated by the government's partial retreat over fuel subsidy cuts and the planned general sales tax in mid-2005.

The early years of the reunified Yemen were characterized by significant positive political developments: The country enjoyed a level of press freedom, political pluralism and popular participation that was remarkable for the region and unique on the Arabian Peninsula (ICG 2003, 4). The constitutional referendum of 1991, in which male and female voters approved a collective presidency and an elected parliament with considerable powers, was followed by the first multi-party parliamentary elections ever held on Yemenite territory (and the Arabian Peninsula as a whole) in April 1993. While the Freedom House Index ranked Yemen as "partly free" at the beginning of the 1990s, the country fell back to "not free" after 1994 (after Glosemeyer 2005, 126f.). This deterioration was mainly due to the increasing

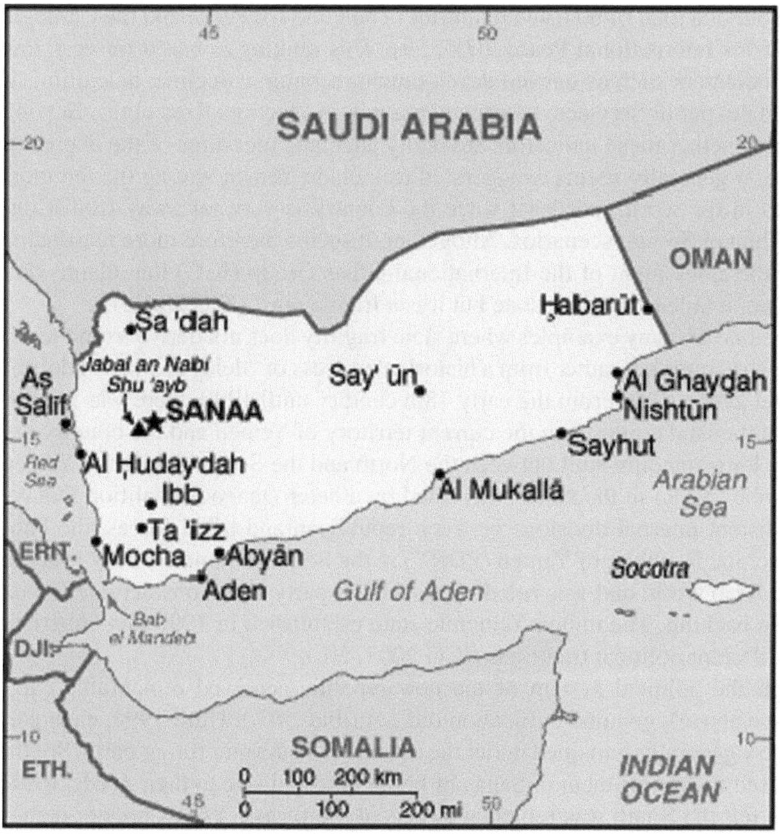

Graph 1 Map of Yemen. Source: CIA (2007)

manipulation of all elections since 1993, the extreme concentration of executive power and restricted press freedom with recurrent attacks on non-governmental journalists. The political pluralism of the early years seems to be eroding. The qualification assessment for the US-funded Millennium Challenge Account, for example, determined that the Yemenite regime had moved backwards from previous assessments: In the 2005 round, Yemen failed all six "ruling justly" indicators and three of the four indicators of "investing in people" (Yemen Times 22 Dec 2005). Also, Transparency International has repeatedly noted widespread and growing corruption, ranking Yemen near the bottom of the corruption scale (ibid.).

1.3 Yemen as a Fragile State

Yemen has figured very prominently in the debate on fragile or failed states. The recently published "Failed States Index", for instance, ranks Yemen as the 8th most failed state out of a total of 60 listed countries (The Fund for Peace and the Carnegie Endowment for International Peace 2005, 59). This ranking is based on very low scores for indicators such as uneven development, economic decline, delegitimisation of the state, public services, security apparatus and factionalized elites. Beyond the question whether these indicators are really adequate measures of the degree of state failure, it generally seems exaggerated to include Yemen among the ten most failed states in the world, not least since the country is very far away from Congolese, Afghan or Somali scenarios. Altogether, it seems therefore more reasonable to endorse the assessment of the International Crisis Group (ICG) that claims that "Yemen is not a failed or failing state but it is a fragile one" (ICG 2003, ii).

Yemen is one of many examples where state fragility does not derive from the decay of state structures but rather from a historical process of "delayed state-building" (Glosemeyer 2005, 122). From the early 18th century until 1990, there was no unified state on the land comprising the current territory of Yemen and the country experienced a long-standing split between the North and the South: While the Yemen Arab Republic (YAR) in the North was ruled by a heterogeneous coalition that reflected persistent internal divisions between republican and tribal forces, the People's Democratic Republic of Yemen (PDRY) in the South became the only Marxist state in the Arab world and was ruled by a socialist party with considerable Soviet and Chinese backing. The united Yemenite state established in 1990 thus inherited strikingly different political traditions (ICG 2003, 2ff.).

Although the political system of the new republic achieved remarkable early progress (see above), genuine unification did not (ibid., 4ff.). Until 1994, each part of the country generally remained under the control of its former ruling party. Southerners accused the government in Sanaa of being unresponsive to their needs, while northerners felt the South was reluctant to accept the former YAR's greater demographic weight. In October 1993, President Saleh's Congress Party (the former ruling party of the YAR) took the initiative to abolish the constitutional principle of collective leadership whereby power had been equally shared between northern and southern leaders. This move sparked increasingly violent clashes between the two parts of the country and culminated into a full-scale civil war in early 1994 that was won by the military domination of Saleh's northern forces by July 1994. Since then, Saleh has retained and consolidated his control of the country. While this has formally ended North–South hostilities, discriminatory state policies continue to threaten the stability of Yemen: Large numbers of Southerners have been marginalized in public life and feel that they receive an insufficient allocation of the wealth produced by the country's natural resources that are mainly located in the South (Yemen Times 22 Dec 2005).

The persistent fragility of the Yemenite state can be best framed in institutional terms.[2] Here, state fragility is a condition where the given institutional arrangements in a state embody and preserve the potential for conflict and the conditions of crisis (Crisis States Programme 2005, 7–8). Fragile states commonly involve a situation

[2] Institutions are defined as "the rules of the game of a society, or, more formally, are the humanly devised constraints that structure human interaction. They are composed of formal rules (statute law, common law, regulations), informal constraints (conventions, norms of behaviour and self-imposed codes of conduct), and the enforcement characteristics of both" (North 1993, 23).

of "institutional multiplicity", which means that formal (or statutory) institutional arrangements are seriously challenged by informal rival institutional systems at all levels of society. In the case of Yemen, the Yemenite state has never been able to establish an effective monopoly of violence over its whole territory and remains unable to control its borders, particularly those shared with Saudi Arabia (Yemen Times 22 Dec 2005). Large areas of the vast and sparsely populated country – primarily the mountains and deserts of the northern and eastern governorates – are beyond the effective control of the central government and important parts of the population continue to live by traditional institutional systems and actively resist stronger government authority. In this context, one has to emphasize the traditional autonomy of Yemen's numerous tribes that provide social services and continue to wield considerable power – a situation that is facilitated by the widespread availability of small and even heavy arms.[3]

Due to the absence of an effective monopoly of violence, Yemen continues to be affected by social unrest in both urban and rural areas that take the form of tribal clashes, kidnapping, military-tribal confrontations, demonstrations and bomb attacks (ICG 2003, 13ff.). The incomplete control over its territory, combined with the existence of a variety of home-grown Islamist movements, has over the last several years attracted increasing international attention: The Yemenite origin of leading al-Qaeda operatives and the assumed presence in Yemen of many al-Qaeda members who fled Afghanistan on the one hand, and, on the other, a series of politically motivated attacks on US and European targets (including, those on the USS Cole in October 2000 and on the French oil tanker Limburg in October 2002), have placed Yemen under the spotlight of the international debate on the war against terrorism (ibid., 1). The Yemenite government has since 11 September 2001 cooperated with the US and taken steps against al-Qaeda. In the light of the widespread popular hostility towards the US, however, this is an extremely delicate balancing act for a fragile country like Yemen (ibid., 23f.).

Altogether, Yemen as a fragile state is rather difficult to categorize. As already indicated above, it seems most convincing to portray Yemen as a country that is still in a difficult process of state-building and continues to be affected by institutional fragility – a situation that may be very pronounced in Yemen but is generally quite common for many low-income countries. If one attempts to apply the typology developed by the DAC Fragile States Group, things become more complicated and blurred. This is largely due to its rather vague underlying concepts of political willingness and capacity that are generally very difficult to apply to countries as a whole. Nonetheless, one may be inclined to argue that the current Yemenite context resembles a situation of (enduring) recovery where a stable government is in place and basic state functions are slowly being established (cf. Meagher 2005, 7). Despite authoritarian tendencies and high levels of corruption, national leadership is demonstrating political will to make progress in developing the infrastructure of a modern state but the performance and capacity of government remains rather weak in terms of policy development and implementation.

[3] Today, men living in both urban and rural areas are likely to own at least one pistol or rifle, often an AK-47. Outside the main cities, overtly carrying weapons in public is accepted as normal (ICG 2003, 1).

2 The Yemenite Water Sector

Yemen is one of the most water-stressed countries in the world and faces a number of serious problems in the water sector, which have to be solved quickly if living conditions are not to deteriorate in many parts of the country.

2.1 Water Resources and Water Use

Yemen has a predominantly arid climate with average annual rainfall ranging from less than 50 mm in the coastal areas and the deserts to more than 1000 mm on the western slopes of the mountains (FAO 2005). The country has no permanent rivers and is generally extremely water scarce (cf. Table 1). In 2000, actual renewable water resources (km^3/year) were estimated at 4 km^3, which amounts to a per capita water availability of 198 m^3/person/year. This is a very low figure not only by global standards (the global per capita average is 8,548 m^3/person/year) but also far below the "water poverty line" of 1,000 m^3/person/year. Even in the generally water scarce Middle East & North Africa (MENA) region, per capita water availability levels still average 1,505 m^3/person/year.

In 2000, total water withdrawals were estimated at 6.6 km^3/year, thereby exceeding by more than half the annual renewable freshwater resources (see Table 1). Notwithstanding the skepticism about the accuracy of the above estimates, the overall picture is clear (Republic of Yemen 2004, 7): Yemen is facing an exacerbating imbalance between water supply and water demand, especially affecting groundwater. The water crisis is starting to take a catastrophic nature in a number of basins where aquifers are depleting very fast, as reflected by rapidly falling groundwater levels at rates reaching 6 m annually in some basins. Even the capital Sanaa is liable to run out of water within the next decade. Declining water tables boost investment cost for pumping, thereby favoring the larger farmers (KfW 2006). Water conflicts are becoming frequent in many basins (see below).

Table 1 Water availability and water use, 2000

	Yemen	MENA
Actual renewable water resources, total (km^3/year)	4	
Actual renewable water resources, per capita (m^3/person/year)	198	1.505
Annual water withdrawals (km^3)	6.6	324.6
Annual water withdrawals, per capita (m^3/person)	368	807
Annual withdrawals, agriculture (%)	95	86
Annual withdrawals, industry (%)	1	6
Annual withdrawals, domestic (%)	4	8

Source: WRI (2005, 208)

Ninty five percent of the annual water withdrawals are used for agricultural purposes, while domestic use and industry account for 4 percent and 1 percent of total water use (see Table 1). Overall water use (or irrigation) efficiency in agriculture is very low (below 40 percent) (ibid.). Most of the cultivated area is still directly rain-fed (55 percent), but the groundwater irrigation share has exploded from 5 to 45 percent since the mid-1970s. 30 percent of the water used for agriculture is going into water-intensive qat production that occupies 11 percent of the cultivated land.[4] Qat can be ascribed a major role in groundwater depletion (Republic of Yemen 2004, 12): If the existing situation continues as it is, then qat farming will in the end deplete rural water and consequently wipe out the rural economy.

With respect to domestic water use, the proportion of unaccounted water (losses) of urban water supply utilities is very high, sometimes approaching 45–50 percent of water production (ibid., 8f.). At the same time, water demand for domestic use is continuously increasing, especially in cities due to rapid population growth and urban migration. It is estimated that an extra 100 million m^3/year are needed for the urban water supply and sanitation (WSS) sector by 2015. Therefore, the re-allocation of water resources from agricultural to domestic and rural to urban use seems inevitable.

Beyond dramatic levels of water scarcity, the pollution of water resources has become a growing problem (ibid.). Cities and industries discharge untreated domestic and industrial wastewater into aquifers in peri-urban areas. While dangers of urban wastewater pollution are more visible, there is also a potential pollution hazard to aquifers from untreated rural wastewater. In addition to environmental impacts, uncontrolled disposal of raw wastewater poses a direct health hazard to urban and rural populations.

2.2 *Water Supply and Sanitation*

Access to improved[5] sources of drinking water and sanitation remains a problem throughout the country (cf. Tables 2 and 3). The major challenge is that service coverage has not kept up with the high population growth.

---

[4] Qat is a non-food crop that has been grown for use as a stimulant for centuries in the Horn of Africa and the Arabian Peninsula. Traditionally, qat has been used as a socializing drug, and this is still very much the case in Yemen where qat-chewing is a predominantly male habit.

[5] The Global Water Supply and Sanitation Assessment 2000 Report (UNICEF and WHO 2000, 4) introduces a definition of coverage that is based on technology type. In past assessments, the coverage figures referred to "safe" water supply and "adequate" sanitation. One of the findings of the 2000 assessment was that there is a lack of information on the safety of the water served to the population and on the adequacy of sanitation facilities. Therefore, the assessment assumed that certain types of technology are safer or more adequate than others and that some of them could not be considered as "coverage." The terms "safe" and "adequate" were replaced with "improved" to accommodate these limitations. The population with access to "improved" water supply and sanitation is considered to be covered. Types of facilities that are considered as improved water sources and improved sanitation facilities are given below.

Table 2 Improved drinking water coverage (percent), 1990–2002

	Yemen (1990)	Yemen (2002)	Western Asia (2002)
Total access to improved drinking water	69	69	88
Total house connections	31	33	63
Urban access to improved drinking water	74	74	95
Urban house connections	64	64	79
Rural access to improved drinking water	68	68	74
Rural house connections	22	22	31

Source: UNICEF and WHO (2004, 31)

In 2002, improved drinking water[6] was available to only 69 percent of the population – 74 percent in urban areas and 68 percent in rural areas. The coverage rate has not improved since 1990 and remains significantly below the Western Asian average of 88 percent (see Table 2). This means that almost one third of the Yemenite population continues to rely on "unsafe" sources of drinking water (e.g. unprotected public wells, rivers, tanker trucks), in particular the poor. It needs to be noted that the water provided through Local Water Supply and Sanitation Corporations (LCs) is generally much cheaper than that provided by private water tankers (e.g. donkey carts, lorries) (Buhl 2005, 13f.). In Zabrid, for example, 1 m^3 of water provided through a donkey cart costs around 200 YR, while the official price ranges from 50 YR in the lowest consumption bracket to 125 YR in the highest.

The Yemenite sanitation coverage rate has risen from 21 percent in 1990 to 30 percent in 2002 but remains significantly (!) below the Western Asian average of 79 percent (see Table 3). While 76 percent of the urban population have access to improved sanitation services percent, this is only the case for 14 percent of the rural

Table 3 Improved sanitation coverage (percent), 1990–2002

	Yemen (1990)	Yemen (2002)	Western Asia (2002)
Total access to improved sanitation	21	30	79
Total house connections	8	11	59
Urban access to improved sanitation	59	76	95
Urban house connections	39	35	82
Rural access to improved sanitation	11	14	49
Rural house connections	0	2	14

Source: UNICEF and WHO (2004, 31)

[6] The following technologies were considered "improved" (UNICEF and WHO 2000, 4): Household connection, public standpipe, borehole, protected dug well, protected spring, rainwater collection. The following technologies were considered "not improved": Unprotected well, unprotected spring, vendor-provided water, bottled water, tanker truck provision of water.

population. This situation has obvious negative implications for the hygiene and health of the population, especially in rural areas.

2.3 *Water Institutions and Policy Development*

For centuries, Yemen remained a society based on agriculture, which was almost entirely supported by rain and surface water resources (Republic of Yemen 2004, 6f.). An elaborate system of formal and informal norms and laws slowly evolved to govern the use of surface water in a sustainable and fairly equitable manner. This began to change from the early 1960s, with the advent of modernization characterized by population increase, rising food demand, rapid urbanization, initial industrialization, and the availability of modern technology such as pumps and drilling rigs. As a consequence, the share of groundwater in irrigation started to increase rapidly and soon reached unsustainable levels. These fundamental changes took place in no more than two to three decades and clearly outpaced the evolution of compensatory policy measures. At present, there are thus inadequate institutional and organizational capacities at all levels of government to regulate water resources development.

While Yemen had for a long time no water legislation, the country has enacted a Water Law in 2002 (ibid., 10f.). The latter provides a legal basis for controlling groundwater abstractions and includes measures such as licensing and registration requirements for wells and rigs, and more strict control regimes in water stressed catchments. It also supports decentralization in the form of encouraging the formation of basin committees and requires working closely with Local Councils in the implementation of water management measures. But for all of this to translate into effective water management, more than just the legal framework is needed. While well-drilling licensing has now become obligatory, its implementation still faces many difficulties (ibid., 7ff.). In the absence of a system of generally accepted and effectively enforced water rights, water allocations remain determined by resource capture. This holds particularly true for groundwater resources where traditional customs grant a land owner the right to exploit whatever water that may exist underground – a classical "common pool resources dilemma" that continues to fuel the unsustainable depletion of groundwater.

At the organizational level, the Yemenite water sector was initially characterized by multiple actors and fragmented responsibilities. In recent years, however, Yemen has been slowly moving towards a more integrated sector governance structure. This began in 1996 with the bundling of water resource management functions under one entity – the National Water Resources Authority (NWRA) – and was consolidated by the establishment of the Ministry of Water and Environment (MWE) in 2003 (ibid., 14f.). For the first time, all agencies dealing with water and environment – with the notable exception of the Ministry of Agriculture and Irrigation (MAI) – are now within the fold of the new ministry.[7] While the MWE is

[7] The respective agencies within the fold of the MWE are introduced and discussed below.

generally responsible for sector investment planning and the coordination between all water sector agencies, it is currently still at a capacity development stage and suffers from a lack of competent staff (GTZ and KfW 2005, 9). As a consequence, it has only the mandate (through NWRA) but not the power to enforce a more sustainable approach to national water resources management, in particular vis-à-vis the influential MAI and vested agricultural interests (Interview 2, 24 Jan 2006; Interview 3, 24 Jan 2006).

The NWRA is responsible for the planning and implementation of national water resources management – a process that is generally very difficult due to a lack of reliable data. Yemen follows an approach of "basin co-management" whereby stakeholders and state institutions forge a partnership for managing water resources at the catchments level (Republic of Yemen 2004, 10f.). Concerted efforts need to be made for translating this from a management model into reality: While a few Basin Committees have been formed (e.g. in the Sanaa, Sadah and Amran basins), efforts to create community-based water organizations on a wider scale are constrained by the fact that NWRA at present lacks sufficient financial and human resources and has only a limited regional and local presence. For the co-management approach to work, NWRA has started to prepare regional water management plans for some areas (e.g. for the Taiz, Hadramawt, Tuban-Abyan and Sadah basins), but the pace of plan preparation remains slow because technical capacities for IWRM are still weak and the necessary infrastructure to collect information (e.g. hydro-meteorological monitoring networks, reliable water quality labs) is either non-existent or has only very limited coverage.

In the field of WSS, the Yemenite government has since 1996 engaged upon a reform program to expand the coverage and quality of services (ibid., 18f., 27ff.; Interview 2, 24 Jan 2006). The general orientation of this reform is the gradual departure of the government from its traditional role as an exclusive investor and service provider to a role of sector facilitator and regulator. In the urban WSS sub-sector, the National Water and Sanitation Authority (NWSA) was initially fully responsible for service provision at the central level. The ongoing reform program foresees a decentralization of service provision by transferring responsibility to autonomous LCs, designed to be self-financing in the end. Implementation of the reform program is proceeding and to date nine LCs have been created. While the decentralization process has generally made good progress and is most advanced by regional standards, it is nonetheless far from being achieved (GTZ and KfW 2005, 6f.). Both NWSA and LCs suffer from a lack of qualified personnel and the manning of key posts within LCs has repeatedly been based on political allegiances rather than on professional competences. Also, the precise mandate of the LCs remains unclear, as the necessary by-laws have yet not been adopted. The envisaged involvement of the private sector, finally, has so far been hampered by insufficient physical and legal security and uncertainties about actual levels of water availability. Nonetheless, there has been some progress: While plans for private sector involvement in Aden have been suspended, LC Sanaa has been authorized for a management contract and LC Taiz has just signed a Public–Private-Partnership Agreement with a Dutch utility operator (Interview 2, 24 Jan 2006). In the rural WSS sub-sector, a reform policy is currently in the Cabinet for approval. Here, the government intends to preserve a stronger role in view of poverty alleviation – a role that is delegated to the General Authority for Water and Sanitation Projects (GARWSP) (Republic of Yemen 2004, 27ff.). However, the actual approval of the rural WSS reform policy remains pending due to a conflict over the sanitation part of the document: The Yemenite government fears that all of the rural population will demand sanitation systems if the strategy is approved as it is (Interview 2, 24 Jan 2006).

Concerning the financial viability of the service utilities, full cost recovery has yet not been achieved (Republic of Yemen 2004, 29). At present, cost recovery practice by each LC varies according to the local situation: The accepted norm is for the tariff to be set such that at least the cost of operation, maintenance and depreciation of electro-mechanical equipment are recovered, while the government pays for new schemes, replacements and expansions. In practice, however, the financial situation of many service utilities remains critical – a situation that is largely due to weak management, insufficient maintenance, overstaffing, motivation problems and rising costs of water production (GTZ and KfW 2005, 7f.). While cost-recovery seems somewhat feasible for drinking water supply, it currently remains wishful thinking for sanitation services. Regarding the poor, there is a degree of pro-poor cross subsidy in the block-tariff structure, where an affordable "lifeline" rate is charged on the first block to benefit the poor (Republic of Yemen 2004, 29). However, the better off currently also benefit from the lifeline rate and the tariff-system needs further pro-poor revision. Connection fees, for instance, are a considerable problem for poor and vulnerable households (Buhl 2005, 14).

At the policy level, Yemen has in late 2004 adopted a consolidated strategy for the water sector as a whole – the National Water Sector Strategy and Investment Program (NWSSIP) that was initiated through a comprehensive multi-stakeholder process spearheaded by the MWE (Republic of Yemen 2004). The NWSSIP adopts the vision of IWRM and develops a coordinated set of policies and objectives for five water sub-sectors: (1) water resources management; (2) urban WSS; (3) rural WSS; (4) irrigation; and (5) environmental aspects. It is targeted for 5 years (2005–2009) and principally oriented towards MDG achievement. In the field of rural WSS, however, MDG goal achievement was considered unrealistic and therefore targeted at 50 percent ("half the MDGs"). In order to achieve these goals, the NWSSIP investment program for 2005–2009 totals about US$ 1.5 billion, of which about US$ 1 billion is committed/pipelined by government and donor funding.

Altogether, Yemen has recently made important progress in the institutional and organizational consolidation of its water sector (Interview 2, 24 Jan 2006). At the policy level, the NWSSIP identifies the key sectoral problems and is clearly operationalized. However, the sectoral budget allocations for the next years remain clearly insufficient to implement the whole array of water policies underlying the NWSSIP and continue to exhibit an urban bias (79 percent of sector investment share) (KfW 2006). Even more importantly, the overall positive trend is mitigated by persistent structural deficits at the operational level (MWE, NWRA, NWSA, LCs), in particular a lack of professional and competent human resources to carry out the huge management and development tasks in the water sector (GTZ and KfW 2005, 3).

As long as key actors such as the MWE and the NWRA remain weak – especially vis-à-vis the MAI and vested agricultural interests – a more sustainable approach to national water resources management will be difficult to implement. Here, it is certainly not enough to develop institutionalized coordination mechanisms between the MWE and the MAI as suggested in the NWSSIP. Instead, institutional and organizational improvements with view to IWRM require policy decisions at a higher political level (Cabinet, President) (Interview 3, 24 Jan 2006).

2.4 The Water Sector and State Fragility

The debate on service delivery in fragile states generally assumes that a fragile environment will likely weaken or disrupt (water) service delivery. To a certain extent, this assumption also holds true for the case of Yemen: As many other low income countries, Yemen is still in the difficult process of state-building and continues to be affected by institutional fragility – a situation that gives rise to weak state institutions and organizations and creates difficult framework conditions for all sectoral policies, including water. Nonetheless, the Yemenite case shows that one has to take a close look at each individual policy sector when assessing the impact of state fragility. While one might expect that the incomplete monopoly of violence hampers the development of a national water policy, this is not always the case: Most tribes recognize the role of the central government in providing WSS services, even in a "trouble spot" like Sadah where government control is traditionally very low (Interview 2, 24 Jan 2006). This situation might be explained with both the high profile of water issues and the characteristic of water as a rather non-ideological service (as opposed to education for example) that requires important investments in infrastructure. This is not to deny that the fragility of the Yemenite state does have a negative impact on the water sector. Nonetheless, one has to raise the question whether these difficulties are really any different from the problems encountered in most low-income countries.

If one admits that the fragility of the Yemenite state has a negative impact on the water sector as a whole, there is furthermore reason to assume that state fragility varies in its impact on different types of water services – depending on the public good characteristic of the respective service. The water sector is generally comprised of different kinds of goods: (a) a container of potable water is in many settings a private good (high rivalry and excludability) that is bought and sold; (b) water from a river or pond is a common-pool resource; (c) water from a communal well is a public good (albeit highly localized); and (d) piped water is a toll good (low rivalry, high excludability) that is provided either by government or by an independent agency under public regulation (Meagher 2005, 16f.). In theory, state fragility is least likely to affect communities without wells or pipe networks, and communities that rely on water vendors – an assumption that is confirmed in the Yemenite case. More vulnerable are publicly-provided and -maintained wells and standpipes, and especially piped water supply networks: Piped water supply occurs within a context defined by legal and regulatory structures set up by districts and central government to govern water rights, real property, and utilities – structures that remain inoperative or inadequate in Yemen. On the other hand, one may argue that the poorest water users – and here in particular women – are most affected by the persistent deficits in the water sector since they continue to rely on unsafe and often expensive water sources.

Furthermore, the case of Yemen illustrates that the relationship between state fragility and the water sector may actually be reciprocal rather than unidirectional. On the one hand, state fragility has a negative impact on the water sector, while on the other hand water-related issues exercise a negative influence on the stability of the Yemenite state. A study conducted by the International Crisis Group (ICG) shows that nearly half the cases of inter-tribal violent conflict – only the tip of the iceberg as there are no official statistics – are, or were initially, related to land or water (ICG 2003, 14). The drilling of new wells, for instance, is particularly controversial in the light of widespread water scarcity and has given rise to recurrent hostilities (GTZ and KfW 2005, 8f.). Along with a shortage in social services and employment, increasing scarcity of water resources has thus been a major cause of tension and instability. If levels of water scarcity continue to worsen, there is reason to expect increasing internal migration towards the less water-scarce coastal areas – a scenario that implies a high conflict potential and threatens to further destabilize the Yemenite state.

3 International Donor Agencies and the Challenge of State Fragility: The Experience of German–Yemenite Development Cooperation

3.1 Donor-Recipient Relations

The Yemenite government generally adopts a very cooperative approach vis-à-vis international donor organizations (Interview 1, 23 Jan 2006). It has actively encouraged far reaching donor participation in the preparation of the NWSSIP and explicitly welcomes initiatives for better donor coordination. The country participates in the OECD (DAC) Learning and Advisory Process on Difficult Partnerships (Fragile States) (DfID 2005, 2), even though it refuses the label "fragile".

German–Yemenite relations are generally described as cooperative, open and trustful (Interview 2, 24 Jan 2006; Interview 3, 24 Jan 2006). The Yemenite government recognizes the importance of German support to its water sector, in particular the German "multi-level approach" (see below). Despite good cooperation at the operational level, German claims for more integrated institutional responsibilities in water management – in particular with view to strengthening the MWE vis-à-vis the MAI – have so far not been taken into account.

3.2 The German Support to the Yemenite Water Sector

Yemen features very prominently in the field of German development cooperation: The country is not only one out of currently 40 priority partner countries but was also selected as one of four pilot countries under the "Action Programme 2015" to halve poverty by 2015. In this context, the German government supports the Yemenite water sector with a comprehensive set of development projects through German Technical Cooperation (GTZ) and German Financial Cooperation (KfW). The German support started as early as in 1980 and amounted to a total of € 268 million for the period between 1980 and 2006 (GTZ and KfW 2005, 1; Interview 2, 24 Jan 2006). This makes Germany the most important international donor in the Yemenite water sector alongside the World Bank.

Yemenite–German development cooperation in the field of WSS is intended to contribute significantly to the overarching goal of poverty reduction as formulated in the German government's "Action Programme 2015" (BMZ 2002, 3f.). Specific sectoral goals include:

- securing the sustainable availability of scarce water resources and crisis prevention through IWRM;
- giving priority to the demand management side;
- protecting water resources from pollution and over-exploitation;
- securing access for all citizens to clean drinking water and sanitation in order to improve the health situation;
- working towards decentralized and commercially oriented management of water utilities;
- promoting private sector involvement in the field of service provision and investment;
- facilitating the active participation and responsibility on the part of all users (women, farmers, small and micro entrepreneurs, industry); and
- fostering water saving in agriculture through the re-use of water and the application of water saving irrigation techniques.

KfW provides the bulk of German financial support to the Yemenite water sector. The current KfW portfolio (€ 198.53 million) includes a total of nine infrastructure projects that all focus on equipping urban areas with adequate water and/ or sanitation facilities (GTZ and KfW 2005, 1; KfW 2005b). While a few measures target the improvement of sewage disposal only (Ibb, Aden), most projects focus on both access to clean drinking water and adequate sanitation services (cf. Box 1). The overarching goal is to provide adequate WSS services and thereby improve the health situation of the population in the target areas. To ensure the sustainability of the infrastructure networks, some of the projects include separate complementary measures that provide training and advisory services (Aden, Sadah, Provincial Towns Programme II). The Aden sanitation project, for instance, includes both an education and training component and a "Utility Support Programme" (USP) designed to strengthen the LC for the Aden Governorate (LWSCA) and prepare its future transformation into an autonomous, commercialized water supply enterprise

Box 1 KfW urban and water sanitation projects

1. Water supply and sanitation in Ibb (€ 26.5 million)
2. Extension of sewage treatment plant in Ibb (€ 4.85 million)
3. Sanitation in Aden (€ 38.35 million)
4. Sanitation in Zabid (€ 6.6 million)
5. Water supply and sanitation in Sadah (€ 14.11 million)
6. Provincial Towns Programme (PTP I) – water supply and sanitation (€ 57.77 million)
7. Anti-terror funds – work-intensive infrastructure measure in the context of Provincial Towns Programme I (Amran & Yarim) (€ 3 million)
8. Provincial Towns Programme (PTP II) – water supply and sanitation (€ 30 million)
9. Provincial Towns Programme (PTP II) – water supply and sanitation (€ 30 million)

Source: GTZ and KfW (2005, 12); KfW (2005b)

(GTZ and KfW 2005, Annex 2). Also, KfW has conducted comprehensive socio-economic baseline and poverty surveys, which have improved the poverty- and target group orientation of German support to the Yemenite water sector and established a basis for poverty-oriented action at the local level (Interview 2, 24 Jan 2006). Beyond the mere focus on WSS, KfW projects also aim at the protection and integrated use of scarce water resources (e.g. Sadah, Provincial Towns Programme II). This goal is to be achieved through complementary measures that foresee securing water availability, equal sharing of water resources and the establishment of conflict resolution mechanisms and thereby contribute to (water) conflict prevention at the local level. Finally, it needs to be noted that KfW is also heavily involved in sectoral/political dialogue with the Yemenite government at the macro level (the German coordinator for the Yemenite water sector is a KfW representative).

German Technical Cooperation (GTZ) currently operates a set of five development projects (with a total volume of € 21.6 million) that provide technical support to the ongoing Yemenite water sector reform and promote institutional and organizational capacity-building (cf. Box 2) (GTZ and KfW 2005, 1). A first project supports the Technical Secretariat for WSS Sector Reform in developing the national water sector framework according to the reform principles of decentralization, corporatization, commercialization and private sector participation (ibid., Annexes 9–15; GTZ 2004a, b; GTZ 2005). More specifically, the project aims at strengthening planning- and policy-making processes at the national level, enhancing the development of adequate institutional, legal and organizational arrangements for the urban WSS sector and developing planning and monitoring instruments. While the first project focuses on national water policy development, a second project provides specific advisory services to the autonomous and commercialized LCs. This involves support to company organization and -management, financial management,

Box 2 GTZ support to the Yemenite water sector

1. Support to the Technical Secretariat for Water Supply and Sanitation Sector Reform (€ 5.11 million)
2. Advisory Services to the Water Supply and Sanitation Sector (€ 10.1 million)
3. Personnel Development in the Water Supply and Sanitation Sector (€ 3.18 million)
4. Introduction of a GIS-based Operations Management System (OMS) in Urban Water Supply and Sanitation Utilities (€ 1.25 million)
5. Integrated Water Resources Management (IWRM) Programme (€ 2 million)

Source: GTZ and KfW (2005, 12)

operation and maintenance and the introduction of a cost-recovering tariff structure. Measures for organizational capacity building are complemented by a public awareness project component where especially trained Community Mobilizing Workers (CMWs) explain the adopted water reform policies to the local population (see below).

A third, highly related project focuses on personnel capacity-building in the water sector whereby the staff of both national agencies (MWE, NWSA) and the local WSS utilities is trained and qualified through a large variety of personnel development instruments. The organizational capacity-building approach underlying the previous projects is complemented by a fourth project that introduces GIS-based Operations Management Systems in selected urban WSS utilities. While the first four projects focus on urban WSS, a final project aims at establishing the organizational structures for IWRM at the basin level. This involves both support to NWRA and the creation of basin committees in Amran and Sadah where involved parties and stakeholders work towards the elaboration of integrated water management plans. In order to (1) strengthen the coordination and alignment of the projects, (2) further emphasize institutional and organizational capacity building and (3) increase the poverty orientation of aid, GTZ has recently decided to shift from a project to a program approach whereby the five projects are integrated into a program entitled "Institutional Development of the Water Sector" (GTZ 2006, 1f.).

The German support to the Yemenite water sector through KfW and GTZ is highly integrated and complementary (Interview 2, 24 Jan 2006). As KfW and GTZ work in close cooperation, the nature and success of their work has to be seen and analyzed in conjunction. While all GTZ projects focus their support on those areas where KfW finances the development of urban WSS infrastructure, KfW provides the GTZ projects with the indispensable (financial) leverage and establishes the necessary basis for the realization of technical cooperation (e.g. through baseline and poverty surveys). The integrated and complementary approach between GTZ and KfW applies to all levels of intervention since the two German agencies assume both the support to the decentralized WSS utilities and the sectoral advice/dialogue in close coordination. On the whole, German support to the Yemenite water sector has so far been very successful, as it has contributed to a number of important structural improvements (ibid.; GTZ 2006, 1ff.). At the macro level, the sectoral dialogue through GTZ and KfW has acted as a key facilitator to the elaboration and revision of the NWSSIP and provided substantial support to the ongoing decentralization of WSS services. Furthermore, the GTZ and KfW advisory services to the urban WSS sector have led to very positive results: The various capacity-building measures in support of the decentralized WSS utilities and their embedding in civil society (through dialogue with water users) have led to significantly improved company management, measured with indicators such as water losses, cost-recovery and consumer satisfaction. The WSS utilities supported by German Development Cooperation now serve about one million people and are probably among the most successful in the entire region. Despite the overall success of German development cooperation, some projects continue to suffer from delays in implementation that can be mainly attributed to the insufficient and declining availability of necessary water resources (e.g. in Amran, Yarim or Ibb), recurrent problems with land acquisition and enduringly modest capacities of project executing organizations despite undeniable improvements (GTZ and KfW 2005, 4, Annexes 1–8; KfW 2005a, Annex 4).

3.3 Service Delivery Adaptations, Trade-offs and Sustainability

As argued above, the Yemenite case resembles most a fragility scenario of (enduring) recovery where a relatively stable government is in place and basic state functions are slowly being established. Despite authoritarian tendencies and high levels of corruption, national leadership is demonstrating political will to make progress in developing the infrastructure of a modern state. This general assessment also holds true for the Yemenite water sector: While the capacities of the government remain weak and critical issues such as water resources vs. agriculture are still not high on the agenda, Yemen nonetheless displays increasing commitment to move towards the institutional and organizational consolidation of its water sector. Altogether, this recovery setting of modest capacities but existing political willingness offers rather favorable framework conditions for the involvement of German donor agencies. In this context, one may even raise the question whether the existing framework conditions in the Yemenite water sector are really more difficult than those encountered in any low-income country.

Against this background, it is generally difficult to discuss the German support to the Yemenite water sector in terms of service delivery *adaptations*: Given the relatively favorable sectoral framework conditions, representatives of GTZ and KfW seem to perceive their activities as an innovative example of support to water sector reform in a water scarce low-income country rather than a deliberate response strategy to a situation of state fragility (Interview 1, 23 Jan 2006; Interview 2, 24 Jan

2006; Interview 3, 24 Jan 2006). Nonetheless, it is plausible to interpret the German approach as an adequate response to the given fragility scenario: In line with the assumptions discussed in the literature on donor engagement in fragile states, German development cooperation addresses the Yemenite recovery setting by adopting a "working with government" approach.[8] This approach is designed as a "multi-level strategy" (GTZ 2006, 3) that combines forms of cooperation with both central and sub-national governments and also introduces elements of community participation.

The first aim of this "multi-level strategy" is to facilitate the establishment of autonomous WSS utilities at the *micro* level that receive multi-facetted support through GTZ and KfW in form of infrastructure and capacity-building measures. At the same time, German development cooperation also intervenes at the *meso* level, e.g. by supporting the development of river basin management plans. Finally, German supports targets the macro level by strengthening processes of water policy development at the national level, in particular the creation of adequate institutional, legal and organizational arrangements for the urban WSS sector. The basic aim behind this multi-level approach is to foster synergies and learning between the different levels of intervention, that is to transfer the experiences and knowledge gained in the context of projects at the micro/meso level to project measures at the macro level (and vice versa) (Interview 3, 24 Jan 2006).

This *scaling up of aid* strategy is followed by both KfW and GTZ and has generally been successful in creating valuable synergies between the different levels of intervention. The new KfW Provincial Towns Programme II, for instance, establishes comprehensive milestones to be achieved at the micro (project executing agencies, local government) and the macro level (MWE, NWRA) and thereby anchors the program in the overall water sector reform (KfW 2005b, Annex 14). In the case of GTZ, far-reaching synergies exist between the two GTZ projects "Support to the Technical Secretariat for Water Supply and Sanitation Sector Reform" (macro level) and "Advisory Services to the Water Supply and Sanitation Sector" (micro level), while the "multi-level strategy" has been less successful in other cases (GTZ 2006, 3). The projects in support of personnel development and IWRM, for instance, remain so far largely limited to the subsectoral (or micro) level. In the context of the shift from a project to a program approach (see above), GTZ has therefore decided to strengthen the "multi-level approach" by systematically applying it to all components of the new program. The "scaling-up of aid" to the macro level is particularly relevant in the case of IWRM since uncertainties in national water resources management tend to have an adverse impact on infrastructure measures and the sustainable operation and management of the service utilities.

One of the main innovations of German support to the Yemenite water sector is that sectoral reform measures are directly communicated to the affected water

[8] The debate on donor involvement in fragile states distinguishes between situations where donors can fully engage with the partner government and those where it may be necessary to avoid the central government as much as possible (cf. Meagher 2005, 17ff.): While the "first best" solution is always to have a willing and capable state, aid donors need to find second-best solutions to public services problems when the state is not willing or is very incapable. This means very limited engagement with government – or avoidance.

users (ibid., 2). The GTZ project-trained Community Mobilizing Workers (CMW), mainly women, do not only provide training in hygiene and water borne diseases but also promote the decentralization and commercialization of service utilities using a wide range of culturally sensitive ways (Buhl 2005, 12). They address schools, mosques, individual households and qat sessions to raise awareness for the tariff system and explain how to make effective use of the new service facilities. CMWs seem to have been successful in promoting the new service delivery approach, especially among poor households that now display an increased willingness to pay for the services. Also, the newly created customer relations departments of the service utilities have employed a number of CMWs – a fact that further underlines the success of the project component. Altogether, the CMW initiative is an innovative example on how to enhance citizen "voice" at the local level and improve the accountability relations between service providers and water users. Ultimately, this may not only improve the quality of service provision but also enhance community participation and thereby contribute to a process of state building from below.

Another important aspect of German support to the Yemenite water sector is a number of measures to prevent water-related conflicts that typically arise over water quantities / quality or the introduction of water tariffs. While the deployment of CMWs as discussed above already exhibits a clear conflict focus with view to minimizing conflicts over water tariffs, German development cooperation also supports the Advisory Committees of the decentralized service utilities that assemble relevant stakeholders and help to institutionalize the existing conflict potential (GTZ and KfW 2005, 8). In the field of water resources management, GTZ and KfW projects provide for the creation of basin committees in Amran and Sadah where involved stakeholders work towards the elaboration of integrated water management plans – a procedure that balances competing user interests and thereby reduces the risk of future water conflicts. The increasingly explicit conflict focus of German water projects can be best illustrated at the example of the new KfW project Provincial Towns Programme II (KfW 2005b, Annex 15). As the Ja'ar / Zinjibar project component displays high conflict potential due to competing water demands between the Abyan branch and LC Aden, conflict-relevant aspects have to a very high degree been taken into consideration during the planning process: The issues of securing and maintaining water availability, equal sharing of water resources and the establishment of conflict resolution mechanisms at the governorate and inter-governorate level are now integral parts of the project concept. Altogether, the German focus on water conflict prevention exercises an important function in the Yemenite context: As water scarcity and resulting water conflicts tend to enhance state fragility (see above), conflict prevention measures may contribute to the long-term consolidation of the Yemenite state. Against this background, it is no coincidence that the Yemenite government has specifically proposed to include "trouble spots" such as Abyan or Sadah (low government control, Islamist activities) among the project areas.

With respect to the sustainability of aid, one may generally expect a trade-off between short-term service delivery in situations of state fragility and medium and longer-term building of local and national governance structures. In the Yemenite recovery setting, however, framework conditions are sufficiently stable to minimize the tensions between the two priorities. German development cooperation has therefore devised an integrated "multi-level strategy" that anchors measures for improved WSS in (more) sustainable governance structures at all levels of government. Nonetheless, the sustainability of German aid should not be taken for granted. The most important obstacle to the long-term sustainability of German support to the Yemenite water sector is the insufficient and declining availability of water resources (Interview 2, 24 Jan 2006). Furthermore, capacity building measures do not always have the desired effects. The Technical Secretariat for WSS Sector Reform, for instance, has received substantial GTZ support but still remains to be integrated into the structure of the MWE (GTZ 2006). This has created a parallel structure that may hinder organizational capacity building and undermine the urgent strengthening of the MWE. Finally, high levels of corruption remain an important concern. While there is little, if any information on corruption in the water sector, it seems plausible to assume that "working with government" approaches can further enrich corrupt networks that feed on sectoral resources. Even though German donor agencies undertake considerable effort to minimize the danger of corruption (e.g. through spending audits, procurement regulations), it is systemic and thus very difficult to reduce to zero (Interview 2, 24 Jan 2006). On the other hand, large-scale corruption is more likely to occur in the oil or gas sectors (Interview 3, 24 Jan 2006).

In sum, German development cooperation addresses the Yemenite recovery setting (high political will, modest capacities) through a "multi-level strategy" that combines "top down" and "bottom up" approaches to state building: German donor agencies provide comprehensive support to sectoral reform at the macro/meso level and to the creation of decentralized and commercialized service utilities at the micro level. This is bolstered by a number of public awareness and conflict prevention measures designed to ease the high conflict potential of water resources in the Yemenite context. While the German support to the Yemenite water sector is clearly beneficial to the ongoing elaboration of a sound sectoral framework, one has to bear in mind that the overall framework conditions for water resources management in Yemen remain largely unsustainable. As long as key actors such as the MWE and the NWRA remain weak – especially vis-à-vis the MAI and vested agricultural interests – a truly sustainable approach to water resources management will be difficult to implement. Here, the recent decision by GTZ to focus its new water program on institutional reform at the macro level can only be a first step in the right direction. Instead, institutional and organizational improvements with view to long-term sustainability require political pressure at a higher political level (Cabinet, President) – a challenge that needs to be assumed by BMZ and coordinated with other donors.

3.4 Donor Coordination

Mutually complementary cooperation on the part of the German implementing agencies is of vital importance for the realization of a sustainable sectoral framework

(BMZ 2002, 14). With view to a German development policy "cast from the same mould", GTZ and KfW do not only work in close collaboration among themselves (see above) but also cooperate with other German organizations such as the German Development Service (ded), the Federal Institute for Geosciences and Natural Resources (BGR) or the Centre for International Migration (CIM) (GTZ 2006, 3f.). Coordination between KfW and GTZ has intensified over time and is now regarded as "excellent" (see above). Since 2005, both organizations are coupled through a "milestones concept", which means that the KfW project cycle establishes milestones that are – with GTZ support – to be achieved by the Yemenite side. While inter-German coordination can generally be regarded as comprehensive and successful, it also involves substantial costs. A cost-benefit analysis has so far not been conducted.

Coordination between international donor organizations is of crucial importance. There are currently many international donors that are active in the Yemenite water sector. The by far most important ones in terms of aid volume include the World Bank, Germany and the Netherlands, while the water-related support provided by UNDP, Japan, France, US AID and the EU is less relevant (BMZ 2002, 13; Interview 2, 24 Jan 2006). Arab funds (Arab Fund, OPEC Fund, Islamic Bank, Abu Dhabi Fund) provide erratic but large-scale support that is not coordinated at all. Donor coordination has so far been ambiguous. While it has generally improved in recent years, there is still little project-related collaboration at the operational level (GTZ 2006, 4f.). Where coordination exists, it takes place between the "large" donor organizations (see above).

Strategic donor coordination, on the other hand, has made substantial progress in recent years (GTZ and KfW 2005, 2; GTZ 2006; 4f.). Here, the elaboration of the NWSSIP – that was itself largely donor-driven – has provided a common reference point and thereby improved the overall prospects for donor coordination. In this context, Germany has started an initiative for better donor coordination that was explicitly welcomed by the Yemenite side and led to a "Joint Donor Declaration" by a total of 10 donor organizations in January 2005. Current discussions among the "core donor group" (Germany, Netherlands, World Bank) focus on the possibility of adopting a Sector Wide Approach (SWAp) (Interview 2, 24 Jan 2006; Interview 3, 24 Jan 2006). While the Netherlands and the World Bank seem to be in favor of a SWAp (albeit at different paces), Germany has yet to define its position on this question. A SWAp may generally be regarded a good opportunity to improve the alignment and harmonization of aid and give important leverage to the donor community, in particular with respect to working towards a more sustainable sectoral framework. On the downside, a SWAp might also lead to problems in the context of weak sector organizations and high levels of corruption.

4 Lessons Learned

Contemporary Yemen is still in a difficult process of state building and continues to be affected by institutional fragility – a situation that might be especially pronounced in Yemen but is generally quite common for many low-income countries. The Yemenite case resembles most a fragility scenario of (enduring) recovery where a relatively stable government is in place and basic state functions are slowly being established. Despite authoritarian tendencies and high levels of corruption, national leadership is demonstrating political will to make progress in developing the (water) infrastructure of a modern state but the performance and capacity of government remain rather weak in terms of (water) policy development and implementation. Altogether, this recovery setting of modest capacities but high political willingness offers rather favorable framework conditions for the involvement of (German) donor agencies. Here, one may even raise the question whether the framework conditions in a Yemenite-like recovery setting are really any more difficult than those encountered in most low-income countries.

Bearing in mind the OECD "Principles for good international engagement in fragile states" (OECD 2005b, 8ff.), the lessons learned from the German donor engagement in Yemen can be summarized as follows:

1. **Take context as a starting point and differentiate between sectors.** The Yemenite case underlines the (only seemingly trivial) need to calibrate analysis and action to particular country/sector circumstances. While Yemen is often considered a particularly fragile or even failed state, a closer look reveals that the framework conditions in the Yemenite water sector are not particularly fragile. Most tribes recognize the role of the central government in providing water and sanitation services, even in "trouble spots" where government control is traditionally very low. Instead, the recovery setting of modest capacities but high political willingness allows for donor strategies that resemble "traditional" support packages to water sector reform. For comparative purposes, it would be interesting to find out whether the framework conditions in other sectors are more difficult and require more "fragility-specific" donor strategies (e.g. in education that is a more ideological and contested service).
2. **Focus on state-building**… The long-term vision for international engagement in fragile environments must focus on supporting the creation of viable sovereign states. A Yemenite-like scenario of (enduring) recovery with modest capacities but high political willingness offers relatively favorable framework conditions for state-building approaches: As there is no need to "avoid" the government, donors can focus their attention on supporting the government in the creation of sustainable (water) governance structures.
3. … **and intervene at different levels.** Donor strategies in recovery settings should combine "top down" and "bottom up" approaches to state building. German donor agencies have devised a "multi-level strategy" that includes (1) support to sectoral reform at the macro level; (2) the strengthening of river basin organization at the meso level; and (3) the creation of decentralized and commercialized service utilities at the micro level. The basic strength of this multi-level approach is that it fosters synergies and learning between the different levels of intervention, which means that the knowledge gained in the context of projects at the micro/meso level is transferred to measures at the macro level (and vice versa). The scaling-up of aid to the macro level is particularly important since proposals

for (water) sector reform are based on practical experience at the micro/meso level – a sequence that will enhance the credibility of the reform.

4. **Enhance dialogue and participation.** State-building strategies in recovery settings should be based on dialogue and participation. The German support to the Yemenite water sector offers an innovative example of how to enhance dialogue with water users: GTZ project-trained Community Mobilizing Workers (CMW) do not only provide training in hygiene and water borne diseases but also promote and explain the decentralization and commercialization of service utilities (e.g. through awareness campaigns on water tariffs). Such initiatives help to strengthen citizen "voice" at the local level and improve the accountability relations between service providers and water users. Ultimately, this may not only improve the quality of service provision but also enhance community participation and state building from below.
5. **Move from reaction to (conflict) prevention.** Preventive action can minimize the risk of future conflict and contribute to long-term stability and development. This is particularly relevant in water-scarce countries like Yemen where water-related conflicts constitute an important source of (future) state fragility. German development cooperation has taken up the challenge by shifting from reaction to conflict prevention: Several WSS projects include risk analyses and support the establishment of conflict resolution mechanisms at the governorate and inter-governorate level. In this context, the concept of IWRM may help to contribute to long-term state consolidation. Altogether, the Yemenite case shows that water can both ease and worsen state fragility.
6. **Align with local priorities.** Where governments demonstrate political will but lack capacity, international donors should fully align assistance behind government strategies. German support to the Yemenite water sector is fully aligned, as the various projects are all based on Yemenite (sector) strategies and rooted in Yemenite structures (no project implementation units).
7. **Coordinate between international donors.** Coordination between international donor organizations is of crucial importance to promote the state building agenda. In the case of Yemen, strategic donor coordination has made important progress in recent years: Here, the elaboration of the NWSSIP has provided a common reference point, followed by a "Joint Donor Declaration" in January 2005. The currently discussed option of a SWAp might help to improve the alignment of aid and give more leverage to the donor community, in particular with respect to working towards a more sustainable sectoral framework. On the downside, a SWAp might lead to problems in the context of weak sector organizations and high levels of corruption.

Acknowledgements This study was originally prepared for Deutsche Gesellschaft für Technische Zusammenarbeit (GTZ) and served as a contribution to the OECD workstream on service delivery in fragile states. The author wishes to thank GTZ for the kind permission to publish the findings in this volume.

References

Berry C, Forder A, Sultan S, Moreno-Torres M (2004) Approaches to improving the delivery of social services in difficult environments, DFID PRDE Working Paper 3, Oct 2004

BMZ (2002) Strategy for Yemeni–German cooperation in the priority area of water and waste management (unpublished draft)

Branchflower A (2004) How important are difficult environments to achieving the MDGs? DFID PRDE Working Paper 2, Sep 2004

Buhl S (2005) Poverty orientation of the Yemeni–German urban water and water sewerage projects, Report prepared on behalf of GTZ, Sanaa, Eschborn

CIA (2007) The World Factbook – Yemen. https://www.cia.gov/library/publications/the-world-factbook/geos/ym.html. Cited 19 Jun 2007

Crisis States Programme (2005) War, state collapse and reconstruction: phase 2 of the Crisis States Programme, Crisis States Working Papers, 1, Crisis States Research Centre, School of Economics, London

DfID (2005) Yemen SFD – Institutional evaluation methodology linked to fragile states and state-building agendas (unpublished draft)

FAO (2005) AQUASTAT Yemen Country Profile. http://www.fao.org/ag/agl/aglw/aquastat/countries/yemen/index.stm. Cited 18 Nov 2005

Glosemeyer I (2005) Jemen: Staatsbildung mit Hindernissen. In: Schneckener U (ed.): States at risk. Fragile Staaten als Sicherheits- und Entwicklungsproblem, Stiftung Wissenschaft und Politik, Berlin, 122–138

GTZ (2004a) Ausbildung im Wassersektor, Fortschrittsbericht an das BMZ, Projekt-/Auftragsnummer 2003.2212.3, Berichtsnummer 4

GTZ (2004b) Support to the Technical Secretariat for water supply and sanitation sector reform, Progress report for the BMZ, Project number 2000.2096.6, Reporting number 3

GTZ (2005) KV-Beratung des Wasserver- und Abwasserentsorgungssektors, Fortschrittsbericht an das BMZ, Projekt-/Auftragsnummer 2002.2508.6-001.00, Berichtsnummer 11

GTZ (2006) Lernerfahrungen aus den bisherigen TZ-Projekten im Wassersektor, KV Programm "Institutionelle Entwicklung des Wassersektors", Jemen, PN: 2005.2185.6

GTZ, KfW (2005) Entwicklungszusammenarbeit mit der Republik Jemen, Gemeinsame Berichterstattung der GTZ und KfW im Schwerpunkt Wasser/Abwasser in Abstimmung mit DED, BGR, InWEnt und CIM

ICG (2003) Yemen: coping with terrorism and violence in a fragile state, ICG Middle East Report No 8, Amman, Brussels

KfW (2005a) Finanzielle Zusammenarbeit mit der Republik Jemen, Wasserver- und Abwasserentsorgung in Provinzstädten (PTP I), Abschlusskontrollbericht, BMZ Nr.: 2002 65 165

KfW (2005b) Finanzielle Zusammenarbeit mit der Republik Jemen, Wasserver- und Abwasserentsorgung in Provinzstädten II – Al Shehr, Ja'ar und Zinjibar, Programmprüfungsbericht, BMZ Nr.: 2002 65 553

KfW (2006) Yemen's water & sanitation sector. Selected key sector issues. Update January 2006, Presentation by Gerhard Redecker, KfW-Office Sanaa

Meagher P (2005) Service delivery in fragile states. Framing the issues, IRIS Center, University of Maryland, College Park, Revised draft, Jul 25, 2005

North DC (1993) The New Institutional Economics and third world development. In: Harris J, Hunter J, Lewis CM (eds.) The New Institutional Economics and third world development, Routledge, London, New York, 17–26

OECD (2005a) Service delivery in fragile states: advancing donor practise, DAC Fragile States Group, Concept note prepared by USAID and UNICEF, DCD(2005)15/REV1

OCED (2005b) Piloting the principles for good international engagement in fragile states, DAC Fragile States Group, Concept note, DCD(2005)11/REV2

Republic of Yemen (2004) National water sector strategy and investment program, 2005–2009, NWSSIP, Sanaa, Yemen

The Economist Intelligence Unit (2006) Yemen country report, Jan 2006
The Fund for Peace and the Carnegie Endowment for International Peace (2005) The failed states index. In: Foreign Policy, Issue 149, July/August 2005, 56–65
UNDP (2005) Yemen Country Sheet, Human Development Report 2005. http://hdr.undp.org/statistics/data/countries.cfm?c=YEM. Cited 27 Nov 2005
UNICEF, WHO (2000) Global water supply and sanitation assessment 2000 report. http://www.who.int/water_sanitation_health/monitoring/jmp2000.pdf. Cited 4 Dec 2005
UNICEF, WHO (2004) Meeting the MDG drinking water and sanitation target. A mid-term assessment of progress. http://www.wssinfo.org/en/40_MDG2004.html. Cited 4 Dec 2005
WRI (2005) World resources 2005. The wealth of the poor: managing ecosystems to fight poverty. http://pdf.wri.org/wrr05_lores.pdf. Cited 12 Dec 2005
Yemen Times (2005) Yemen: failure or democracy, Issue 905, Vol. 14, Dec 22–25, 2005. http://www.yementimes.com/article.shtml?i= 905&p=report&a=1. Cited 14 Jan 2006

EU Water Initiative – A (non-) Innovative Form of Development Cooperation

Lena Partzsch

Abstract Facing the reality of a global economy and an increasingly interdependent world, conventional forms of pure state-to-state development cooperation are being brought into question. At the Johannesburg summit in 2002, the European Union announced the EU Water Initiative (EUWI) with 'partners' from the private sector and from civil society organisations. This multi-stakeholder partnership is a key strategy in tackling the global water crisis. The article argues that, on the one hand, the EUWI follows an innovative approach in terms of coordinating state actors and 'new' actors from private sector and civil society, as well as from different policy fields. On the other hand, the partnership must be considered non innovative because it is de facto dominated by European actors, while actors in the partner countries are under-represented.

1 Introduction

> The water related MDGs and WSSD targets cannot be achieved by conventional means. They require a great mobilization of partners ranging from government, IFIs and other donors, to civil society organizations, water users and the water industry, both in Europe and in partner countries. (European Commission 2005)

Facing the reality of a global economy and an increasingly interdependent world, conventional forms of pure state-to-state development cooperation are being brought into question. Power has been redistributed among states, markets and civil society. Foreign direct investment (FDI) by private corporations brings much more change on the ground in developing countries than official development aid (ODA) has ever been able to. However, essential problems have not been solved and in some cases have even been aggravated. The global water crisis can definitely be considered a

Lena Partzsch
e-mail: lena.partzsch@ufz.de

W. Scheumann et al. (eds.), *Water Politics and Development Cooperation*,
doi: 10.1007/978-3-540-76707-7_17,

persistent problem as hardly any improvements have been made, despite promises and claims to the contrary (Postel 2006, 112; Simonis 2001, 15).[1]

More than ever before, the world faces serious challenges in terms of quantity and quality of freshwater. The hydrological cycle yields a limited quantity of water per time period, and this quantity is currently over-exploited. Global water consumption is increasing at more than twice the rate of population growth, and within the next two decades water availability per capita is therefore expected to drop by a third (Gilles et al. 2005, 5). Meanwhile, more than a billion people lack access to potable water and twice as many do not have basic sanitation. The UN Millennium Development Goals (MDGs) and the Johannesburg Plan of Implementation (JPOI) aim to halve those numbers by 2015. Furthermore, national plans for improved water efficiency[2] and integrated water resource management have to be elaborated.[3]

At the World Summit on Sustainable Development (WSSD) in Johannesburg 2002, the European Union (EU) announced the EU Water Initiative (EUWI) which is a key strategy in tackling the global water crisis. The EUWI represents a new approach in development cooperation for two main reasons. First, it encompasses not only state actors, but 'partners' from the private sector and from civil society organizations as well. This is a clear indicator of the withering away of the nation-state as the one and only player in the field of development cooperation. Second, the EUWI shows that the dichotomy between internal and external politics which has dominated our understanding of politics so far is going to be transformed by EU water strategies.

This article, first, examines how non-state actors, and especially women, have been considered in EU water policy in the past. Special emphasis is put on the EU Water Framework Directive (WFD) and the Communication from the European Commission on EU water strategies for developing countries. Against this background, one can assess the extent to which the EUWI can be considered innovative or not. Evidence shows that, in fact, the partnership builds on both internal and external EU water strategies but goes further than both in terms of stakeholder participation.

In the second part of this chapter, theoretical assumptions behind the new partnership approach are elaborated on and an analytical framework is developed in order to examine the inclusion of non-state actors in the EUWI. New modes of governance such as partnerships are meant to enhance not only effectiveness (Witte and Reinicke 2005); scholars of deliberative democracy argue that they increase legitimacy as well because of the inclusion of new non-state actors (Nanz and Steffek 2005, 79). The EUWI is divided into four regional components which are investigated in the third part of this article.

[1] For more detailed information on progress in access to water (not resource protection) see UN Millennium Development Indicators: http://millenniumindicators.un.org/unsd/mi/mi_worldmillennium_new.asp. Cited 18 Jul 2006.

[2] Increasing water efficiency means same or better yields with less water volume.

[3] Millennium Development Declaration: www.un.org/millennium/declaration/ares552e.pdf; Johannesburg Plan of Implementation: www.un.org/esa/sustdev/documents/WSSD_POI_PD/English/WSSD_PlanImpl.pdf. Cited 1 Mar 2006.

The data which has been used comes from documentary material, a participatory observation of the EUWI annual meeting in Stockholm in August 2005 and 2006, and the transcripts of 25 qualitative interviews with people connected to the partnership. For reasons of a balanced sampling, not only proponents, but also opponents of the EUWI and of partnerships in general have been interviewed. Since policy cycles are small in the field of water, all interviewees preferred to remain anonymous. Therefore only the actor category and date of interview are given as reference.

The article argues that, on the one hand, the EUWI follows an innovative approach in terms of coordinating state actors and 'new' actors from private sector and civil society, as well as from different policy fields. On the other hand, the partnership must be considered non innovative because it is de facto dominated by European actors, while actors in the partner countries are under-represented.

2 Participation in European Union's Water Policy

A participative approach in water management and development was endorsed at the international level by the Dublin Conference in 1992. Special emphasis was thereby put on women to participate at all levels in water resources programs, including decision-making. Stakeholder participation has been an essential part of EU water policy.[4] Both internal and external EU water strategies therefore encourage the essential role of non-state actors. Participation is anchored in the WFD and in the Communication from the European Commission on water management in developing countries. However, the EUWI goes one step further; it is not only about participation but partnership with non-state actors.

Dublin Guiding Principles

Principle No. 2 – *Water development and management should be based on a participatory approach, involving users, planners and policy-makers at all levels.*

The participatory approach involves raising awareness of the importance of water among policy-makers and the general public. It means that decisions are taken at the lowest appropriate level, with full public consultation and involvement of users in the planning and implementation of water projects.

Principle No. 3 – *Women play a central part in the provision, management and safeguarding of water.*

This pivotal role of women as providers and users of water and guardians of the living environment has seldom been reflected in institutional arrangements

[4] Dublin Declaration: http://www.wmo.ch/web/homs/documents/english/icwedece.html. Cited 1 Jun 2006.

for the development and management of water resources. Acceptance and implementation of this principle requires positive policies to address women's specific needs and to equip and empower women to participate at all levels in water resources programmes, including decision-making and implementation, in ways defined by them.

2.1 The European Water Framework Directive

The 'Directive 2000/60/EC of the European Parliament and of the Council of 23 October 2000 establishing a framework for Community action in the field of water policy'[5] – in short: the Water Framework Directive (WFD) – is the heart of EU water policy. The document has caused a general revision of European legislation on water. The WFD unifies the 'patchwork' of different, partly contradictive directives and separate regulations, and creates a legal framework for the administration and the protection of water resources in the EU.

The Directive was formulated from a primarily ecological point of view (Mostert 2003, 525). The preamble states: 'Water is not a commercial product like any other but, rather, a heritage which must be protected, defended and treated as such' (preamble, no. 1). Environmental pollution shall be prevented and an ecologically sustainable water policy shall be established through the WFD. The aim is to keep European waters of 'good status'. Water quality shall even be improved by 2015 (preamble no. 26, 28 und 33; art. 1, 2, 4 and 6).

The Directive was originally adopted as EU internal policy, which means that it applies only to the territory of EU member states. However, it is also considered a worldwide model for 'Integrated Water Resource Management' (IWRM) (Rahaman et al. 2001). Outside the EU, the WFD is promoted by programs such as the 'Joint Framework Directive/EU Water Initiative Process' – in short: 'Joint Process' (Dimas 2005, 3; European Commission 2003a, 5). Of course, it is still not legally binding outside the EU.

A key element of the WFD, along with the IWRM concept, is the participation of non-state actors, e.g. environmental organizations (Mostert 2003, 526). To quote from the preamble (no. 14): 'The success of this Directive relies on close cooperation and coherent action at Community, Member State and local level as well as on information, consultation and involvement of the public, including users.' The involvement of stakeholders and the broader public is forecasted in various ways (preamble no. 14, 46; art. 14; Annex VII, A9).

According to article 14, 'member states shall encourage the active involvement of all interested parties in the implementation of this Directive, in particular in the production, review and updating of the river basin management plans.' In order to

[5] http://ec.europa.eu/environment/water/water-framework/index_en.html. Cited 2 Jun 2006.

guarantee public information and consultation, the Directive even prescribes contact points and procedures for obtaining the background documentation and information (Annex VII, A.11) (Mostert 2003, 526). Thereby, an open and participative process is promoted as a goal in itself to support democratic procedures.

However, in contrast to most IWRM frameworks, the WFD does not prescribe a gender-inclusive strategy (Rahaman et al. 2001, 5). This goes alongside an apparent know-how and research deficit on gender-specific needs of women and men with regards to water resources in the EU. Almost a hundred percent of households in the EU are connected to water systems, and women do not share the burden of carrying water like, for example, they do in Africa. Therefore, the need for gender-inclusive strategies is not evident and was obviously not considered relevant for the WFD.

However, gender-inclusive strategies go far beyond general access to water. For example, they are also about equal employment and salaries in the water service sector, equal access to information and equal representation in water decision-making. Considering this broader approach to understanding gender, a need for progress on gender issues does not only exist in developing countries, but in the EU as well (Braunmühl and Winterfeld 2003; Kunst et al. 2002).

In summary, the WFD encompasses various possibilities for the participation of stakeholders and the broader public, which do not directly discriminate against either women or men. However, the WFD does not follow a gender-inclusive strategy.

2.2 *The European Communication on Water Management in Developing Countries*

While the WFD was originally meant to be EU legislation limited to the territory of the EU member states, water strategies for outside the EU, namely developing countries, were prepared too. The key document in this respect is the 'Communication from the Commission to the Council and the European Parliament. Water Management in Developing Countries. Policy and Priorities for EU Development Cooperation' (COM 2002, 132).[6] The Communication was published in March 2002 before the Johannesburg summit where the EUWI was initiated in September 2002.

The Communication points in a different direction from the WFD, in that while the WFD is environmental legislation, the Communication is clearly settled in a context of development cooperation. Indeed, the document focuses mainly on increasing access to water supply and sanitation, and pays less attention to resource protection. It has no binding character and is only a guideline for EU support on water resources management in developing countries (European Commission 2002a, 3).

In May 2002, the Council of Development Ministers implicitly referred to the Communication with a Resolution[7] which points to the relevance of water scarcity

[6] Communication: http://europa.eu.int/eur-lex/en/com/cnc/2002/com2002_0132en01.pdf. Cited 1 Jun 2006.

[7] Council Resolution: http://register.consilium.eu.int/pdf/fr/02/st08/08958f2.pdf. Cited 1 Jun 2006.

and decreasing water quality especially in developing countries. In this document, the need for a 'strategic partnership' with non-state actors is declared. This became reality three months later with the initiation of the EUWI. The EUWI partners interpret the Resolution as the Council's mandate for the EUWI (EUWI 2005a, 7).

The European Parliament also stated their position towards the Communication in a Report[8] and Resolution[9] but this was actually in 2003 when the EUWI had already been created. Paul A.A.J.G. Lannoye was appointed Rapporteur by the Committee on Development and Cooperation of the European Parliament.

The Communication from the Commission emphasizes the need for the participation of all stakeholders 'in a spirit of balanced partnership at all levels and in all contexts, and including a gender balance'. Participation of 'representatives of civil society such as NGOs and user associations, and the private sector' and their ownership are considered 'a key to success of policies and activities' (European Commission 2003a, 16). Non-state actors are regarded as relevant in two ways: First, in terms of mobilizing new private funding. In this context, water-related services, according to the Commission, shall be made more attractive for investment 'to bridge the gap between current and required levels of expenditures' (European Commission 2002a, 16). Second, stakeholder participation is regarded as useful in terms of knowledge and information exchange. Furthermore, their involvement is seen as a significant contribution to institutional sustainability and conflict prevention (European Commission 2002a, 3 and 18). In this respect, 'participation by all stakeholders (and especially women)' (European Commission 2002a, 3) is not only recommended as a goal in itself, and for democratic concerns, but it is considered essential for implementation, for example in order to establish water pricing (European Commission 2002a, 15).

This objective is highly criticized in the Lannoye Report from the European Parliament. In fact, the Report lists 'the privatization of water services' as one of the main reasons for the 'developing world's affliction by impoverishment in general and water shortage in particular' (European Parliament 2003, 12). The Report questions whether water pricing and the privatization of water services benefit the poor and enquires whether only French and German corporations' benefit from it (European Parliament 2003, 13). However, the European Parliament agrees with the participative approach, although it does point to disparities in participation and disapproves of the fact that the involvement of citizens and the advancement of women are barely mentioned in the Communication (European Parliament 2003, 12).

However, under the title of 'Horizontal and other aspects', a paragraph of the Commission's Communication deals with 'Gender balance'. The Commission asserts that 'mainstreaming gender equity should give special attention to water-related policies and programmes' (European Commission 2003a, 11). In concrete terms, nonetheless, the paragraph refers only to 'the heavy burden imposed on women in terms of their time and energy in collecting domestic water supplies'

[8] Report: http://www.europarl.europa.eu/registre/seance_pleniere/textes_deposes/rapports/2003/0273/P5_A(2003)0273_EN.doc. Cited 1 Jun 2006.

[9] Parliament Resolution: http://www.europarl.europa.eu/registre/seance_pleniere/textes_adoptes/definitif/2003/09-04/0377/P5_TA(2003)0377_EN.doc. Cited 1 Jun 2006.

(European Commission 2002a, 11). There is no reflection on social construction of gender identities and the relations between women and men. Furthermore, impacts of development programs, in terms of the reproduction of inequalities are not dealt with. For example, there is no mention of the impacts of the replacement of subsistence farming, which is mainly done by women, by modern irrigation farming, which is mostly run by men (Deutscher Bundestag 2002, 361; Braunmühl and Winterfeld 2003, 49).

In conclusion, the Communication differs from the WFD in that it encompasses not only the involvement of stakeholders, but it also emphases the need for gender balance. Nevertheless, the gender-inclusive strategy is limited. Furthermore, while in the WFD the participation of non-state actors is first and foremost treated as a democratic goal, the Communication puts stronger emphasis on its role in the enhancement of implementation.

2.3 The EU Water Initiative

At the World Summit on Sustainable Development in 2002, the EU announced the Water Initiative (EUWI) which can be considered to be a key strategy in tackling the global water crisis. This so called 'partnership' has encompassed private sector and civil society from the very beginning. Indeed, non-state actors along with the European Commission, EU member states and partner governments from the South initiated the EUWI (European Commission 2002b). This is clearly different from the processes which led to the adoption of the WFD and the publication of the Communication from the Commission. Furthermore, the European Parliament was not involved in the EUWI process.

The EUWI is structured in four regional multi-stakeholder partnerships: the EU Water Initiative for Africa (A-EUWI); the EU Water Initiative for Eastern Europe, Caucasus and Central Asia (EECCA-EUWI); the EU Water Initiative for the Mediterranean (MED-EUWI); and the EU Water Initiative for Latin America (LA-EUWI). Apart from the regional components the EUWI includes two crosscutting components on research and financing. The initiative is expected to continue until 2015 which is when the MDG and the JPOI targets are due (European Commission 2002b).

As the EUWI is a political initiative, it does not follow a clearly determined institutional procedure with documents originating from it in a constitutive order, like, for example, as was the case with the legislative procedure which led to the adoption of the WFD. Moreover, there is no fixed amount of money which is supposed to be spent, something which one would normally expect to form the basis of any development cooperation program. Instead, the initiative represents an open process which brings together state and non-state actors from the sectors of development, environment and research, as well as from inside and outside the EU. The aim is to implement the water related MDGs and JPOI targets.

When the EUWI was initiated in Johannesburg, a working document was available called 'EU Water Initiative: Water for Life Health, livelihoods, economic development, peace and security'[10] which summarized the main ideas behind the initiative (European Commission 2002b). During a first 'design phase', institutional structures and procedures were then developed. Those were finally adopted at the EUWI annual meeting, the so called 'Multi-Stakeholder Forum' (MStF), in Stockholm in 2004. Goals and structures, however, depend on an open process in which any change is possible at any time (European Commission 2002b, 10; EUWI 2005a, 7; Interview with state actor on 29 November 2004). Nonetheless, the following goals seem to endure throughout (Dimas 2005, 3):

1. **Reinforce political will and commitment to action**: The initiative seeks to give the global water crisis a higher priority on the political agenda as water is considered essential for poverty eradication and sustainable development.
2. **Improve coordination and cooperation**: The initiative seeks to provide an umbrella for a range of existing state and non-state activities in order to create synergy effects in the field of water supply and sanitation. Furthermore, regional and sub-regional water cooperation is aimed to be enhanced whereby the WFD is promoted as best practice ('Joint Process').
3. **Increase the efficiency of existing EU aid flows**: The initiative does not seek to increase the amount but the efficiency of ODA, especially by establishing water tariffs and private funding for water supply and sanitation.

Even if there is some overarching structure (annual Multi-Stakeholder Forum, common website etc.), the four regional partnerships of the EUWI each have developed an independent structure on their own. They also have their own websites, for example, which provide much broader documentation and more detailed information than that available on the common website. Some have also established their own regional secretariats.

3 A New Governance Mode

We have learnt that the WFD and the Communication both encourage the participation of non-state actors but that the EUWI goes much further, whereby non-state actors are not only meant to participate, but are indeed meant to act as partners, in the whole political process, from the outset (see Table 1).

The inclusion of non-state actors in political processes is often referred to by the term 'new governance'.[11] Participation of non-state actors is a step towards 'new governance' while partnership in fact represents a new mode of governance (Witte

[10] EUWI document: http://europa.eu.int/eur-lex/en/com/cnc/2002/com2002_0132en01.pdf. Cited 1 Jun 2006.

[11] Governance is thereby not synonymous with government. While "government suggests activities that are backed by formal authority, by police powers to insure the implementation of duly constituted policies, (...) governance refers to activities backed by shared goals that may or may not derive from legal and formally prescribed responsibilities and that do not necessarily rely on police powers to over-come defiance and attain compliance" (Rosenau 1992, 4).

Table 1 Participation in EU Water strategies

	Scope	Dublin principle no. 2: participatory approach	Dublin principle no. 3 – Women's participation
EU Water Framework Directive (WFD)	EU member states	Involvement of all interested parties encouraged by art. 14; see also preamble no. 14, 46; annex VII, A9	Not considered
Communication from the European Commission on Water Management in Developing Countries	Guideline for EU support to developing countries	Participation of all stakeholders is required, including economic and social actors, representatives of civil society such as NGOs and user associations, and the private sector	Gender Mainstreaming
EU Water Initiative (EUWI)	(1) Africa; (2) Eastern Europe, Caucasus and Central Asia; (3) the Mediterranean; and (4) Latin America	Partnership with private sector and civil society	Considered only in partnership with Latin America

et al. 2003; Witte and Reinicke 2005). At the same time, some scholars argue that the inclusion of non-state actors also bears a new source of legitimacy (Nanz and Steffek 2005). Let me elaborate on these aspects through a small theoretical excursion where we will first explore the shift from participation to partnership, before going on to discuss aspects of legitimacy.

3.1 The Shift from Participation to Partnership

Diverse transnational non-state actors have appeared during the last two decades. There are multinational corporations on the one hand and actors from global civil society on the other. Often, the non-profit 'civil society sector' is thereby considered counterpart to the profit-oriented 'private sector'. This leads to a trinity of state actors, private actors and actors from civil society (Hummel 2001, 32).

Governance is characterized by the inclusion of non-state actors – whereas conventional regulation is exclusively performed by state actors. New modes of governance which encompass non-state actors are meant to be more efficient, flexible and effective than conventional 'top down' regulation. Hence, it is assumed that they may have a greater impact on problems which traditional nation-state activity has failed to solve, such as climate change, loss of biodiversity or global water crisis (Messner and Nuscheler 2003, 9f.; Rosenau 1992, 8ff.).

When analyzing partnerships, most scholars turn to network theory (Stewart and Gray 2006, 363; Witte et al. 2003, 64). Indeed, network theory helps to understand the institutional setting comprising state and non-state cooperation. Scharpf (2000), for example, uses the term 'network' in order to describe a specific institutional context in which relevant actors interact. Referring to game theory, he distinguishes four forms of institutional context. For the purpose of this article, his distinction between two of these, network and hierarchical organization is significant.

Hierarchical organization refers to the institutional setting of the nation-state (and the European Union), and represents conventional regulation 'from the top down to the bottom' by state actors backed by formal authority. Implementation is ensured by the State's monopoly on (police) power (Scharpf 2000, 169 and 230).

In contrast, networks do not have a hierarchical structure, but instead are horizontal (there is no leadership, there is no (own) legal system backing exchange between actors). Relations are semi-permanent, based on resource exchange, mutual support and common norms between network members. Membership is voluntary. Hence, decisions cannot be enforced against a member and must thus be based on consensus. If actors contribute resources to the network by, for example, delivering a service to another member, they have no guarantee for receiving anything in return. Trust among network members is therefore essential and comparable to power as the main element of hierarchy (Börzel 1998, 254; Scharpf 2000, 231).

The WFD is clearly settled in the hierarchical structure of the EU. Stakeholder participation might contribute additional resources (for example extra information for the river basin management plans), and thereby influence political output, but state actors have the final say; they are accountable to the people for decisions made under the auspices of the WFD. The situation is even clearer in the case of the European Communication on water management in developing countries. Here, participation is meant to enhance implementation of what has been previously agreed by state actors. Participation is therefore established in the hierarchical system of the nation-state or the EU, respectively.

In contrast, the EUWI has included non-state actors from the very beginning. Non-state actors were indeed among the initial partners of the EUWI. This is obviously different from the processes which led to the adoption of the WFD and the Communication. Instead of following a legislative procedure, the EUWI began with a design phase in which all participating state and non-state actors had to agree on an over-all structure.

Hence, there is a shift from a hierarchical to a network structure through leaving behind conventional procedures and including new actors. At the same time, the institutional context of the network overcomes sector divisions and the traditional dichotomy of what is considered internal and external affairs ('thinking out of the box'). This is noticeable in two ways: Firstly, former EU internal policies are now also applied outside the EU. Namely, the WFD is promoted as a model to implement IWRM in third countries by the EUWI ('Joint Process') (Dimas 2005, 3; European Commission 2003a, 5). Secondly, not only actors from the field of development cooperation or external affairs take part in the EUWI, but also state and non-state actors from different policy fields, e.g. DG Environment and DG Research on behalf of the European Commission (LA-EUWI 2005b, 1). As such, competencies of different resorts are recognized as overlapping and boundaries have blurred, not only in terms of state versus non-state spheres, but also in terms of territory and sector policies.

Thus, the EUWI represents a new mode of governance for two reasons: (1) the inclusion of non-state actors, and (2) the overlapping of spheres in terms of sectors (e.g. environment and development) and territory (inside and outside the EU). After explaining the shift from participation to partnership, I will now discuss its implications in terms of legitimacy.

3.2 Stakeholder Inclusion as a Source of Legitimacy

Legitimacy means to accept a political order as binding. Today it is oriented on democratic norms and fair procedures (input legitimacy) and on an effective and just performance (output legitimacy) (Scharpf 2000, 349). According to liberal theories, state actors alone are entitled to prescribe behavior for others because only state actors can be held accountable by the political institutions; in democracy, they are elected representatives. Hence, while performance of certain services is transferable, the transfer of legitimacy is not as easy. New modes of governance thus demand new sources of legitimacy. There are various normative aspects which must be considered such as the reasons given to justify authority, who or what confers legitimacy and why legitimacy is accepted by others or not (Hummel 2001, 30).

Already the transfer of responsibilities from the nation-state to international intergovernmental organizations has caused a debate on legitimacy beyond the nation state (Zürn 2000, 183). About 300 detailed international regimes and conventions such as those on climate, biodiversity, and human rights exist. Repeatedly, governments and national parliaments are thereby forced to 'take it or leave it' in terms of decisions on international agreements of some hundred pages. Compromises which were achieved at the international level are then hard to communicate to citizens and may provoke conflicts (Mathews 1997, 65). Hence, legitimating international politics has never been unproblematic. The emergence of new non-state actors at the global level opens, however, a new chapter in this debate.

In contrast to national governments, non-state actors are not entitled to act 'in the name of the people' because they are not authorized by the people and, thus, cannot be held accountable. Managers of companies and associations are indeed not accountable to the whole of society but only to their individual members – at least in a legal and financial sense. According to liberal theory of international law, only states and their governments are legitimate givers and takers of international legislation (Cutler 2002, 32).

As the 'new' actors are not foreseen by the state-centered concept of international relations, which only acknowledges sovereign territorial states and their representatives, they must consequently be considered non-legitimate actors by definition. The question of how to legitimize non-state actors at the global level, therefore, means leaving behind the conventional understanding of international relations as pure state-to-state affairs.

After all, deficits in conventional input legitimacy are accepted by proponents of new governance approaches, such as partnerships, in favor of a more effective and just performance (output legitimacy) (Witte et al. 2003, 59). Romano Prodi, when he was president of the European Commission, formulated it this way: '[A]t the end of the day, what interests them (the citizens) is not who solves these problems, but the fact that they are being solved' (Magnette 2001, 1).

From this very pragmatic output-oriented view, decisions by any actor can be considered legitimate if, and because, they effectively contribute to solving a problem and increase social welfare. However, there are some scholars such as Braunmühl and Winterfeld (2003, 18) who criticize this view from a neo-gramscian perspective, arguing that certain questions are faded out such as 'for whom what and why a problem is considered as such, who is able to get attention for his – or her – problems in the public sphere and who is under which mechanisms included in problem solving strategies and who in which form of implementation'. They point the finger to power asymmetries and formal or informal hierarchies among diverse actors. Referring to water they ask: 'Who actually seeks to provide whom with water?' (Braunmühl and Winterfeld 2003, 48).

Indeed, from a democratic standpoint, legitimacy by output alone is weak and can hardly be considered sufficient. Theories of deliberative democracy thus argue that output legitimacy should be balanced through stakeholder inclusion on the input side. The deficit in conventional 'input legitimacy' is meant to be overcome by 'throughput legitimacy', in other words, stakeholder involvement (Dingwerth 2004, 86; Nanz and Steffek 2005, 79). Besides the democratic benefits seen in the representation of diverse interests, special emphasis is put on stakeholders' expertise. Stakeholders enhance legitimacy to a process from a throughput oriented perspective, if their knowledge flows into the process which generates the output (Nanz and Steffek 2005, 79 and 85). Hence, legitimacy can be increased by the participation and partnering of non-state actors. Participation is accordingly even considered essential for 'good governance' (e.g. European Commission 2001).

Nevertheless, the question of who is represented by the participating stakeholders and who is excluded from the partnership is important. All interested parties and water users are potential stakeholders, but some take part in partnerships and others are excluded. Attention is normally paid only to those who are able to make themselves heard and seen – to the cost of those who are heard and seen less (or not). Of course, state actors and big corporations benefit in this regard from greater resources available to them. However, ordinary citizens are relevant actors too. In a study on water supply in South Africa (Nelspruit and Dolphin Coast) Lombard (2000, 49), for example, indicates the relevance of water users who get attention through non-payment of water bills.

4 The EU Water Initiative: Empirical Findings

It has become clear that analyzing partnerships must go beyond naming the actors de facto involved and must also examine relations of power and authority. This, I will now do with reference to empirical findings from the four regional components of the EUWI: the A-EUWI, the EECCA-EUWI, the MED-EUWI and the LA-EUWI. The aim of which is to show (a) who is actually involved and (b) whether power asymmetries exist. Of course, both foci of analysis overlap.

4.1 The EUWI for Africa

(a) The actors who are most actively involved in the EU Water Initiative for Africa (A-EUWI) are state actors, namely the European Commission, EU member states and African States. This has been evident from the very beginning when these actors signed the initiative at the WSSD. Official leadership of this component was initially with Denmark and France, later Germany joined in. The official counterpart to the EU member states is the African Ministers' Council on Water (AMCOW). The International Council for Local Environmental Initiatives (ICLEI) is also involved which means state actors from the partner countries' local level also participate in the partnership (European Commission 2003c, 5; EUWI 2005b).

All initial non-state actors come from Europe, though. The private sector is involved in the partnership with Suez, its (previous) subsidiary Northumbrian Water, RWE Thames Water and Vivendi/Veolia. Civil society is represented through NGOs such as Greencross, Tearfund, WaterAid, World Wide Fund for Nature (WWF) and Global Water Partnership (GWP) (European Commission 2002b, i; EUWI 2006).

(b) State actors dominate the EUWI over non-state actors as leadership is with state actors only, and the EUWI Secretariat is based at the European Commission. Thus the Commission sets the rules of the game (EUWI 2005b; interview with state actor on 2 December 2004) what aggravates existing inequalities, especially, between North and South. The EU is in the donor position while African partners are receiving ODA. The same is true for multinational corporations and FDI.

Stakeholders involved in the A-EUWI are not all independent. The civil society partners cooperated already in the past with the private sector. For example, WWF runs a Corporate Club consisting of corporate members such as Unilever, Coca-Cola, and Nestlé (WWF 2006). Moreover, for example, GWP encompasses state members and members from the private sector. The current chair Margaret Catley-Carlson is former head of the Canadian International Development Agency and also leads the Water Resources Advisory Committee which was founded by Suez (Rekacewicz 2005). Suez executive, René Coulomb, sits on the GWP steering committee (Barlow 2006).

Furthermore, it is questionable if NGOs originating from the North are really supposed to represent people from the South. Independent African civil society has different political standpoints concerning water policies than European civil society involved. The Free Water Movement, for example, opposes water commodification while GWP promoted water being an economic good (Bond 2004, 144; GWP 2006). Besides, the A-EUWI has no gender-inclusive strategy to support women's involvement.

As a consequence, there are power asymmetries and internal hierarchies among A-EUWI partners, among state and non-state actors and among partners from the North and South. The decision-making procedures at work might somehow crack persistent problems at the level of paper statements. Nevertheless, above and beyond aspects of democracy, there may be difficulties encountered when decisions come to be implemented if there is no sense of ownership on the side of the partner countries.

4.2 EUWI for Eastern Europe, Caucasus and Central Asia

(a) The EU Water Initiative for Eastern Europe, Caucasus and Central Asia (EECCA-EUWI) focuses on 12 countries, namely Armenia, Azerbaijan, Belarus, Georgia, Kazakhstan, Kyrgyzstan, Moldavia, Russia, Tadzhikistan, Turkmenistan, the Ukraine and Uzbekistan. It is part of the intergovernmental process 'Environment for Europe' and was initiated at the WSSD. Thus, the partnership has been driven mainly by state actors, especially, from Western Europe which are the old EU member states, Switzerland, Norway and the European Commission. Initially, leadership was with Denmark and Russia whereby Russia's role is often described as 'assisting' Denmark or even 'passive'. Since October 2004, the European Commission took over the lead from Denmark. EECCA countries assured to give support to the partnership (EUWI 2005b, 2006; UNECE 2003, vi).

Still, private companies and actors from civil society are among the partners. Companies which participate are not listed or named anywhere in public documents. At the Multi-Stakeholder Forum in 2005, a Suez representative gave a presentation on behalf of the water suppliers involved in the EECCA-EUWI. Civil society actors involved in the partnership are GWP, WWF and the Ukrainian NGO MAMA-86. In partnership documents, moreover, some international organizations are listed as important partners such as OECD, UNECE, UNDP, UNEP and WHO Regional Office for Europe as well as IFIs (European Commission 2002b, i; UNECE 2003, vii).

(b) There is no clear procedure for partner selection. In general, the countries and partners that are involved in the EECCA partnership are those that the EU has had contact with in the past, or which actually turned to the Secretariat to ask to participate (EUWI 2005b).

As the EECCA-EUWI is part of the intergovernmental 'Environment for Europe' process, some meetings include only state actors and exclude private sector and civil society. This exclusivity results in an internal hierarchy among partners which gives priority to state actors. However, state actors invite non-state actors to stakeholder forums which in principal can be attended by any interested parties or concerned citizens (Government of Kazakhstan et al. 2005, 2). In this regard, the EECCA-EUWI structure is open and actors who can afford to participate are welcome. GWP prepared a study on IWRM in the region what exemplifies the openness of the partnership to non-state actors' input (UNECE 2003, vi). However, at the Multi-Stakeholder Forum in 2005, only five representatives from the EECCA region were present which might indicate further (indirect) mechanisms of exclusion. Moreover, there is also no strategy to support women's involvement in the EECCA-EUWI (EUWI 2005b[12]).

Again, GWP and WWF can only hesitantly be considered antagonists to the private sector. MAMA-86 is an environmental NGO which supports water pricing and does not oppose water privatization in general. This is a major difference to African civil society which is engaged in water issues for often only social objectives (increasing access). Hence, MAMA-86 is a NGO involved in the EECCA-EUWI which obviously stems from the region. So it can more convincingly claim to represent local people than international NGOs like GWP and WWF. Nevertheless, also WWF and GWP have been working on ecological and conservation issues like water protection in the EECCA region, and they have offices on the ground in these countries (WWF 2006; GWP 2006).

What is true for the A-EUWI in terms of North–South dimensions is true for the EECCA region too: Asymmetries exist in this case between the West and the East. State actors from the West dominate the process as donors. Furthermore, Western companies from the EU are considered potential investors in EECCA countries (European Commission 2003a, 8). Actors from the EECCA region in return are in the position of (dependent) recipients. Then again, in contrast to Africa, there is less geographical distance from EECCA countries to the EU. Water over-use and pollution in EECCA countries, e.g. sewage in the Baltic Sea, ultimately affect EU member states. Countries such as Denmark have a direct benefit from other riparian States pumping less and properly treated waste water into the Baltic Sea. In this regard, EU member states are also dependent on EECCA countries. Furthermore, EU member states and investors have to rely, of course, on EECCA governments to stick to agreements concerning such things as water tariffs. Consequently, power asymmetries exist but there are dependencies in both ways.

4.3 The EUWI for the Mediterranean

(a) When analyzing the EU Water Initiative for the Mediterranean (MED-EUWI), the role of the Greek government is striking. The MED-EUWI was initiated during the Greek Presidency of the European Council (after the WSSD). Furthermore, the European Commission plays an important role (European Commission 2004a, 20; 2004b, 7). All riparian States of the Mediterranean Sea are included, thus riparian EU member states on the one hand and from South-East Europe and South-East Mediterranean on the other hand (European Commission 2004a, Annex 2).

Moreover, international organization such as UNEP and International Finance Institutions are involved (EUWI 2005b). The MED-EUWI aims for coordination

[12] Request by European Commission for showing of hands according to home region.

with state-to-state programs such as the Centre for Environment and Development for the Arab Region and Europe (CEDARE), UNDP and the World Bank as well as with the other EUWI partnerships (European Commission 2004a, 3).

In terms of private sector participation, not only water suppliers are supposed to contribute, but also 'new' sectors and decision making cycles besides traditional and well known water actors (European Commission 2004b, 21). From the part of the stakeholders, GWP-Mediterranean (GWP-Med) plays a major role in the MED-EUWI, e.g. in running the regional secretariat on behalf of the government (European Commission 2004b, 8; MED-EUWI 2005a, 4, 11). Other civil society actors involved are the Mediterranean Wetlands Initiative (MedWet), Euro-Mediterranean Information System on the know-how in the Water sector (EMWIS), Mediterranean Water Institute (IME), Mediterranean Network of Basin Organisations (MENBO), Programme Solidarité Eau (Pseau) and WWF International (MED-EUWI 2005a).

(b) The Greek government is responsible for inviting actors to partnership meetings such as the WG meetings, but there are no concrete selection criteria in place. Instead selection based vaguely on geographical representation and degree of engagement in the partnership and it is thus largely up to the Greek government to decide who participates (MED-EUWI 2005a, 4). GWP-Med is in a powerful position, too, because it runs the secretariat. A main partnership objective is the improvement of knowledge on water, especially IWRM (European Commission 2004a, 21). Civil society has the potential to contribute in this regard. Their expertise flows into the process (EUWI 2005b) and thereby increases throughput legitimacy. But again, as in the A-EUWI and the EECCA-EUWI, civil society does not function as counterpart to either state actors or private sector what becomes obvious in the GWP mandate to run the secretariat on behalf of the Greek government. Moreover, there is no gender-inclusive strategy to support women's involvement in the MED-EUWI.

Because Greece as a donor has the MED-EUWI leadership, existing power asymmetries between donors and recipients are enhanced by the partnership setting (European Commission 2004a, Annex 2; EUWI 2005a, 6). Whether or not an activity is implemented, depends on finance available (European Commission 2004a, 29; 2004b, 11). Decision-making structures are consequently focused on donors.

Besides financial dependencies, existing knowledge asymmetries among partners are mentioned in the documents. Such asymmetries can generate one-sided benefit and cause mistrust among partners (European Commission 2004a, 26). The MED-EUWI does however seek to provide access to additional information and capacity building. Thus, in as far as, it achieves this, it avoids asymmetries and contributes to solving the water crisis in the Mediterranean. In generating publicity, furthermore, recipients are able to pressure for the financial aid needed (European Commission 2004b, 21).

4.4 The EUWI for Latin America

(a) The EU Water Initiative for Latin America (LA-EUWI) consists mainly of state actors. This partnership was announced at the WSSD in 2002 by Portugal and Spain but was not initiated before the World Water Forum in March 2006 (LA-EUWI 2005b, 2). Portugal, Spain and the European Commission are clearly interested in the LA-EUWI while other EU member states do not take an active part (LA-EUWI 2004a, 1). Leadership is thus with Portugal and Spain from the European side and Mexico from the side of the Latin-American partner countries (European Commission 2005, 11; LA-EUWI 2005a).

Partner countries are Mexico in North America; Costa Rica, El Salvador, Guatemala, Honduras, Nicaragua and Panama in Central America; Cuba and the Dominican Republic in the Caribbean; Argentina, Bolivia, Brasilia, Chile, Columbia, Ecuador, Paraguay, Peru, Uruguay and Venezuela in South America. However, it is not clear if all of them are really involved in the partnership process. At the MStF in Stockholm 2005 and 2006, for example, there were far less Latin Americans present, than would be expected if you consider the list of countries officially involved in the LA-EUWI (European Commission 2003b, 11; EUWI 2005b).

Civil society actors which are involved in the partnership are the Water Centre for the Humid Tropics of Latin America and the Caribbean (CATHALAC), Centre for Research, Education and Development (CIED) and Latin America Network of Basin Organizations (LANBO/RELOC) (European Commission 2005, 25). The Freshwater Action Network (FAN) takes part too. It is based at Water Aid in London, but claims to represent 400 member organizations worldwide, e.g. FAN-CA in Latin America (FAN 2006). At the MStF in Stockholm 2005, someone from 'Platform de Agua', a FAN-CA member organization in Honduras, represented civil society from Latin America (EUWI 2005b).

GWP as well as divers international organizations have been invited to participate (LA-EUWI 2005b: 2). The private sector would then be involved, at least, indirectly because private companies are GWP members. Apart from this, there are no private companies mentioned as LA-EUWI partners so far (European Commission 2002b, i; EUWI 2005b).

(b) Participation is generally open to all actors while selected actors are explicitly invited to participation (European Commission 2003b, 29). Unlike the other partnerships, the LA-EUWI seeks to integrate gender issues and the needs of indigenous people into its work (European Commission 2003b, 10). In order to get more actors from local civil society involved, LA-EUWI plans to hold regional multi-stakeholder forums (LA-EUWI 2005, 1).

However, some LA-EUWI meetings have been exclusive to state actors, and some even open only to the lead countries Portugal, Spain and Mexico (LA-EUWI 2005a). The inclusion of Mexico in this grouping demonstrates the intention of the LA-EUWI to share political responsibility for future implementation among not only EU member states, but also Latin-American countries (LA-EUWI 2005b, 1).

Even though Mexico (a target country) is included in this leadership grouping, power asymmetries still exist between donors and recipients, just as in the other EUWI partnerships. Similar consequences can be drawn concerning FDI in a sense that Latin American countries try to be attractive to European investments. However, foreign corporations also depend on the acceptance of local people and state actors. This is especially true for future implementation of partnership agreements. Cases such as Cochabamba where protest against private water corporations became violent make this fact obvious (Gómez and Terhorst 2005, 121). As long as civil society is not included in the partnership, the risk is obvious that these actors will oppose the output of the LA-EUWI. If so, the partnership will not contribute to solving water problems in Latin America, but will in fact, turn out to be a waste of time.

5 Conclusion

Stakeholder participation has always been an essential part of EU water policy. The WFD (internal strategy) and the Communication from the European Commission on water management in developing countries (external strategy) have encouraged the essential role of non-state actors. The EUWI goes further with non-state actors' involvement because participation is meant to turn into partnership. I examined the inclusion of non-state actors in the four regional EUWI partnerships considering aspects of legitimacy. The aim was to see who is de facto involved and if asymmetries exist among 'partners'.

Can the EUWI actually be considered innovative or not? From the above analysis of the EUWI, two main conclusions can be drawn. First, with regard to the partnership approach, one can see that the involvement of non-state actors is fruitful in terms of 'thinking out of the box'. It has the potential to change awareness of the actors involved, and, therefore, opens the ways for new solution.

The second conclusion which can be drawn is, however, less positive for the EUWI. Although each regional partnership claims to have a multi-stakeholder approach, state actors are in fact dominant in all of them. Sometimes, even exclusive structures for state actors exist, especially in the EECCA-EUWI which is part of the intergovernmental 'Environment for Europe' process. Only in MED-EUWI is a civil society actor, namely GWP, permanently involved as core actor.

Furthermore, actors from the North dominate. Leadership is mainly with either the European Commission or an EU member state. Partners from the private sector encompass only the big water corporations from the EU and their subsidiaries. Even civil society actors do not, for the most part, stem from the target regions. GWP, for example, which besides the MED-EUWI is also involved in the A- and EECCA-EUWI and was explicitly invited to join the LA-EUWI, has its headquarters in Stockholm. Besides, civil society actors cannot be considered antagonists to the private sector. GWP is representative again because it has members from the private sector. So it can hardly claim to represent local consumers.

Critical NGOs and activists are almost completely absent from the EUWI, whereby regional disparities exist. The EUWI promotes water commodification that contradicts the objectives of, for example, the Free Water Movement in Africa. In contrast, environmental NGOs which are more present in the EECCA region, e.g. MAMA-86, favor water pricing as an instrument for water saving and resource protection. In this regard, multi-stakeholder partnerships with the private sector are more likely to work where civil society is first of all interested in environmental protection and not necessarily against water commodification.

The disproportionate weight of (Western) European actors in the partnerships means however, that their interests are much more prevalent than those of partners from the South and the East. Moreover, the LA-EUWI is the only partnership among the four which encompasses a gender-inclusive approach and that seeks to integrate indigenous needs.

Concerning power asymmetries between countries, donor versus recipient disparities are most obvious. Regional differences have been examined. Indeed the EU is directly affected by pollution of common waters in the EECCA and MED regions (Baltic Sea, Mediterranean Sea). In contrast, pollution in Africa and Latin America has no direct or only long-term impacts on the EU (migration etc.). Besides, in all regions, the EU obviously depends on the partners on the ground for implementation. Actors from the EU might be able to run the EUWI without active engagement from the partner regions but the EU depends on partner countries cooperation, if it seeks to have any impact on the ground. The EUWI might effectively produce paper statements. However, when it comes to reality on the ground, the output must be accepted as legitimate, otherwise, it will not be implemented by the target countries. Hence, the EU dominance in the partnerships is very problematic – in terms of effectiveness as well as legitimacy.

In summary, the EUWI started with an innovative ambition: The new partnership approach seeks to coordinate state and non-state actors from different policy fields and beyond national boundaries as partners. However, power asymmetries, rather than being overcome, are actually aggravated by the EUWI structures. Hence, in this regard, the partnership cannot be considered innovative. Target countries are under-represented in the partnership process. This is likely to have major impacts when it comes to the reality on the ground. The effective and just performance of the EUWI will hence be hindered by deficits in legitimacy. Thus, the EUWI needs to be reformed in a way that redresses this deficit balance. A starting point would be to highlight the innovative aspects of the partnership approach and their value for solving global water crisis.

References

Barlow M (2006) The failure of governments from the booklet blue gold. The global water crisis and the commodification of the world's water supply. A special report issued by the International Forum on Globalization (IFG), www.thirdworldtraveler.com/Water/Failure_ Governments_BG.html. Cited 31 Jul 2006

Bond P (2004) Talk left walk right. South Africa's frustrated global reforms, University of KwaZulu-Natal Press, KwaZulu-Natal

Börzel T (1998) Organizing Babylon – On the different concepts of policy networks. In: Public Administration 76, 253–273

Braunmühl C von, Winterfeld U von (2003) Global Governance. Eine begriffliche Erkundung im Spannungsfeld von Nachhaltigkeit, Globalisierung und Demokratie, Wuppertal Papers Nr. 135/2003, www.wupperinst.org/Publikationen/WP/WP135.pdf Cited 1 Jun 2006

Cutler CA (2002) Private international regimes and interfirm cooperation. In: Hall RB, Biersteker TJ (ed) The emergence of private authority in global governance, Cambridge University Press, Cambridge, 23–40

Deutscher Bundestag (2002) Enquête-Kommission Globalisierung der Weltwirtschaft – Herausforderungen und Antworten, Schlussbericht, Drucksache 14/9200, 12-06-2002. www.bundestag.de/gremien/welt/glob_end. Cited 1 Jun 2006

Dimas S (2005) EU water for life initiative. The European Commission side event. Meeting the MDGs and the JPOI goals on water and sanitation, European Union Delegation of the European Commission to the United Nations, New York, Apr 21, 2005

Dingwerth K (2004) Effektivität und Legitimität globaler Politiknetzwerke. In: Brühl T et al. (eds) Unternehmen in der Weltpolitik. Politiknetzwerke, Unternehmensregeln und die Zukunft des Multilateralismus, Dietz, Bonn, 74–95

European Commission (2001) European Governance. A white paper, COM (2001) 428 final, 25-07-2001, Brussels

European Commission (2002a) 'Communication from the Commission to the Council and the European Parliament. Water management in developing countries. Policy and priorities for EU development cooperation' (COM 2002-132)

European Commission (2002b) EU Water Initiative: water for life health, livelihoods, economic development, peace and security, working document, Draft 21-08-2002. www.johannesburgsummit.org/html/sustainable_dev/p2_managing_resources/2508_eu_water_initiative.pdf. Cited 1 Jun 2006

European Commission (2003a) Water for life. EU Water Initiative. International cooperation – from knowledge to action, EUR 20612 (booklet). http://ec.europa.eu/research/water-initiative/pdf/water-for-life_en.pdf. Cited 1 Jun 2006

European Commission (2003b) The European Union Water Initiative. Latin American component. Programme document version 031014/LAWI. Proposal of the conference of Ibero-American water directors (Cancun, 24–25 September 2003) to the III Forum of Ibero-American Environment Ministers

European Commission (2003c) Initiative européenne pour l'eau. Composante "GIRE en Afrique". Document de travail 1: identification de bassins transfrontaliers pour mise en œuvre de l'Initiative (Document de Travail)

European Commission (2004a) Mediterranean component of the EU Water Initiative (MED EUWI) Activity Plan 2004–2006, DRAFT 1

European Commission (2004b) Mediterranean component of the EU Water Initiative (MED EUWI). Strategic partnership on water for sustainable development. MED EUWI Progress Report February 2003 – May 2004

European Commission (2005) EU Water Initiative. Water for life (Update), KH-66-05-701-EN-C, Brussels

European Parliament (2003) Report on the Commission communication on water management in developing countries and priorities for EU development cooperation. Rapporteur Paul A.A.J.G. Lannoye, A5-0273/200322, July 2003

EUWI (2005a) EU Water Initiative. Strategy for development of the EUWI, Revised Draft, 06-07-2005

EUWI (2005b) Participant observation at multi-stakeholder forum 22–25 August 2005, Stockholm

EUWI (2006) www.euwi.net. Cited 1 Feb 2006

FAN (2006) www.freshwateraction.net/fan-ca/index.asp. Cited 1 Feb 2006

Gilles I, Slaymaler T, Grimm S, Newborne P (2005) Relations between the EU development policies and the ongoing EU position in the WTO/GATS negotiations on the liberalisation of water services. Report. Draft, 03-05-2005, http://www.ec-prep.org. Cited 30 Sep 2005

GWP (2006) www.gwpforum.org. Cited 1 May 2005

Gómez LS, Terhorst P (2005) Cochabamba, Bolivia: Public-collective partnership after the water war. In: Belanya B et al. (ed) Reclaiming public water. Achievements, struggles and visions from around the world, Transnational Institute (TNI)/Corporate Europe Observatory (CEO), Amsterdam, 121–130

Government of Kazakhstan/UNDP/Global Water Partnership/Government of Norway (2005) Progress report on project for a national integrated water resources and water efficiency plan for Kazakhstan (Project No. 00034982), Period covered: January–June 2005

Hummel H (2001) Die Privatisierung der Weltpolitik. Tendenzen, Spielräume und Alternativen. In: Brühl T et al. (ed) Die Privatisierung der Weltpolitik. Entstaatlichung und Kommerzialisierung im Globalisierungsprozess, Dietz, Bonn, 22–56

Kunst S, Kruse T, Burmeister A (eds) (2002) Sustainable water and soil management, Springer, Berlin, Heidelberg

LA-EUWI (2005a) Conferencia de Directores Generales Iberoamericanos del Agua. Meeting of the 'Comité de Enlace' (Network Committee) of the Conference of Ibero-American Water Directors, Madrid, 30 March 2005

LA-EUWI (2005b) Work Programme 2005–2006 – Draft 2, March 10, 2005. http://www.euwi.net. Cited 7 Jun 2005

Lombard CS (2000) Stakeholder assumptions in public-private partnerships in water services, Master Thesis, University of Witwatersrand, Johannesburg

Magnette P (2001) European governance and civic participation: can the European Union be politicised? Jean Monet Working Paper No. 6, Harvard Law School, New York

Mathews JT (1997) Power shift. In: Foreign Affairs 76 (1), 50–66

MED-EUWI (2005a) Mediterranean component of the EU Water Initiative (MED EUWI). Conclusions of the 4th meeting of the Med EUWI Working Group, 17 January 2005

Messner D, Nuscheler F (2003) Das Konzept Global Governance. Stand und Perspektiven. INEF-Report, Duisburg, http://ineuni-due.de/page/documents/report67.pdf. Cited 1 May 2005

Mostert E (2003) The European water framework directive and water management research. In: Physics and Chemistry of the Earth 28, 523–527

Nanz P, Steffek J (2005) Legitimation durch Deliberation? Die Rolle der Zivilgesellschaft in der supranationalen Politik. In: Knodt M, Finke B (ed) Europäische Zivilgesellschaft. Konzepte, Akteure, Strategien, VS, Wiesbaden, 79–102

Postel S (2006) Die Süßwasserökosysteme schützen! In: Worldwatch Institute (ed): Zur Lage der Welt 2006. China, Indien und unsere gemeinsame Zukunft, Westfälisches Dampfboot, Münster, 112–147

Rahaman MM, Varis O, Kajandar T (2001) EU Water Framework Directive vs. international principles concerning IWRM: The seven mismatches, Water Resource Laboratory, Helsinki University. http://www.euwi.net/file_upload/niki_tmpphp23UK4Y.pdf. Cited 1 Jun 2005

Rekacewicz P (2005) Wasser als Wirtschaftsgut. http://www.wasser-symposium.ch/meldungen/mehr.php?id=44. Cited 31 Jul 2006

Rosenau JN (1992) Governance, order, and change in world politics; in: Czempiel EO, Rosenau JN (eds) Governance without government: order and change in world politics, Cambridge University Press, Cambridge, 1–29

Scharpf Fritz (2000) Interaktionsformen. Akteurszentrierter Institutionalismus in der Politikforschung, Leske + Budrich, Opladen

Simonis, Udo E (2001) Wasser als Konfliktursache – Plädoyer für eine internationale Wasserstrategie, WZB FS II, Berlin, 01–406

Stewart A, Gray T (2006) The authenticity of 'type two' multistakeholder partnerships for water and sanitation in Africa: when is a stakeholder a partner? In: Environmental Politics 15 (3), 362–378

United Nations Economic Commission for Europe (UNECE) (2003) Fifth ministerial conference. Environment for Europe. Kiev, Ukraine, 21–23 May 2003. Environmental partnership in the UN ECE region: environment strategy for countries of Eastern Europe, Caucasus and Central Asia. Strategic framework (KIEV.CONF/2003/INF/33; United Nations Economic and Social Council ECE/CEP/105/Rev.1, 27 June 2003), http://www.unece.org/env/documents/2003/ece/cep/ece.cep.94.rev.1.e.pdf. Cited 1 Jun 2006

Witte JM, Streck C, Benner T (eds) (2003) Progress or peril? Partnerships and networks in global environmental governance. The Post-Johannesburg agenda, Global Public Policy Institute, Berlin

Witte M, Reinicke W (2005) Business as unusual. Facilitating United Nations reform through partnerships, United Nations Global Compact Office, New York
WWF (2006) www.wwf.org. Cited 1 Jul 2006
Zürn M (2000) Democratic governance beyond the nation-state: the EU and other international institutions. In: European Journal of International Relations 6, 183–221

Index

MIX
Papier aus verantwortungsvollen Quellen
Paper from responsible sources
FSC® C105338

If you have any concerns about our products,
you can contact us on
ProductSafety@springernature.com

In case Publisher is established outside the EU,
the EU authorized representative is:
Springer Nature Customer Service Center GmbH
Europaplatz 3, 69115 Heidelberg, Germany

Printed by Libri Plureos GmbH
in Hamburg, Germany